Camille Flammarion, James Glaisher

The Atmosphere that which Gives Life to the Earth and by which Everything Has its Being

An Ethereal Sea that Covers the Whole World

Camille Flammarion, James Glaisher

The Atmosphere that which Gives Life to the Earth and by which Everything Has its Being
An Ethereal Sea that Covers the Whole World

ISBN/EAN: 9783744727631

Printed in Europe, USA, Canada, Australia, Japan

Cover: Foto ©berggeist007 / pixelio.de

More available books at **www.hansebooks.com**

A. Marie pinx'

Eug. Ciebri chromolith'

HALO

Imprimerie A. Pierre et Sœur Sy Paris

THE ATMOSPHERE THAT WHICH GIVES LIFE TO THE EARTH AND BY WHICH EVERYTHING HAS ITS BEING AN ETHEREAL SEA THAT COVERS THE WHOLE WORLD BY CAMILLE FLAMMARION TRANSLATED BY C B PITMAN AND EDITED BY JAMES GLAISHER F.R.S.

"EVERYTHING THAT TAKES PLACE AROUND US IS INTERESTING WHEN INSTEAD OF REMAINING AS ONE BORN BLIND A MAN HAS LEARNED TO KEEP HIMSELF IN INTELLIGENT COMMUNICATION WITH NATURE"

ILLUSTRATED WITH EIGHTY SIX WOODCUTS AND TEN CHROMO LITHOGRAPHS

NEW YORK PUBLISHED BY THE DRALLOP PUBLISHING COMPANY 166 SIXTH AVENUE MDCCCXCVI

PREFACE BY THE EDITOR.

THE following work is translated and abridged from M. Flammarion's *L'Atmosphère*, Paris, 1872. That some curtailment of the text of the original work was requisite will be apparent when it is stated that the French Edition contains 824 large pages of closely printed matter, and is of more than twice the extent of the present volume. Not only was some compression necessary in order to bring the work within a reasonable compass, but, independently of this, one or two chapters, such as that on the Respiration and Alimentation of Plants, appeared to have so remote a connection with the subject of the work—the Atmosphere—that their omission would in any case have been desirable.

Every one who has any acquaintance with French popular works on Science is aware that very many exhibit a tendency to imaginative, or, to express my meaning colloquially, "fine" writing, which ill accords with the precision and accuracy that ought to be a characteristic of scientific information, even when expressed in language free from technicalities. There is a good deal of this exalted kind of composition in M. Flammarion's book, which—even in the French not very agreeable to an English reader—becomes, when translated, intolerable. I have, therefore, omitted these rhapsodies very freely, though traces enough of them will be found here and there to betray the French origin of the work.

I may add that the task of editing has not been a light one; besides the necessity for compression and the consequent selection of the matter to be included, I have been obliged to exercise some sort of censorship over the facts contained in the work. It is impossible for any one man to have a complete knowledge of so great a variety of subjects as are treated of by M. Flammarion, and the compiler of such a book must include many things taken from others, of the accuracy of which he is not fully competent to judge. In cases where a statement contained in the original work appeared to me clearly erroneous, I have corrected it,

appended a note, or omitted it altogether; and in cases where I have
been doubtful of the accuracy of a passage, or have differed in opinion
from the author, I have not considered myself justified in making an
alteration, so long as there was no strong *primâ facie* presumption that
the original was incorrect. In spite of obvious blemishes, inseparable
from a translation, and a certain want of continuity in a few places,
which is due to the omission of portions of the book as originally
written, I believe the volume will be found to be readable, popular, and
accurate, and it covers ground not occupied by any one work in our
language.

The work treats on the form, dimensions, and movements of the
earth, and of the influence exerted on meteorology by the physical
conformation of our globe; of the figure, height, color, weight, and
chemical components of the atmosphere; of the meteorological phenomena induced by the action of light, and the optical appearances
which objects present as seen through different atmospheric strata; of
the phenomena connected with heat, wind, clouds, rain, and electricity,
including the subjects of the laws of climate: the contents are therefore of deep importance to all classes of persons, especially to the observer of nature, the agriculturist, and the navigator.

The whole is explained in a very popular manner, and as free as possible from all technicalities; the object having been to produce a work
giving a broad outline of the causes which give rise to facts of every-
day occurrence in the atmosphere, in such a form that any reader who
wished to obtain a general view of such phenomena and their origin
would be readily enabled to do so. The great number of subjects
treated of will thus, to the majority of readers, who merely desire an
insight into the general principles that produce phenomena which
every one has seen or heard of, be found to be rather an advantage, as
the whole range of atmospheric action is thus displayed in the same
volume in moderate compass, without so much detail being anywhere
given as to make the book other than interesting to even the most
casual reader.

The translation was made by Mr. C. B. PITMAN.

January, 1873.

PREFACE.

In eâ vivimus, movemur et sumus.

Of all the various subjects which invite a studious examination, it is impossible to select one possessing a more direct, a more permanent, or a more real interest than that which forms the subject of this work. The Atmosphere gives life to earth, ocean, lakes, rivers, streams, forests, plants, animals, and men; in and by the Atmosphere every thing has its being. It is an ethereal sea reaching over the whole world; its waves wash the mountains and the valleys, and we live beneath it and are penetrated by it. It is the Atmosphere which makes its way as a life-giving fluid into our lungs, which gives an impulse to the frail existence of the new-born babe, and receives the last gasp of the dying man upon his bed of pain. It is the Atmosphere which imparts verdure to the fertile fields, nourishing at once the tiny flower and the mighty tree; which stores up the solar rays in order to give us the benefit of them in the future. It is the Atmosphere which adorns with an azure vault the planet in which we move, and makes us an abode in the midst of which we act as if we were the sole tenants of the infinite —the masters of the universe. It is the Atmosphere which illuminates this vault with the soft glitter of twilight, with the waving splendors of the aurora borealis, with the quivering of the lightning and the multiform phenomena of the heavens. At one moment it inundates us with light and warmth, at another it causes the rain to pour down in torrents upon the thirsty land. It is the channel by which the sweet perfumes descend from the hills, and the vehicle of the sound which permits human beings to communicate with each other, of the song of the birds, of the sighing of the wind among the trees, of the moaning of the waves. Without it, our planet would be inert and arid, silent and lifeless. By it the globe is peopled with inhabitants of every kind. Its indestructible atoms incorporate themselves in the various living organisms; the particle which escapes with our breath takes refuge in a plant, and, after a long journey, returns to other human bodies; that which we breathe, eat, and drink has already been inhaled, eaten, and drunk millions of times: dead and living, we are all formed of the same substances. . . . . What study can possess a vaster or more direct interest than that of the vital fluid to which we owe the manner of our being and the maintenance of our life?

The study of the Atmosphere, of its physical condition, of its movements, of its functions, and of the laws which regulate its phenomena, forms a special branch of human research. This science, which since the days of Aristotle has been designated *Meteorology*, belongs in part to Astronomy, which shows the movements of our planet around the sun—movements to which we owe day and night, season, climates, solar action, or, in a word, the basis of the subject. On the other hand, it appertains to Natural Philosophy and Mechanics, which explain and measure the forces brought into play. As it exists in the present day, Meteorology is a new science, of recent establishment, scarcely as yet fixed in its elementary principles.

We are assisting at its elaboration, at its struggling into life. The present generation has seen the establishment of meteorological societies throughout the different nations of Europe, and of special observatories for the exclusive study of the problems relating to the Atmosphere. The analysis of climates, seasons, currents, and periodical phenomena is scarcely terminated. The examination of atmospheric disturbances, of tempestuous movements, and of storms, has been made, so to speak, before our own eyes. The science of the Atmosphere is the question of the day. We are just now, in regard to this study, in an analogous situation to that of modern Astronomy in the days of Kepler. Astronomy was founded in the seventeenth century. Meteorology will be the work of the nineteenth.

I have endeavored to collect in this work all that is at present positively known about this important subject, to represent as completely as possible the actual state of our knowledge about the Atmosphere and its work—that is, about the air, the seasons, the climates, the winds, the clouds, the rain, the hurricanes, the storms, the lightning, the meteors—in a word, the phenomena of time, and above all, the general upholding of terrestrial life. It is, in fact, a synthesis of the research effected during the last half century (especially during the latter portion of it) as to the great phenomena of terrestrial nature, and the forces which produce them. The great majority of us, inhabitants of the earth, no matter to what nation we belong, pass our lives without attempting to form an idea of our actual position, without asking ourselves what is the force which prepares for us our daily bread, ripens for us the grapes that give the wine, presides over the change in the seasons, and alternates the exhilarating blue sky with the rains and cold of inhospitable winter. Yet, why should we live in such a state of ignorance? I venture to hope that after perusing this work there will be no difficulty in understanding the life and movements of the globe. Every thing which takes place around us is interesting when, instead of remaining as one born blind, a man has learned to appreciate external things and to keep himself in intelligent communication with Nature.

I could have wished to keep this work, destined as it is for the general public, free from scientific terms and figures which constitute its basis. I have done so as far as possible, but without in any point sacrificing accuracy and precision in respect to observed facts. It seems to me, too, that what is termed the public (that is, every one) has become somewhat scientific itself, since so many excellent works have popularized ideas previously reserved for a small circle of the elect.

Camille Flammarion.

Paris: *November*, 1871.

CONTENTS.

BOOK FIRST.

OUR PLANET AND ITS VITAL FLUID.

CHAP.   PAGE

I. THE TERRESTRIAL GLOBE.................................. 17

II. THE ATMOSPHERIC ENVELOPE........................... 23

III. THE HEIGHT OF THE ATMOSPHERE 28

IV. WEIGHT OF THE TERRESTRIAL ATMOSPHERE—THE BAROMETER
AND ATMOSPHERIC PRESSURE...... 38

V. CHEMICAL COMPONENTS OF THE AIR..................... 57

VI. SOUND AND THE VOICE................................ 75

VII. AERONAUTICAL ASCENTS 85

BOOK SECOND.

LIGHT AND THE OPTICAL PHENOMENA OF THE AIR.

I. THE DAY.. 103

II. EVENING.. 113

III. THE RAINBOW.. 121

IV. ANTHELIA: SPECTRE-SHADOWS UPON MOUNTAINS—THE ULLOA
CIRCLE—CIRCLE SEEN FROM A BALLOON.................. 127

V. HALOS: PARHELIA — PARASELENES — CIRCLES SURROUNDING
AND TRAVERSING THE SUN — CORONAS — COLUMNS—VARI-
OUS PHENOMENA..................................... 137

VI. THE MIRAGE.. 149

CHAP.   PAGE

VII. Shooting-stars—Bolides—Aerolites—Stones falling from
        the Sky.. 163

VIII. The Zodiacal Light....................................... 174

BOOK THIRD.

TEMPERATURE.

I. Heat: the Thermometer—Quantity of Heat received—
        Temperature of the Sun—Temperature of Space...... 181

II. Heat in the Atmosphere................................... 190

III. The Temperature of the Air: Its mean Condition—Daily
        and Monthly Variations of the Temperature — Tem-
        perature of each Summer, Winter, and Year at Paris
        and at Greenwich since the last Century—Daily and
        Monthly Variations of the Barometer................ 202

IV. Remarkable Summers—The highest known Temperatures 218

V. Autumn—Winter: Winter Landscapes—Cold—Snow—Ice
        —Hoar-frost, Rime, etc. — Remarkable Winters—The
        lowest known Temperatures........................... 229

VI. Climate: Distribution of Temperature over the Globe—
        Isothermal Lines — The Equator — The Tropics — The
        Temperate Regions — The Poles — The Climate of
        France... 245

BOOK FOURTH.

THE WIND.

I. The Wind and its Causes: General Circulation of the
        Atmosphere — The Regular and Periodical Winds —
        Trade-winds—The Monsoon—Breezes.................. 269

II. The Sea Currents: Meteorology of the Ocean—Maritime
        Routes—The Gulf Stream............................. 284

CHAP.   PAGE

III. THE VARIABLE WINDS—THE WIND IN OUR CLIMATES—MEAN
DIRECTIONS IN EUROPE AND IN FRANCE—RELATIVE FRE-
QUENCY OF DIFFERENT WINDS—RISE OF THE WINDS ACCORD-
ING TO THE TIMES AND PLACES—MONTHLY AND DIURNAL VA-
RIATION IN INTENSITY.................................... 297

IV. RESPECTING CERTAIN SPECIAL WINDS: THE BISE—THE BORA—
THE GALLEGO—THE MISTRAL—THE HARMATTAN—THE SI-
MOOM—THE KHAMSEEN—THE SIROCCO—THE SOLANO....... 318

V. THE POWER OF THE AIR: THE HURRICANE—THE CYCLONE—
THE TEMPEST.. 327

VI. TROMBES, WHIRLWINDS, OR WATER-SPOUTS.................... 337

BOOK FIFTH.

WATER—CLOUDS—RAIN.

I. THE WATER UPON THE SURFACE OF THE EARTH AND IN THE AT-
MOSPHERE: THE EARTH—VOLUME AND WEIGHT OF THE WA-
TER THROUGHOUT THE GLOBE—PERPETUAL CIRCULATION—
VAPOR OF WATER IN THE ATMOSPHERE—ITS VARIATIONS AC-
CORDING TO THE HEIGHT, THE LOCALITY, AND THE WEATHER
—THE HYGROMETER—DEW—WHITE FROST.............. 355

II. THE CLOUDS: WHAT A CLOUD IS—THE MANNER OF ITS FORMA-
TION—MIST—OBSERVATIONS TAKEN FROM A BALLOON AND
FROM MOUNTAINS—DIFFERENT KINDS OF CLOUDS—THEIR
SHAPES—THEIR HEIGHTS 363

III. RAIN: GENERAL CONDITIONS OF THE FORMATION OF RAIN—ITS
DISTRIBUTION OVER THE GLOBE—RAIN IN EUROPE......... 381

IV. HAIL: PRODUCTION OF HAIL—COURSE OF HAILSTORMS—VARY-
ING DISTRIBUTION OF HAILSTORMS IN DIFFERENT PARTS OF
THE COUNTRY—HEAVIEST HAILSTORMS KNOWN—NATURE,
SIZE, AND SHAPE OF HAILSTONES—PERIODS OF THEIR OCCUR-
RENCE.. 390

V. PRODIGIES: SHOWERS OF BLOOD—OF EARTH—OF SULPHUR—
OF PLANTS—OF FROGS—OF FISH—OF VARIOUS KINDS OF AN-
IMALS.. 401

BOOK SIXTH.

ELECTRICITY, THUNDER-STORMS, AND LIGHTNING.

CHAP.   PAGE

I. ELECTRICITY UPON THE EARTH AND IN THE ATMOSPHERE: ELECTRIC CONDITION OF THE TERRESTRIAL GLOBE—DISCOVERY OF ATMOSPHERIC ELECTRICITY—EXPERIMENTS OF OTTO DE GUÉRICKE, WALL, NOLLET, FRANKLIN, ROMAS, RICHMANN, SAUSSURE, ETC.—ELECTRICITY OF THE SOIL, OF THE CLOUDS, OF THE AIR—FORMATION OF THUNDER-STORMS. 423

II. LIGHTNING AND THUNDER. 431

III. THE SAINT ELMO FIRES AND THE JACK-O'-LANTERNS. 441

IV. AURORÆ BOREALES. 445

ILLUSTRATIONS.

CHROMO-LITHOGRAPHS.

FIG.   PAGE
1. Halo...*Frontispiece.*
2. Sunset at Sea...*To face* 119
3. The Rainbow.." 121
4. Lunar Rainbow seen at Compiègne.......................... " 126
5. Sunrise from the Righi... " 127
6. African Mirage... " 149
7. Summer Landscape.. " 218
8. Winter Landscape.. " 229
9. The Storm.. " 423
10. Aurora Borealis seen at Paris, May 13, 1869.......... " 445

WOOD-CUTS.

1. Mathematical Limit of the Shape of the Atmosphere............................... 29
2. Measure of the Height of the Atmosphere, according to the Length of Twilight...... 32
3. Thickness of the Earth's Crust, of our Atmosphere, and of a higher Atmosphere..... 34
4. Suction-Pump.. 39
5. Suction and Forcing Pump.. 40
6. Torricelli inventing the Barometer.. 41
7. Barometer Tube full of Quicksilver... 43
8. The Tube in the Basin... 43
9. Otto de Guéricke's Experiment... 45
10. The Magdeburg Hemispheres.. 46
11. Atmospheric Pressure. Rupture of Equilibrium..................................... 47
12. Atmospheric Pressure under an inverted Glass...................................... 47
13. Diagram showing the Decrease of atmospheric Pressure, according to Height....... 51
14. Variation in the atmospheric Pressure at the Level of the Sea...................... 52
15. Lavoisier analyzing atmospheric Air.. 56
16. Matrass or Glass Vessel.. 58
17. The Apparatus for Analysis of Air.. 58
18. Mercury-Eudiometer, for analyzing Air.. 59
19. Apparatus for analyzing Air by the Method of Weight............................... 60
20. Apparatus for obtaining the Proportion of carbonic Acid in Air.................... 61
21. Apparatus for separating the Oxygen from the Nitrogen............................ 62
22. Vibrations of a Blade... 75
23. Vibration of a Cord... 76
24. Illustration of Hawksbee's Experiment.. 78
25. Baroscope.. 86
26. Soap-bubbles inflated with Hydrogen.. 88
27. Distribution of Kinds of Birds according to Height of Flight....................... 97
28. Lunar Day.. 111
29. Atmospheric Refraction.. 114
30. Simple Reflection of Rays in a Drop of Rain.. 121

FIG.		PAGE
31.	Formation of the Rainbow	123
32.	Double Reflection of Rays in a Drop of Rain	124
33.	Theory of the two Arches of a Rainbow	124
34.	Triple Rainbow	125
35.	The Spectre of the Brocken	129
36.	The Ulloa Circle	132
37.	Theory of the Halo	140
38.	Halo seen in Norway	142
39.	Corona formed around the Moon by Diffraction	147
40.	Explanation of the ordinary Mirage	152
41.	Mirage seen at Paris in 1869	158
42.	Lateral Mirage seen on the Lake of Geneva	160
43.	La Fata Morgana	162
44.	Shooting-stars	165
45.	Fall of a Bolide in the Daytime	170
46.	The Caille Aërolite, weighing 12½ cwt.	172
47.	The Pyrheliometer	183
48.	Relative Intensity of the calorific, luminous, and chemical Rays of the Sun	192
49.	Inequality of the Thickness of Air traversed by the Sun	196
50.	Regular Diurnal Oscillation of the Barometer	213
51.	Regular Monthly Oscillation of the Barometer	216
52.	Snow Crystals	231
53.	Winter.—The Seine full of floating Ice	235
54.	Comparative Temperatures of Rome, London, Paris, Vienna, St. Petersburg	251
55.	The last human Dwelling-places. Esquimaux of the Polar Regions	262
56.	Ice at the Pole	264
57.	Section of the Atmosphere, showing its general Circulation	272
58.	Average annual Prevalence of different Winds at London	305
59.	Average annual Prevalence of the different Winds at Brussels	305
60.	Monthly Intensity of the Winds	307
61.	Diurnal Intensity of the Winds	307
62.	The Simoom	323
63.	Whirlwind	346
64.	Sand Whirlwind	348
65.	Water-spout at Sea	350
66.	Intense Fog in one of the Islands of the Antipodes	368
67.	Intense Fog in the Spitzbergen Mountains	369
68.	Formation of a Thunder-cloud	377
69.	Above and below the Rain-cloud	380
70.	Diminution in the Rain-fall from the Tropics to the Poles	383
71.	Increase of Rain, according to the Undulations of the Soil	384
72.	Comparative Depths of Rain-fall	385
73.	Section of Hailstones, showing their ordinary interior Structure	398
74.	Section of a Hailstone, enlarged	399
75.	Different Forms of Hail	400
76.	Rain of Blood in Provence, July, 1608	404
77.	Shower of Locusts	417
78.	Shower of Cock-chafers	418
79.	Experiments of Franklin and Romas	424
80.	Richmann, of St. Petersburg, struck by Lightning during an electrical Experiment	426
81.	Harvesters killed by Lightning	438
82.	Curious Freak of Lightning	440
83.	Saint Elmo Fire over the Spire of Notre-Dame, Paris	442
84.	An Aurora Borealis over the Polar Sea	447
85.	Aurora Borealis observed at Bossekop (Spitzbergen), January 6, 1839	449
86.	Aurora Borealis observed at Bossekop (Spitzbergen), January 21, 1839	451

BOOK FIRST.

OUR PLANET AND ITS VITAL FLUID.

THE ATMOSPHERE.

CHAPTER I.

THE TERRESTRIAL GLOBE.

BORNE forward in space, in obedience to the mysterious laws of universal gravity, our globe travels therein with a rapidity that our closest study can scarcely conceive. Let us imagine a sphere absolutely free, isolated on all sides, without any prop or stay, placed in the midst of space. If this sphere were alone in the immensity, it would remain thus suspended, motionless, without power to incline to this side or to that. Eternally fixed, it would constitute in itself the whole of creation; astronomy and physics, mechanics and biology, would all be included in its conception. But the earth is not the only world existing in space. Millions of celestial bodies have been formed, like itself, in the infinite heavens, and their co-existence establishes between them relations inherent in the very constitution of matter. The earth, in particular, belongs to a system of planets analogous to itself, having the same origin and the same destiny, situated at various distances around the same centre, and governed by the same motive power. Our planetary system is composed essentially of eight worlds, made to revolve in successive orbits, the exterior one of which is seven thousand million leagues in extent. The sun, a colossal star nearly a million and a half times larger than the earth, and 350,000 times as heavy, occupies the centre of these orbits; or, to speak more accurately, a focus of one of the nearly circular ellipses which they describe. It is around this gigantic star that take place the revolutions of the planets, which are performed with an indescribable speed on account of the length of the circumference to be traversed. Far from being motionless, as it appears to us, the globe which we inhabit revolves at an average distance of ninety-one and a half millions of miles from the sun, and over an orbit which does not measure less than 587 millions of miles. These are traversed in 365 days and six hours — that is to say, that we move through space with a speed of more than one and a half million of miles per day, or more than 66,000 miles an hour. The most rapid of express trains can scarcely accomplish more than twenty-five leagues an hour. Upon the invisible roads of the heavens the earth moves with a speed eleven hundred times greater. The difference is so enormous, that it is impossible to express it in this work by a geometrical figure. If the distance traversed in an hour by a locomotive was represented by one tenth of an inch, it would be necessary to trace a line more than nine feet long to indicate the comparative advance made by our planet during the same space of time. I will add, as a point of comparison, that the movement of the tortoise is about eleven hundred times less rapid than that of an express train. Consequently, were an express train to be sent in pursuit of the earth, it would be as a tortoise in pursuit of an express train.

Situated as we are about the globe, infinitely small mollusks, made to adhere to its surface by its central attraction, and carried away with it, we are unable to appreciate this movement or form a direct idea concerning it. It is only by the observation of the corresponding change of position in the celestial perspectives, and by calculations based thereon, that we have been able—and this only during the last few centuries— to acquire a knowledge of its nature, its form, and its importance. From the deck of a ship, from a railway-carriage, or the car of a balloon, we are alike unable to form an idea of the movement that is transferring us from one place to another, because we participate in it; and without some object of comparison not partaking of the motion, it is impossible for us to appreciate it. To form an idea of the rapidity of the earth's motion, we must imagine ourselves placed not upon the earth's surface but outside it, in space itself, not far from the course along which it hurries so impetuously. Then we should see far in the distance—to our left, I will suppose—a little star shining amidst the rest in the gloom of space. Then this little star would seem to grow larger, and to draw nearer to us. Soon there would be perceptible a disk like that of the moon, upon which we should also recognize spots formed by the optical difference between continents and seas, by the polar snows and the cloudy bands of the tropics. We should endeavor to distinguish upon this gradually swelling globe the principal geographical shapes visible athwart the vapors and clouds of the atmosphere, when suddenly, standing out against the sky and covering the immensity of its dome, the globe would meet our affrighted gaze, as if it were a giant emerging from the abysses of space. Then, rapidly, without giving us time to recognize it, the colossus would rush away to our right, quickly diminishing in size, and silently burying itself in the dark depths beyond. So moves the globe we inhabit, and we are borne along by it like so many grains of dust adhering to the whirling surface of a cannon-ball projected into space.

How great a difference there is between this truth and the ancient fallacy which represented the earth as the support of the firmament! During the reign of illusion—so old, and yet so difficult to dispel, even in our epoch, from certain minds—the earth was believed to form in itself alone the living universe, and to represent the whole of nature. It was the centre and objective of all creation, while the rest of space was but a vast and silent solitude. There was a higher region in the universe—viz., the heavens, the empyreum; a lower region—viz., the earth, hell. Mysticism had created the world for terrestrial humanity alone, as being the centre of Divine Will. In the present day we know that the heavens are but boundless space, and that the earth is in the heavens just as the other stars; we contemplate in the firmament worlds similar to our own, and the starry night addresses itself to our minds with a new eloquence. The terrestrial globe, with its humanity, is no longer more than an atom cast into the infinite—one of the countless fly-wheels which, in tens of thousands, constitute the mysterious mechanism of the physical world. Our planetary system, despite its vastness, compared to the microscopical volume of this earth, is, sun and all, eclipsed in the presence of the extent and number of the stars, which are solar centres of systems distinct from ours. The astonished gaze encounters distant suns whose light takes hundreds and thousands of years to reach us, notwithstanding its wondrous speed of 186,000 miles a second; farther still the eye may contemplate pale masses of stars which, seen nearer, would resemble our Milky Way, and would be found to be composed of millions of suns and systems; beyond these, again, the eye and the mind still seek to discover more distant creations, but the sweep of our fatigued conceptions soon falls to a lower level, worn out and lost by this interminable flight into the regions of infinity.

An invisible star, lost in the myriads of stars, the earth is borne along in the heavens by various movements far more numerous and peculiar than most people would be inclined to suppose. The most important

one is that of revolution, which we have noticed above, and by virtue
of which the earth moves round the sun at the rate of one and a half
million of miles a day. A second movement, that of *rotation*, causes it
to turn round its own axis in the course of every four-and-twenty hours.
It may be at once seen, in examining this movement of the globe, that
the different points of the terrestrial surface have a different speed, according to their distance from the axis of rotation. At the equator,
where the speed is greatest, the terrestrial surface has to traverse 25,000
miles in twenty-four hours; that is, more than 1040 miles an hour, or
about seventeen a minute. In the latitude of London, where the circle
is perceptibly smaller, the speed is eleven miles a minute. At Rekia-
witz, one of the towns almost in the heart of the polar region, the speed
is seven and a half miles a minute; and finally, at the poles themselves,
it is nil. A third movement, that which constitutes the *precession of the
equinoxes*, causes the terrestrial axis to accomplish a slow rotation, which
occupies not less than 25,868 years, and in virtue of which all the stars
of heaven annually seem to change their position, to return to the same
point only at the close of this great secular cycle. A fourth movement
gradually makes a change in the position of the perihelion, which makes
the circuit of the orbit in 20,984 years, so that in this other cycle the
seasons successively take the place the one of the other. A fifth movement causes the plane of the earth's orbit, which it describes around the
sun, to oscillate, and diminishes the obliquity of the ecliptic at present,
to increase it in the future. A sixth movement, due to the action of
the moon, and called *nutation*, causes the pole of the equator to describe
upon the celestial sphere a small ellipse in eighteen years and eight
months. A seventh movement, caused by the attraction of the planets,
and principally by the gigantic world of Jupiter and our neighbor Venus, occasions perturbations, calculable beforehand, in the curve described by our planet around the sun, swelling or flattening it, according
to the variations of distance. An eighth movement, more considerable
and less exactly measured than the preceding ones, though its existence
is incontestable, is the transport of the whole planetary system in space.
The sun is thus not motionless, but traverses an immense orbital line,
the direction of which is at present toward the constellation of Her-
cules. The speed of this general movement is estimated at 487,000
miles a day. The laws of motion would incline one to believe that the
sun gravitates around a centre as yet unknown to us. If so, how vast
must be the extent of the circumference of the ellipse which it describes,

since for the last century it has followed, as far as we can judge, a perfectly straight line!

These different movements, which cause the earth to travel in space, are ascertained with certainty, thanks to the vast number of the observations of the stars made for more than 4000 years, and to the definite nature of the modern principles of celestial mechanics. The knowledge of these constitutes the essential basis of the highest and most substantial of sciences. The earth is henceforth inscribed in the ranks of the stars, in spite of the evidence of the senses, in spite of secular illusions and errors, and, above all, in spite of human conceit, which had for a long time complacently formed a creation for man alone. Drawn here and there by these diverse movements—some of which, such as that of the *perturbations*, are extremely complicated—the terrestrial globe travels onward, whirling along, balancing itself under the influence of varied forces, rushing with an incomprehensible rapidity toward an unknown goal. Since the beginning of the world, the earth has not twice passed the same spot, and the place which we occupy at this very moment is rapidly sinking behind into our track never to return. The very terrestrial surface, too, undergoes changes every century, every year, every day, and the conditions of life change throughout eternity as throughout space. After having thus examined the movement of the earth in space, we must join to it, in order to complete its astronomical aspect, the motion of the moon round the earth in twenty-nine days and a half. The moon is only $\frac{1}{49}$ of the size, and $\frac{1}{81}$ of the weight of the earth. Its action upon the ocean and the atmosphere is, nevertheless, comparable with that of the sun, and is even more important as regards the production of tides: it is as useful to know its movement about us as to know that of our planet about its primary. The revolution of the moon around the earth takes place really in twenty-seven days and eight hours, but during these twenty-seven days the earth has not been motionless, but, on the contrary, has advanced a certain distance. The moon employs about two days more to complete its revolution and to return to the same point in relation to the sun, which gives twenty-nine days and thirteen hours for the lunation or the cycle of phases. The revolution in twenty-seven days is called the *sidereal* revolution, because in that time the moon returns upon the celestial sphere to the same position in relation to the stars. We see that to return to the same position in relation to the sun, and to accomplish its synodical revolution, our satellite must make more than a circle upon the celestial sphere, and pass over in addition the distance which the earth has traveled during that time. If we suppose the earth motionless, the movement of the moon round it may be nearly represented by a circle. In reality, it is a sinuous line, resulting from the combination of the two movements.

Three stars thus command our attention in the general history of nature—the sun, the earth, and the moon. They are held up, isolated, in space in a manner dependent on their respective weights. The sun weighs two quadrillions of tons (two followed by twenty-four zeros). The sun is 355.000 times heavier than the earth, the latter eighty times more so than the moon. The sun holds the earth at arms-length, so to speak, ninety-one and a half millions of miles distant; the earth holds the moon—also by the influence of its mass—at a distance of 237,000 miles.

In gravitating around our luminary, the earth, constantly immersed in its rays, brings the different portions of its surface successively into its fertilizing emanations. Morning succeeds evening, and spring autumn. Night, like winter, is but the transition from one light to another. The solar heat keeps in continual work the mighty factory of the terrestrial atmosphere, forming the currents, the winds, the tempests, and the breezes; preserving the water liquid and the air gaseous, raising water from the inexhaustible wells of the ocean, producing the mists, the clouds, the rains, and the storms; organizing, in a word, the permanent system of the vital circulation of the globe.

It is this system of circulation, with the varied phenomena of the atmospheric world, which we are about to study in this work. The subject is vast and grand, for upon it depends all terrestrial life. In studying it we learn, therefore, the very organism of existence upon the planet we inhabit.

CHAPTER II.

THE ATMOSPHERIC ENVELOPE.

OUR globe, the motions of which we have been explaining, is encircled by a gaseous film which adheres to its entire spherical surface. This layer of fluid extends with uniform thickness all round the globe, covering it on every side. We have already compared the earth in the midst of space to a cannon-ball launched into the air; by imagining this cannon-ball surrounded by a thin ring of smoke not more than $\frac{1}{200}$ of an inch thick, we may form some idea of the position of the atmosphere around the terrestrial globe. It is, indeed, from this position that the atmosphere derives its name ('Ατμός, vapor; and Σφαῖρα, sphere). being, as it were, a second sphere of vapor concentric with the solid sphere of the globe itself. As a rule, sufficient importance is not attached to the functions of this atmospheric envelope. It is from it that we draw our being. Plants, animals, and men imbibe therefrom the first elements of their existence. The earth's organization is so ordered that the atmosphere is sovereign of all things, and that the *savant* can say of it as the theologian said of God: "In it we live and move, and have our being."

The air is the first bond of society. Were the atmosphere to vanish into space, an eternal silence would be the lot of the terrestrial surface. We may not think of the fact with our forgetfulness of nature, but none the less the air is the great medium of sound, the liquid channel in which our words travel, the vehicle of language, of ideas, and of social communication.

It is also the first element of our bodily tissues. Breathing affords three-quarters of our nourishment; the other quarter we obtain in the aliment, solid and fluid, in which oxygen, hydrogen, nitrogen, and carbonic acid are the chief component parts. Further, the particles which are at the present moment incorporated in our organism will make their escape either in perspiration or in the process of breathing; and. after having sojourned for a certain time in the atmosphere, will be re-incorporated in some other organism, either of plant, animal, or man.

With the unceasing metamorphoses in beings and in things, there is

at the same time going on a continuous exchange between the products
of nature and the moving flood of the atmosphere, by virtue of which
the gases of the air take up their abode in the animal, the plant, or the
stone, while the primitive elements, momentarily incorporated in an
organism, or in the terrestrial strata, effect their release and help to re-
compose the aërial fluid. Each atom of air, therefore, passes from life
to life, as it escapes from death after death; being in turn wind, flood,
earth, animal, or flower, it is successively employed in the composition
of a thousand different beings. The inexhaustible source whence every thing that lives draws breath, the air is, besides, an immense reservoir into which every thing that dies pours its last breath; under its
action, vegetables and animals and various organisms are brought into
existence, and then perish. Life and death are alike in the air which
we breathe, and perpetually succeed the one to the other by the exchange of gaseous particles; thus the atom of oxygen which escapes
from the ancient oak may make its way into the lungs of the infant in
the cradle, and the last sigh of the dying man may go to nourish the
brilliant petal of a flower. The breeze which caresses the blades of
grass goes on its way until it becomes a tempest that uproots the forest-
trees and strews the shore with shipwrecks; and so, by an infinite concentration of partial death, the atmosphere provides an unfailing supply of aliment for the universal life spread over the surface of the
earth.

It is this unceasing activity of the aërial envelope of gas which forms,
nourishes, and sustains the vegetable carpet that extends over the surface of the dry land. From the meanest blade of grass to the colossal
Baobab, this rich and diversified covering draws all its sustenance from
the air.

And while it keeps up the vital circulation of the earth by incessant
exchanges of which it is the vehicle, the atmosphere is also the aërial
laboratory of that splendid world of colors which brightens the surface
of our planet. It is owing to the reflection of the blue rays that the
sky and the distant heights near the horizon assume their lovely azure
tint, which varies according to the altitude of the spot and the abundance of the exhalations; and to it also we owe the contrast of the
clouds. It is in consequence of the refraction of the luminous rays, as
they pass obliquely across the aërial strata, that the sun announces its
approach every morning by the soft and pure melody of the glowing
dawn, and makes its appearance before the astronomical hour at which

it should rise; it is owing to a similar phenomenon that, toward evening, it apparently slackens the speed of its descent beneath the horizon, and, when it has disappeared, leaves floating upon the western heights the fantastic fragments of its blazoned bed. Without the gaseous envelope of our planet, we should never have that varied play of light, those changing harmonies of color, those gradual transformations of delicate shades which lighten up the world, from the gleaming brightness of the summer sun down to the shadows which cover, as with a veil, the forest depths.

The study of the atmosphere embraces also the general conditions of terrestrial existence. The notion of life is so bound up in all our conceptions with that of the forces which we see ever at work in nature, that the myths of the early inhabitants of the world always attributed to these forces the generation of plants and animals, and imagined the epoch anterior to life as that of primitive chaos and struggle of the elements. "If we do not consider," says Humboldt, "the study of physical phenomena so much as bearing on our material wants as in their general influence upon the intellectual progress of humanity, it will be found that the highest and most important result of our investigation will be the knowledge of the intercommunication of the forces of nature, and the certainty of their mutual dependence upon each other. It is the perception of these relations which enlarges the views and ennobles our enjoyment of them. This enlargement of the view is the result of observation, of meditation, and of the spirit of the age in which all the directions of thought concentrate themselves. History teaches him who can travel back through the strata of preceding centuries to the farthest roots of knowledge how, for thousands of years, the human race has labored to grasp, through ever-recurring changes, the fixity of the laws of nature, and to gradually conquer a large portion of the physical world by the force of intelligence."

The most important result of a rational examination of nature is, that it leads one to comprehend unity and harmony in this immense assembly of things and forces, to embrace with equal ardor what is due to the discoveries of past ages and to those of our own time, and to analyze the details of phenomena without succumbing beneath their weight. It is thus that it has been given to man to show himself worthy of his high destiny, by penetrating into the meaning of nature, unveiling its secrets, and mastering by thought the materials collected by observation.

We may now contemplate our planet traveling in space, and keeping about it the aërial envelope which adheres to its surface. Our imagination can easily comprehend the general shape of this gaseous sphere which encircles the solid globe, and which is comparatively thin and of slight bulk.

The exterior surface of the atmosphere is therefore curved like that of the sea, for, like water, the external layer of air tends to a level, all points of which are at equal distances from the centre. To the eyes of novices, it seems difficult to reconcile the idea of the *spherical* surface of the ocean with what is commonly termed a *level;* the idea that the air has a horizontal level like water, and that, like an aërial ocean, this level is always tending to an equilibrium, seems at first sight somewhat obscure. Nevertheless, not only does the air possess to an unlimited degree all the properties of elasticity and mobility of a fluid seeking equilibrium, but, different in this respect from water and other liquids, it is extremely capable of compression and, consequently, susceptible of extreme expansion. These are facts which must always be kept in mind, for they will assist in the understanding of a great number of atmospheric conditions explained in future chapters of this work.

What, then, is the thickness of this gaseous stratum which envelops our globe? This is the point which we shall examine in the next chapter.

To ascertain the height to which the atmosphere extends, it would be necessary to calculate the density of the air at different elevations in the average state, leaving out of consideration accidental disturbances. This can be done when we know the temperature of the air, its pressure, and the tension of the vapor of water which it contains. It would, further, be necessary, in order to obtain an exact determination, to take account, first, of the gradual diminution in weight as the distance from the centre of the earth is increased; secondly, of the variation in the centrifugal force according to the latitude. These variations are, however, slight, and scarcely affect the calculation, in consequence of the coat of air being of such insignificant thickness as compared to the radius of the globe.

The height of the atmosphere has its limits, which, as we shall see, are somewhat confined. If the air had no elasticity, its limit would be at a distance where the centrifugal force was in equilibrium with the weight; but as this condition does not exist, its elasticity must necessarily be counterbalanced by a force of some kind, and this force is the weight of the strata of air which are above the particular one we are considering. But the higher we ascend the more rarefied does the air become, and when the last strata are reached there is nothing to keep them down. Nevertheless, the atmosphere being limited, as we shall presently see, these strata can not be lost in space; and it is probable that in consequence of their rarefaction and the great decline in their temperature, their physical condition is so modified that the elastic force becomes nil. Laplace has pointed out this indispensable condition; Poisson has specified it, by showing that the equilibrium would still be possible with a very considerable limiting density, provided that the fluid was not capable of expansion; and Biot, who has summed up these conditions, clearly indicates the state of these external inexpansible strata in his remark that they must be like "a liquid which does not evaporate." We will now examine the mechanical and physical conditions of this aërial envelope, estimate its exterior shape, and measure its height.

CHAPTER III.

THE HEIGHT OF THE ATMOSPHERE.

As the earth travels in space with enormous swiftness, carrying along with it, adhering to its surface, the gaseous body that encircles it, it naturally follows that this latter does not extend indefinitely into space, but ceases to exist at a certain distance from the surface. How far can it extend? Carried along by the rotation of the globe in its daily movement, we may conclude that at a certain height above the ground the movement of the atmosphere is so rapid that the centrifugal force which it acquires would hurl into space the outside particles of air, which would then cease to adhere to the surface and, for the same reason, to form part of the atmosphere.

Certain inventors of methods of aërial navigation have vaguely imagined that the atmosphere does not entirely turn round with the earth, so that, by rising to a certain height, we could see the globe moving around beneath our feet, and should only have to wait until the meridian, where we wished to alight, passed under the balloon, to find ourselves transported there by the rotation of the globe. Such an idea is, of course, absurd, as the atmosphere and all that it contains partake equally with the earth in the rotation of the latter.

The centrifugal force increases as the square of the velocity, and at the equator its amount is $\frac{1}{289}$ part of that of gravity, so that a body at the equator weighs less than the same body at either of the poles by $\frac{1}{289}$ of its weight. If, therefore, the earth rotated on its axis seventeen times as fast as it does, since seventeen times seventeen is equal to two hundred and eighty-nine, a body at the equator would not have any weight. A stone, for instance, detached from the ground by the action of the hand, would not fall down again; we should become so feather-like, that, in dancing upon the surface, we should resemble aërial nymphs displaced by the wind. As the circumferences of circles vary as their radii, at seventeen times the distance from the surface to the centre of the earth—that is to say, at a height of about sixteen times the radius of the earth, or about 63,000 miles—if the other quantities involved remained unchanged, the atmosphere would cease to rotate with the earth; but, in point of fact, the weight does not remain unchanged, but diminishes as the distance from the centre of attraction is increased.

By combining this diminution with the increase of centrifugal force, we find that at a distance of about 6·61 times the radius of the earth from its centre, which corresponds to a height above its surface of about 21,000 miles, the centrifugal force is equal to the weight, and consequently the aërial particles which might happen to be in these regions must of necessity escape. This is the distance at which a satellite would gravitate in exactly twenty-three hours fifty-six minutes, the time occupied by our planet in its rotation. It is, *theoretically*, the *maximum limit* of the atmosphere, which, however, as a matter of fact, is far from extending to so great a height, as we shall see; but, mathematically, it might do so, and it is only at this enormous distance that the centrifugal force would be sufficiently great to prevent the atmosphere from existing as such.

Such is the extreme and maximum limit of the atmosphere; but it is at a far lower elevation that the air we breathe really ceases. Thus, at the height of 10,000 feet—the height of Mount Ætna—there is beneath the mountaineer nearly a third of the aërial mass; at 18,000 feet, which is less than that of the peaks of many mountains, the column of air which presses upon the soil has already lost half its weight, and consequently at this point the whole gaseous mass, which reaches far up into the sky, does not weigh more than the strata which are compressed into the region below.

In consequence of the forces that act upon it, the shape of the atmosphere is not absolutely spherical, but swollen out at the equator, where it is much higher than at the poles. The maximum limit of this figure, in the case where the flattening is greatest, has been given by Laplace. The diameter of the atmosphere at the equator is a third greater than at the poles.[*] It is the mathematical limit, beyond which the terrestrial

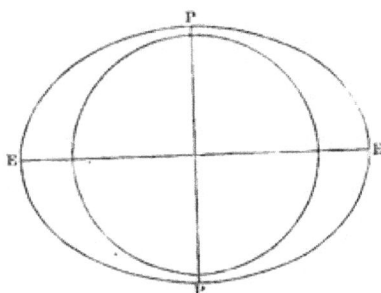

Fig. 1.—Mathematical limit of the shape of the atmosphere.

---

[*] [This is inaccurate. Laplace proved that the ratio of the least (the polar) diameter to the greatest (the equatorial) diameter could not be less than $\frac{2}{3}$ (not $\frac{3}{4}$, as in the text). Fig. 1 is

atmosphere can not pass. But it has not this exaggerated shape, though in reality it is perceptibly denser at the equator than at the poles. It may be remarked that it is probable that a detached train of the lighter gases remains constantly in the rear of the globe during its rapid revolution around the sun. It need scarcely be added that the shape of the atmosphere undergoes further change, owing to the atmospheric tides, which are due to the varying attraction of the sun and the moon.

The decreasing weight of the atmospheric strata affords us the first means of calculating a minimum limit of the height of the atmosphere. Mechanics have given us the maximum limit, and it is in this instance to physics that we shall have recourse.

Consider a vertical column of air, then the pressure at any point must be equal to the weight of air above; or, in other words, any portion of the column measured from the ground supports all the rest of the column above; the lower strata of the atmosphere are, therefore, more pressed down (and consequently denser), because they have a greater weight resting on them. The barometer, which measures this pressure of the air, is higher at the foot than at the summit of a mountain; and the relation which exists between the pressure and the height is so close, that the difference in level between two points may be deduced from the difference in the heights of the barometrical columns simultaneously placed at these two stations. The smaller the pressure the more dilated is the air; so that, at first sight, it would seem as if the atmosphere must extend to an immense distance.

A celebrated natural philosopher, Mariotte, first determined the law of the compression of gases; and the result of his researches shows that the quantity of air contained in the same volume—or, in other words, the density of the air—is proportionate to the pressure to which it is subjected. Until within the last few years this law was considered entirely accurate; but recently it has appeared most difficult to conceive why the terrestrial atmosphere does not extend very far into space; while other considerations indicate that it is necessarily limited, and

therefore incorrectly drawn, as the protuberance should be considerably greater. It may be mentioned that one consequence deduced by Laplace from his result is, that the Zodiacal light can not be produced by reflection on the atmosphere of the sun, as the former always appears in the form of a thin lens, the ratio of the polar to the equatorial diameter being much less than $\frac{2}{3}$. Laplace's investigation is given in vol. ii., pp. 194-197 of the *Mécanique Céleste* (National Edition).—Ed.]

ceases at a short distance above the ground. This apparent contradiction was the result of a too extensive generalization of Mariotte's law, which is simply relative instead of rigorously definite; and Regnault has studied the differences which exist between the theoretical law and the facts of the case.

Subsequently to these investigations, M. Linis has ascertained, by introducing very small portions of air into a large barometrical instrument made for the purpose, that the differences between the results of observation and the theory usually adopted are still greater. By diminishing sufficiently the quantity of air, it has been possible to find a limit at which the particles, far from separating from each other, as would happen were the gases capable of indefinite dilatation, seem, on the contrary, to have a mutual tendency to adhesion similar to that of the molecules in a viscous liquid. The elasticity of the air, producing expansion, ceases, therefore, at a certain degree of dilatation, from which point this gas assumes the character of a liquid, but a liquid out of all comparison lighter than those with which we are acquainted.

By means of this decrease in the density of the air in proportion to its height, Biot has, by an examination of the physical conditions of equilibrium and a complete discussion of the observations obtained at different degrees of altitude by Gay-Lussac, Humboldt, and Boussingault, demonstrated that the minimum height of the atmosphere is 160,000 feet, or about thirty miles. At that height the air must be as rarefied as beneath the exhausted receiver of an air-pump; that is to say, as rarefied as the air in the nearest approach to a vacuum that we can make.

Thus the minimum height of the atmosphere is thirty miles, and the maximum 21,000. Hence we have two defined limits, but with a great distance between them. There are, however, other methods by which we can get nearer to the truth. Efforts have been made to measure the height of the atmosphere optically, by studying the length of the twilight and the length of time during which the solar rays continue to reach the aërial regions when the luminary himself has sunk below the horizon.

If the atmosphere were unlimited, the phenomenon of night would be entirely unknown to us; the light of the sun, reaching the strata of air which are sufficiently distant from the earth, would be continuously sent on to us by reflection from these strata. On the other hand, the absence of any aërial envelope would cause the night to begin exactly

at sunset, and the light of day to burst upon us immediately the sun
rose. As it is, every one knows that the twilight of evening and the
morning dawn prolong the time during which we enjoy the solar light.
It will be readily imagined that the observation of these phenomena at
once suggested the idea of seeking to resolve, by their agency, the
height to which the atmosphere extended.

Suppose the earth to be represented by the circle, radius O A, and
that its atmosphere is limited by the circumference F G H I C. It is
evident that, when the sun has sunk beneath the horizon F A C B of the

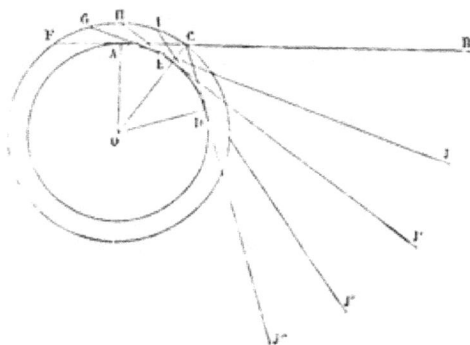

place A, it will only give
light to a portion of the
atmosphere. Thus, when
the sun arrives at J, if
we imagine a tangent
cone to the earth, hav-
ing the sun for its sum-
mit, all those parts of
the atmosphere situated
below J G will be de-
prived of light, and the
part C I H G will alone
be illuminated. Later
on, when the sun reaches

Fig. 2.—Measure of the height of the atmosphere, according to
the length of twilight.

J', the portion bounded by C I H will alone be subject to its light;
later still, only from C to I; and finally, when the sun gets to J''', upon
the tangent line from C, the intersection of the plane of the horizon
F A C B and the limiting sphere of the atmosphere, the twilight ceases.
From the moment, therefore, that the sun sets, we ought to see a sort
of arc appear on the opposite side of the horizon, rising gradually until
it reaches the zenith, and then slowly descend until it finally disappears. Such is the theory that the earliest astronomers conceived as
to the phenomenon of twilight. In the optics of Alhasen (in the tenth
century) we find that the angle of the sun's declivity for the close of the
twilight or the break of dawn was taken as 18°, and this estimate is
still adopted by modern astronomers as the average amount.

In our climate it is difficult to distinguish with accuracy the limit of
separation between that part of the atmosphere which is lighted by the
sun and that which does not receive its rays directly. But Lacaille, in
his voyage to the Cape of Good Hope, recognized all the phases which

have been enumerated theoretically. He says: "Upon the 16th and 17th of April, 1751, while at sea and in calm weather, the sky being extremely clear and serene, at the point where I could distinguish Venus at the horizon as a star of the second magnitude, I saw the twilight terminated in the arc of a circle as regularly as possible. Having regulated my watch by the exact hour, according to sunset, I saw this arc lost in the horizon, and I calculated, by the hour at which I made this observation, that the sun had descended below the horizon, on the 16th of April, 16° 38′, and on the 17th, 17° 13′."

Other observations have since been made, as we shall see further on.

It is easy to understand that, once having ascertained the apparent daily circle described by the sun upon a certain date, and the position of the observer upon the earth, we can calculate, by the time that has elapsed between the hour of sunset and the moment of the crepuscular arc's disappearance, the angle traversed by the sun below the horizon. It will also be understood that, according to the time and place, there will be found a difference both in regard to twilight and dawn, since the variations in the relative position of the sun and the state of the air must necessarily influence the direction and quantity of the light which, after countless reflections and refractions, reaches the observer.

We will study, in the second book, the optical effects of twilight; at present we are only concerned with the relation existing between its duration and the height of the atmosphere.

Now, the time during which the sun, after sinking below the horizon of a particular spot, continues to give light directly to part of the atmosphere visible from this place, depends upon the thickness of the aërial strata which envelop the earth. Let us suppose, for instance, that we pass a plane (Fig. 2) through the place A, the centre, O, of the earth and the centre of the sun; this plane will cut the earth in the circle O A. Let F A B be the intersection of the horizon of the spot A with this same plane; from C draw the tangent C D to the earth; all that part of the atmosphere visible at A will cease to be illuminated by the sun when, in its apparent diurnal movement, it has sunk below C D J‴. Now we have seen that, from the duration of twilight, it was concluded that it came to an end when the angle B C J‴ of descent below the horizon was 18°. As the angle O A C is a right angle, and as O A is the radius of the earth, we know one side and the angles of the triangle O A C, and consequently are enabled to calculate the other parts. O C

may therefore be regarded as known, and thence it results that we have
the height, E C, of the atmosphere, for E C = O C — O E.

Such is the method devised by Kepler for deducing the height of the
atmosphere from the phenomena of twilight. The results which it has
furnished agree with the preceding, and give our atmosphere a height
of from thirty to thirty-seven miles.* The average radius of the earth
being 3908 miles, it will be seen that this height is but a little more
than the 130th part of this radius; that is to say, that if the earth were
represented by a sphere about twenty-two feet in
diameter, the atmosphere would be like a coat of
vapor adhering to the surface, with a thickness of
about one inch.

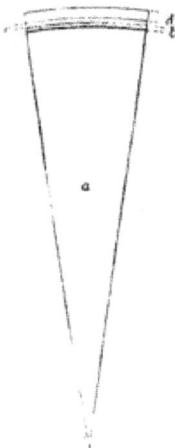

Figure 3 represents exactly this relation. It
shows — firstly, the incandescent interior of the
globe, which is a; secondly, the solid crust, b, on
which we live (it is but twelve leagues, or thirty
miles, thick, as, in consequence of the increased temperature of one degree (Fahrenheit) for fifty feet,
minerals fuse at this depth);† thirdly, the thickness
of the aërial layer which we breathe, and which is
represented by c; and, fourthly, the probable height
of a very light atmosphere, d, over and above ours,
of which we are about to treat.

Fig. 3.— Section showing
the relative thickness of
the earth's crust, of our
atmosphere, and of a
higher atmosphere.

It may be further mentioned, in reference to the
measurement of the height of the atmosphere by
the duration of twilight, that certain observers have
obtained, as the result of similar researches, an elevation much greater than that given above, affording a clear proof that
the twelve leagues actually represents the minimum only. M. Liais has
made a direct calculation of this height by observing the duration of

* [It is to be noted that different methods give different heights for the atmosphere, but
there is no discrepancy, as different things are meant. Thus, if experiments on twilight give
forty miles as the height, this implies that the air above this elevation reflects no appreciable
amount of light: while, if we define the height to be to the point where the friction will not
set light to a meteor, we have about seventy miles; but, of course, there is no reason why there
should not be some air at much greater heights. — Ed.]

† [This is the observed rate of decrease at the surface of the earth, but it is not true that
the thickness of the crust must be as stated in the text. It follows, from several considerations of other kinds, that the thickness of the crust is in all probability not less than 600
miles. — Ed.]

twilight and of the crepuscular curve, which colors the sky with that lovely rose tint which is so remarkable, especially in southern countries. These observations have been made both on the Atlantic, during a voyage from France to Rio Janeiro, and in the bay upon the shores of which the last-named city stands. They give, as a minimum, 180 miles, and, as a probable height, 204 miles.

By observing, from the summit of the Faulhorn, the course of the crepuscular arcs, Bravais obtained a height of seventy-one and a half miles. The height, however, varies according to the temperature of the seasons, and remains always greatest at the equator. Another method, different from the preceding, consists in measuring the thickness of the penumbra which surrounds the earth's shadow on the moon during lunar eclipses, as well as the phenomena of refraction produced. This measurement gives from fifty to sixty miles as the thickness of the terrestrial atmosphere, the influence of which is felt under this special aspect.

The observations which accord the atmosphere a height far greater than the theoretical thirty-eight miles have been for many years the object of special discussion. Quételet, director of the Brussels Observatory, has, after much research on this head, arrived at the conclusion that it does indeed extend much higher than had been supposed, but that the upper strata are not quite of the same nature as those nearer the earth.

This addition is supposed to be due to an *ethereal* atmosphere, very rarefied and differing from the *lower* atmosphere in which we live. It is the region where are mostly seen the shooting stars, which afterward disappear when they reach the terrestrial atmosphere.

The upper atmosphere* is still, the lower in continual motion. The special movements caused by the action of the winds and tempests are limited in their height by the effect of the seasons. Thus, as regards our climate, the agitated portion, in the vicinity of the earth, would not be more than from seven to ten miles high during the winter, while its height must be almost double in summer. All that part of the atmosphere which is above the latter would only experience a very slight and scarcely sensible movement, arising from the movable basis upon which it reposes.

The continual disturbances going on in the lower regions cause the air in the inferior atmosphere to be very much alike in its chemical

* [The existence of such an atmosphere seems to me very uncertain.—ED.]

components. No difference has been discovered at the various elevations which it is possible to attain for the purpose of collecting air and
submitting it to analysis.

In the upper atmosphere the phenomena, of which we are scarcely
able to form an idea by judging them from the surface of our globe,
take place. There, also, appear the shooting stars; descending from
a still greater height, the aurora borealis, and those mighty luminous
phenomena which we often witness without having the power to sub-
mit them directly to the test of experiment. All these facts do not escape us altogether, especially as regards the aurora borealis and the
magnetic phenomena. If we can not determine the cause, we can at
least feel the effect with sufficient force to be in a position to appreciate them.

Sir John Herschel, De la Rive, and Hansteen seem to share upon this
point the opinion of Quételet. We can quite admit that, above our atmosphere of oxygen, nitrogen, and vapor of water, there exists an atmosphere excessively light, which may extend two hundred miles in
height, and which is naturally composed of the very lightest gases.

The terrestrial globe being about 8000 miles in diameter, this total
thickness represents the fortieth of the globe's diameter. The simultaneous existence of these two atmospheres is, therefore, the general conclusion at which we will, momentarily at least, stop.

As to the basis of the atmosphere, we may now inquire if it ceases
at the surface of the ground, and does not descend into the interior of
the globe itself.

Pressing upon all bodies upon the surface of the earth, it tends to
penetrate in all directions between the molecules of liquids as into the
interstices of the rocks. It is to be found in water as in all vegetables
and all organic structures; the earth and the porous stones are impregnated with it, and that in proportion to the force with which it presses.
It will be seen, therefore, that the air is not limited to the part which
is, so to speak, a gaseous envelope, and that a sensible fraction of its
constituent elements penetrates the waters of the ocean and the interstices of the ground. Certain *savants* have imagined that the air of
which the atmosphere is composed is but the continuation of an interior atmosphere; but the rise in the temperature, due to the central
heat, would prevent the condensation of gases, and must limit the presence of air in the under strata.

A rough estimate of the quantity of air which is thus introduced into the waters of the ocean may be formed by measuring the absorption of gases by various liquids. Under ordinary pressure, sea-water absorbs from two to three per cent. of its volume, only the proportion of oxygen is much greater than in the ordinary air. The result of the calculation is, that the quantity of air absorbed by the ocean is not above a three-. hundredth part of the atmosphere.

We thus have a tolerably complete determination both as to the height and shape of this terrestrial atmosphere.

CHAPTER IV.

WEIGHT OF THE TERRESTRIAL ATMOSPHERE—THE BAROMETER AND ATMOSPHERIC PRESSURE.

WHILE treating of the height of the atmosphere, we have already seen that the air is denser in the lower regions of the aërial ocean—that is to say, near the surface of the earth—than in the higher regions. The air, light and unsubstantial as it may appear to us to be, has consequently a positive weight. Each square foot of the earth's surface sustains a considerable pressure, the amount of which we shall presently attempt to estimate, corresponding to the height and density of the column of air above it.

Our ancestors were not able to *measure* the atmospheric pressure; but we must not conclude from this that they were ignorant of the effects which it exercised, especially when the wind was violent. Yet this force, which every one felt without being able to measure, was not rendered determinate until the middle of the seventeenth century.

In 1640, the Grand Duke of Tuscany having ordered the construction of fountains upon the terrace of the palace, it was found impossible to make the water rise more than thirty-two feet. The duke wrote to Galileo in reference to this strange refusal of the water to obey the pumps. Torricelli, the pupil and friend of Galileo, gave the true explanation of the fact, and proved, as we shall see, that this column of water of thirty-two feet was in equilibrium with the weight of the atmosphere.

The celebrated invention of Torricelli has sometimes been erroneously attributed to Pascal. The French philosopher himself alludes to the mistake, and shows how much of the merit is due to him in the following terms: "The report of my experiments having been spread abroad in Paris, they have been confounded with those made in Italy; and, thanks to this misunderstanding, some, according me an honor to which I can lay no claim, attributed the Italian experiment to me, while others unjustly deprived me of the credit of those to which I was really entitled. To give to others and to myself the justice due to us, I published, in 1647, the experiments which I had made the year before in Normandy; and that they might not be confounded with one made in Italy, I gave the latter separately and in italics, whereas mine were printed in Roman letters. Not content with giving it these distinctive marks, I have stated in so many words that I am not the inventor of the barometer; that it was made in Italy four years previously, and was the cause of my making similar experiments."

It was, then, the refusal of the water to rise more than thirty-two feet, in obedience to the pumps, which revealed to Torricelli the fact that the atmosphere had weight, and that its whole weight was balanced by a column of water thirty-two feet in height. Let us then examine for a moment the mechanism and action of the pump.

Every one knows that these simple and old-fashioned contrivances serve to raise water either by suction or pressure, or by both combined. Hence their classification as *suction-pumps*, *forcing-pumps*, and *suction and forcing pumps*. Before Galileo's day, the ascension of water in the suction-pump was ascribed to the fact of nature abhorring a vacuum; but it is, in reality, merely an effect of atmospheric pressure.

Take a tube, at the lower extremity of which is a piston, and place this lower end in water. If the piston is drawn up, a vacuum is created below, and the atmospheric pressure, acting upon the surface of the liquid external to the pump, makes it rise in the tube and follow the movement of the piston.

Herein lies the principle of the suction-pump, which is essentially composed of the body of the pump, in which a piston moves, communicating by a tube with a reservoir of water (see Fig. 4). At the point where the body of the pump and the suction-tube join is placed a valve, opening upward, and in the body of the piston there is an opening formed by a similar valve.

For water to reach the body of the pump, the suction-valve must be less than thirty-two or thirty-three feet above the level of the water in the well, otherwise the water would cease to rise at a certain point in the tube,

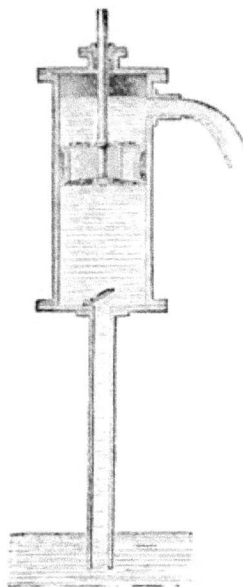

Fig. 4.—Suction-pump.

and the motion of the piston would be unable to raise it any farther.

In addition, to insure raising at each ascent of the piston a volume of water equal to the volume of the body of the pump, the spout must be placed at a less height than thirty-two feet above the reservoir.

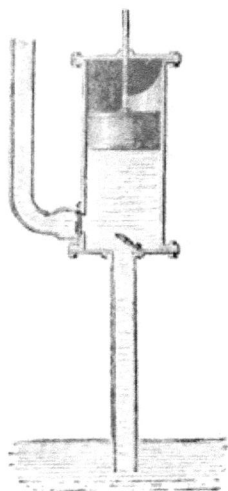

Fig. 5.—Suction and forcing pump.

Thus the suction-pump will not raise water to a height of more than thirty-two feet; but the water having once passed above the piston, the height to which it can then be raised depends solely upon the force which drives the piston.

The suction and force pump (see Fig. 5) raises water both by suction and pressure. At the base of the body of the pump, over the orifice of the suction-pipe, is, as before, a valve opening upward. Another valve, also opening upward, closes the aperture of the bent tube, which runs into a receptacle called the air-vessel.* Then from this reservoir there starts a pipe which serves to raise the water to the required height. Finally, the *force-pump* only acts mechanically, and does not utilize atmospheric pressure. It differs only from the other in that it has no suction-pipe, its body going right into the water which is to be drawn up.

In reference to this elevation of the water only to a certain height, Torricelli, throwing aside, like his master, all idea of a hidden cause, *showed that the pressure of the air compels the water to mount up into the pipe from which the air is withdrawn*, until the weight of water raised into the pipe is equivalent to that of the air which presses upon an equal section of the reservoir from which the water is being raised. By the aid of this principle he was led to invent the barometer. To exercise equal pressures, the liquid columns must be of heights inversely proportional to their density. Thus, a liquid twice as heavy as water would, with a column of sixteen feet, be in equilibrium with the atmosphere; and quicksilver, which is nearly thirteen and a half times as heavy as water, would be in equilibrium if the height of the column were diminished in this proportion—that is, to about twenty-nine inches.

* [The air-vessel is not essential to the principle of the pump: if it were not used the supply of water would be intermittent, as in the common suction-pump, but the effect of the elasticity of the air in the air-vessel is to render the stream of water continuous.—ED.]

Fig. 6.—Torricelli inventing the Barometer.

This conclusion is easily verified. Take a glass tube, three feet in length, and open only at one end; fill it with quicksilver, and then, placing the finger on the open end (see Fig. 7), put the lower portion of the tube into a basin filled with the same liquid, with the end closed by the fin-

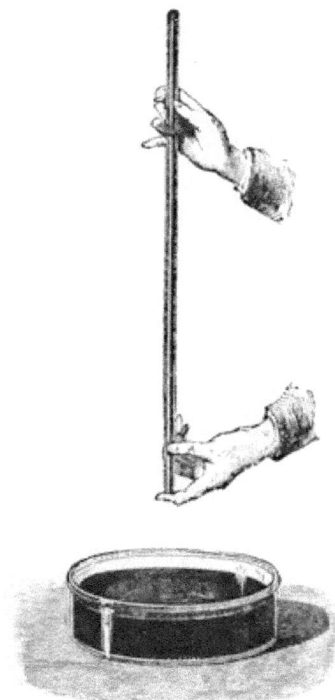

Fig. 7.—The tube full of quicksilver. Fig. 8.—The tube in the basin.

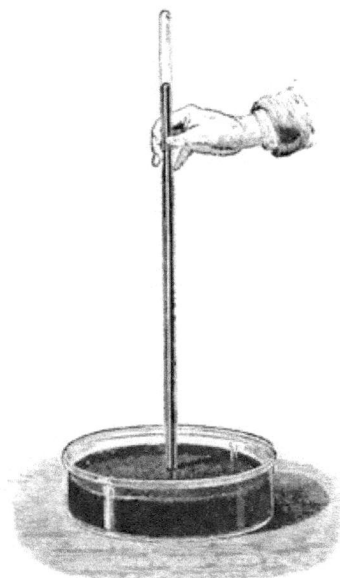

ger downward. Immediately the finger is removed, the quicksilver inside will descend several inches and then stop (see Fig. 8). The equilibrium is established, and the liquid column which remains suspended in the pipe is a true balance, for the weight of the column of mercury is exactly in equilibrium with the atmospheric pressure.

Torricelli gave to this tube of quicksilver, thus placed vertically in a basin of quicksilver, the name of Barometer; that is to say, a contrivance to indicate the weight of the air, from the Greek βάρος, weight, and μέτρον, measure. Its invention by Torricelli dates from 1643. Three years later, Pascal repeated the experiment in France with a water-barometer, and even a wine-barometer. This was at Rouen. His tube was forty-nine feet long, and to avoid the difficulty, insurmounta-

ble in that day, of exhausting the air in it directly, he had it sealed at
one end, filled it with wine, and closed the other end with a cork.
Then, by means of cords and pulleys, the tube was placed upright and
the lower end put into a vessel full of water. As soon as the cork that
kept it closed was removed, the whole liquid column in the tube fell,
until its surface was about thirty-three feet above the level of the water
in the vessel. The remaining sixteen feet above were destitute of air.
Consequently, the liquid column itself formed an equilibrium to the atmospheric pressure, and from this he drew the conclusion that a column
of water (or of wine of the same density) thirty-two feet high weighs as
much as a column of air on the same base.

The surface of the earth is pressed upon as if it was covered with
a body of water thirty-two or thirty-three feet deep, and we who live
upon the bed of this ocean of air undergo the same pressure.

If it is the pressure of the air which causes the elevation of the quicksilver or the water, as we ascend into the atmosphere, the weight of the
column of quicksilver raised, and consequently the height of this column, must gradually diminish in a manner dependent on the strata of
air left beneath it. The experiment was made on the Puy-de-Dôme, according to the instructions of Pascal, by his brother-in-law, Florin Pé-
rier, upon the 19th of September, 1648, and repeated by Pascal himself
on the Tour St. Jacques at Paris. The results were decisive, and the
barometer became an easy and accurate means of measuring the total
weight of the atmosphere, and the variations in the pressure which it
exerts at different times and places upon the surface of the globe. We
thus see that it was between 1643 and 1648 that the atmospheric pressure was demonstrated by the construction of the barometer and the experiments which its discoverers at once entered upon.

By a coincidence not at all unusual in the history of science, while
the indications of the barometer were being studied in Italy and
France, experiments were being made in Holland to ascertain the precise weight of the air, but by quite a different process.

In 1650, Otto de Guéricke, burgomaster of Magdeburg, invented the
air-pump, by which the air may be exhausted from any receptacle and
a nearly absolute vacuum created.

The ingenious inventor conceived in the same year the idea of weighing a globe of glass, first leaving in it the air which it contained, and
then weighing it again when the air had been removed by the air-
pump. The globe, when emptied of air, was found to be less heavy

by about one-third of a grain for every cubic inch of the globe's capacity.

Aristotle had long before suspected that air had weight, and to make sure of the fact, he weighed a leather bottle, first empty and afterward when inflated with air; for, he remarked, if the air has weight, the leather bottle will be heavier when weighed the second time than it was the first time. The experiment not confirming his supposition, he concluded that the air had no weight. Nevertheless, several of the ancient philosophers admitted the material nature of air as a fact. Thus the Epicureans compared the effects of the wind with those of water in motion, and considered the elements of the air as invisible bodies. During the reign of the peripatetic philosophy, however, it was assumed that air was without weight, and there were but few philosophers who did not share this erroneous opinion.

We have seen that, by repeating judiciously the experiment of Aristotle, Otto de Guéricke demonstrated the real weight of air. If Aristotle's experiment led to a contrary result, it must be attributed to the change in the volume of the leather bottle during his two trials, for every body, when weighed in a fluid, loses in weight a quantity equal to the weight of the fluid displaced. The leather bottle made use of by Aristotle would have shown an increase of weight if weighed in a vacuum. Let us suppose that about 1835 cubic inches of air were introduced into it by inspiration; its weight would have increased by about 550 grains, but at the same time the bottle would become inflated, and its volume, being increased by 1835 cubic inches, would have displaced a volume of air of equal weight, so that its loss in weight would be also 550 grains, and the weight of the air and bottle together would consequently remain the same as before. But in the experiment of Otto de Guéricke the globe was always of the same size, whether empty or full of air, and its

Fig. 9.—Otto de Guéricke's experiment.

loss in weight through the displacement of the air being in each case the same, there was, of course, a difference, which proved that air had weight. Otto de Guéricke, at the same time, conceived the idea of the Magdeburg Hemispheres, so called from the town in which he invented them, and which consist of two hollow hemispheres of copper, with a diameter of from four to five inches. The hemispheres fit each other hermetically. One of them has attached to it a cock that screws on to the plate of an air-pump, and the other a ring which acts as a handle to move it backward or forward. As long as the two hemispheres, when in contact, contain air within them they can easily be separated, for there is equilibrium between the expansive force of the interior air and the outside pressure of the atmosphere, but when once a vacuum is formed by the exhaustion of the air, it requires a considerable effort to draw them apart.

In one of these experiments, the learned burgomaster had each hemisphere pulled by four strong horses without succeeding in separating them. The diameter was more than two feet, which gives a total of more than three and a quarter tons as the atmospheric pressure brought to bear in the way of resistance.

The pressure of the atmosphere on a square inch is equivalent to the weight of a column of quicksilver with a volume of 29·92 cubic inches, viz., about fifteen pounds.

Fig. 10.—The Magdeburg Hemispheres.

It is easy and interesting to draw from this the conclusion that, as the superficies of an average human body is sixteen square feet, we may each of us be said to be subject to a pressure of about fifteen tons.

That we are not crushed by this enormous pressure, is because it does not all press vertically down on us. As the air surrounds us on all sides, its pressure is transmitted over our body in all directions, and, in consequence, becomes neutralized. Air penetrates readily and with full pressure into the profoundest cavities of our organism; hence we have the same pressure inside and outside, and thus these weights become exactly balanced. This is easily proved by the experiment of bursting a bladder under the receiver of an air-pump. Take a cylindrical glass vessel, hermetically closed at the upper end by a piece of gold-beater's skin, with the other end placed (see Fig. 11) on the plate of an air-pump; as soon as the air begins to be exhausted from the vessel, the gold-beater's skin becomes depressed under the influence of the atmospheric pressure upon it from above, and soon bursts. The opposite result occurs if the pressure from outside is lessened. If a bird is placed in the vacuum of an air-pump, its body will be seen to swell, its blood to spurt out with violence, and in a short time it perishes, a victim to a kind of explosion the inverse of that just described.

Fig. 11.—Atmospheric pressure: rupture of equilibrium.

This fact is confirmed, as we shall see farther on, by the ascents that have been made to great elevations. Upon reaching the regions where the air is much rarefied, the limbs swell, and the blood has a tendency to force its way through the skin, in consequence of the want of equilibrium between its own tension and that of the external air.*

Any one can show the effect of atmospheric pressure by a very simple experiment. This consists in filling a glass quite with water and laying over the top a sheet of paper. It can then be turned over without spilling any of the liquid, a fact which must be attributed to the pressure which the atmosphere exercises upon the sheet of paper.

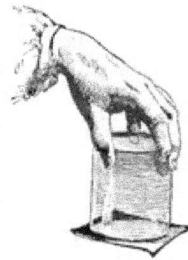

Fig. 12.—Atmospheric pressure under an inverted glass.

It was stated above that, where a vacuum is created, the atmospheric pressure is about fifteen pounds to the square inch. It is this pressure which causes the limpet to adhere to the rock, when this mollusk has by contraction created a vacuum under its shell. The fly, excluding the air from between its feet and the ceiling, is enabled, apparently, to violate the laws of gravity. Cupping-glasses, when applied to the body, act on this same principle, and we can not take a step without observing some fact which is founded on the effects of atmospheric pressure. Such are the general facts and experiments which demonstrated that the air had weight, and gave birth to the instrument wherewith this weight was to be determined, viz., the barometer. It now remains to apply these ideas to the whole atmosphere, the extent of which we endeavored to explain in the preceding chapter.

* [I have neither experienced any of these symptoms myself, nor have I observed them in others.—Ed.]

At the level of the sea the pressure, upon the average, sustains the barometrical column at a height of about 29·92 inches.

Experiments frequently repeated by physical philosophers—and the accuracy of which has been verified—have proved that the weight of the air at 32° (Fahr.) of temperature, and under a pressure of 29·92 inches of mercury, is to the weight of an equal volume of quicksilver in the proportion of unity to 10,509—that is to say, that 10,509 cubic inches of air have the same weight as one cubic inch of mercury. If the density of the strata of air were everywhere the same, it would be easy to deduce from the above result not only the height of a given spot by the aid of the barometer reading there, but also the total height of the atmosphere. It is, indeed, evident that if a fall of an inch in the height of the barometer corresponded to a change of height of 10,509 inches, a fall of 29·92 inches, which is the total height of the barometer, would correspond to 29·92 times 10,509 inches—that is, about five miles. Such would be the height of the atmosphere if its density remained the same from top to bottom, but we have seen that its lower strata are denser than the higher. It follows, therefore, that, to procure a fall of an inch in the mercury of the barometer, it is necessary to traverse a greater distance above the level of the ground or the sea.

Halley was the first to deduce a formula by which heights might be obtained by means of the barometer.

We have seen in the previous chapters that, since the experiments of Mariotte, it has been recognized that air becomes compressed in proportion to the weight above, or to the pressure exerted upon it. Thence it is inferred that, in rising vertically in the atmosphere to successive elevations, increasing in arithmetical progression, the density of the corresponding strata of air would diminish in geometrical progression. This would be accurate if the temperature were everywhere the same, and the difference in height would scarcely be any more complicated than if the density were constant. But the temperature of the air diminishes with increased height, so that the variation in density is not so simple, as the upper strata are more condensed by their lower temperatures than those below.

The relation between temperature and height is rather complicated, as we shall see farther on; and this, of course, renders more difficult the process of measuring heights by the barometer. At the same time, the atmospheric strata always contain a certain quantity of aqueous vapor, the weight of which must be added to that of the air.

Furthermore, the weight of any body, and consequently that of a stratum of air, is proportionately less as the body in question is farther removed from the centre of the earth. And as the weight of bodies varies also according to the latitude on account of centrifugal force, it becomes evident that, for a single formula to be in general use for observations made at different points of the globe, it is indispensable that it should include the latitude of the place of observation.

Laplace has given, in the "Mécanique Céleste," the corrections rendered necessary by these different causes in measuring height, and has deduced from theory alone a formula the accuracy of which has been confirmed by numerous experiments.

To determine the height of a mountain it is necessary that two persons take simultaneous observations of the readings of the barometer, one at its foot, the other at its summit. They must be careful, at the same time, to read the thermometers attached to the barometers, as well as others to determine the temperature of the surrounding air. Two observations will be sufficient; but it is better to have several.

A single observer can also ascertain the difference in level between two stations, not very distant the one from the other, with very fair accuracy, if he takes care to observe the thermometer and barometer at the lower stations, both when he leaves it and returns to it, and infers, from the difference, the reading at the lower when taking that at the higher station.

When, by a long series of observations, the average readings of the barometer and thermometer at a given place have been determined, they may be employed to calculate the absolute elevation of the place above the level of the sea by taking corresponding observations at the level of the ocean. Sufficient barometrical observations have already been made at various elevations for us to be in a position to represent this decrease of atmospheric pressure, with increase of elevation, no longer theoretically, but from direct observation.

From a series of observations, made at very different elevations, the table on the following page has been formed.

This satisfactory series of barometrical observations, which we are able to establish by means of numerous ascents, either in the balloon or up the mountain path, and by researches of several observers in inhabited regions far above the level of the sea, enables us also to endeavor to represent, by a curve and a tint, this rapid decrease in the weight of the atmosphere. In Fig. 13, the horizontal line which forms

4

	Height above the Sea.	Mean Reading.
	Feet.	Inches.
Level of the Ocean...	0	29·92
Mean barometric reading at Greenwich Observatory..........	159	29·74
" " Paris "	213	29·68
" " Strasburg "	472	29·57
" " Toulouse "	650	29·37
Dijon (Perres)...	804	29·21
Geneva Observatory (Plantamour)............................	1,339	28·58
Rodez (Blondeau)...	2,067	27·91
Summit of Vesuvius (Palmieri)................................	3,937	25·98
Guatemala (R. P. Camudas)...................................	4,856	25·24
Guanaxuato (Humboldt)..	6,837	23·62
The Monastery of the Great St. Bernard.....................	8,150	22·17
The Summit of the Faulhorn (Bravais)......................	8,773	21·85
Town of Quito (Fompé)...	9,541	21·02
Summit of Ætna (Elie de Beaumont).........................	10,866	20·08
In several aeronautical ascents (Flammarion)..............	13,124	18·70
Summit of Mont Blanc (Ch. Martins)........................	15,748	16·69
On the Chimborazo (Humboldt and Bonpland)..............	20,014	14·17
The summit of Ibi-Gamin (the highest mountain that has been climbed) (Schlagintweit)......................................	22,113	13·39
In an aeronautical ascent (Gay-Lussac).....................	22,966	12·79
" " (Bixio and Barral)......................	22,966	12·60
In several aeronautical ascents (Glaisher).................	26,247	10·79
In an aeronautical ascent (Glaisher)........................	29,000	9·75
In the highest ascent (Glaisher).............................	37,000	7·00

the base represents the mean state of the barometer at the level of the
sea (29·92 inches). Each other horizontal line indicates the reading of
the barometer corresponding to the elevation which is shown by the
vertical line. In this way, or by the aid of the tinted portion, it will
be noticed that at 8200 feet the pressure is diminished by one-quarter,
at 18,000 feet by one-half, and at 31,168 feet by three-quarters.

The reading of the barometer diminishes, therefore, rapidly as we
rise above the level of the sea. But even there it is not the same all
over the globe's surface. It is lower at the equator than between the
tropics; at the equator it is about 29·84 inches; it then increases up to
the 33d degree of latitude, where it is 30·16 inches; then decreases until the 43d degree (30·00 inches), toward which point it becomes stationary, and so remains up to the forty-eighth degree. Thence it continues to decrease so far as sixty-four degrees, where it stands at 29·65
inches. Lastly, it again increases from that point as far as the remotest
latitudes—at Spitzbergen (seventy-fifth degree), where the height of the
barometer is 29·84 inches. Between the pressures at the thirty-third
degree and the sixty-fourth degree of latitude, there is, therefore, a difference of half an inch. I have laid down these results on a diagram,
and traced the following curve (see Fig. 14, p. 52):

These variations in the atmospheric pressure are probably caused by the trade-winds and upper currents of air, which slightly raise the whole mass of the atmosphere.

It is easy to conceive that the latitude may exercise some influence upon the pressure of the air, inasmuch as the conditions of temperature, pressure, and rotary movement vary with it. It is less easy to explain

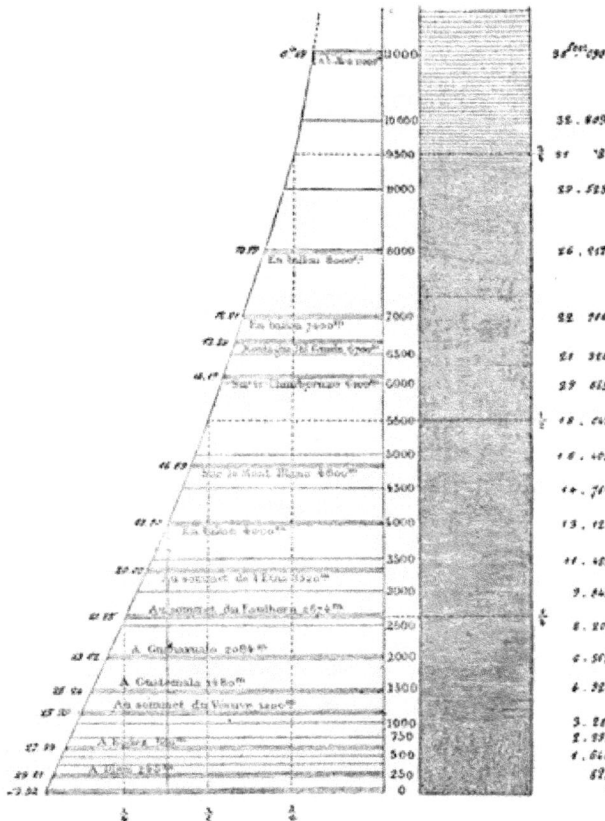

Fig. 13.—Diagram showing the decrease of atmospheric pressure, according to height.

why the longitude should exercise any, but it seems, nevertheless, to do so. In the same latitude, the average pressure of the atmosphere is 0·14 inch greater in the Atlantic than in the Pacific Ocean.

The readings of the barometer are continually changing; but, notwithstanding this, by a careful determination of the mean atmospheric pressure at many places, a map showing the lines of equal barometrical pressure (isobaric lines) can be drawn over the surface of our planet.

The lines of equal pressure—or isobaric lines, as they are technically termed—are at first pretty equally distributed from N. to S., running from W.S.W. to E.N.E. The isobaric line of 29·96 inches passes through the south of England and Holland; that of 30·02 inches near Tours and Nancy; but the centre of France shows a very remarkable

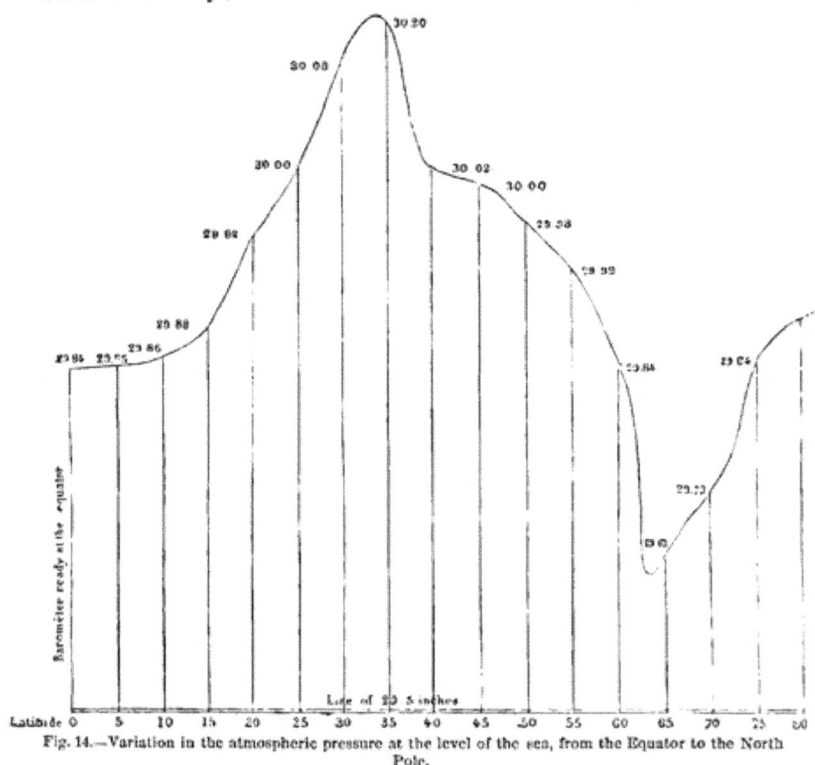

Fig. 14.—Variation in the atmospheric pressure at the level of the sea, from the Equator to the North Pole.

line of pressure, for the isobaric line of 30·04 inches crosses France diagonally, passing close to Strasburg, Chaumont, Dijon, Clermont, and Toulouse. On the other side, toward the S.E., the pressure diminishes, and attains a minimum not less remarkable in the Gulf of Genoa, where the pressure is about 29·98 inches.

The curve of 30·00 inches is formed, and its path pretty well known, in consequence of the numerous points at which observations have been made. The isobaric line of 30·08 inches, which passes close to Oran, and somewhat farther from Algiers, necessarily continues toward the west, nearly parallel with the above. A maximum of pressure in the Atlantic is in thirty-five degrees of north latitude; a minimum of pressure is met with at five degrees north of the equator; a maximum at sixteen degrees south latitude, near St. Helena; and the lowest pressure existing in the world is to the south of Cape Horn, where it does not exceed 29·33 inches. Upon the Asiatic continent the distribution is quite different, and Siberia shows a maximum of about 30·24 inches between Nertchinsk and Bernaoul.

The chief difficulty in calculating altitudes is in reference to the mean level of the sea. Equilibrium upon the surface of the sea is not absolute; its level is affected by various causes, such as centrifugal force in the zone of the equator, the wind, barometrical pressure, and temperature. To these may be added the configuration of the sea-board, which gives a varying effect to the action of the winds and tides. It is well known that the sea rises quicker than it recedes, and when the gulfs are landlocked this effect is more decided. Along the coast the sea must rise higher than it does farther from shore.

The level of the sea at Marseilles is 31·5 inches lower than the average level of the ocean upon the French coast. The Mediterranean must be an inclined plane, falling from the Straits of Gibraltar to the coast of Syria. The last level taken in Egypt, from the Mediterranean to the Red Sea, showed that the latter is higher than the Mediterranean. It is easy to comprehend that these seas, receiving much less water than evaporates from them, must have a tendency to become shallow, and that they are only kept up by the straits that unite them with the ocean.

This first general description of the weight of the air and its pressure upon the spherical surface of the globe will answer our present purpose. It explains in some degree the statics; and we shall soon reach the dynamics. The atmosphere is unceasingly in motion, with its displacements, horizontal, vertical, and oblique. From this cause it results that the weight of air upon a given place, or the height of the barometer, is always changing. Solar heat gives rise to regular *diurnal* and *monthly* variations, the intensity of which differs according to the latitude. The change in the position of the great currents gives rise to extensive variations upon a vast scale. Changes of weather are heralded by these fluctuations, which are bound up with the general pressure.

Under the title of "*Combien pèse la masse entière de tout l'air qui est au*

Monde," Pascal wrote, at the epoch when he devoted himself to his celebrated experiments on atmospheric pressure, a small treatise as simple as it is curious, the first sketch of all that has since been written on this subject, and containing from the outset the absolute reply to the question which forms its title. "We learn," he says, "by these experiments that the air which is over the sea-level weighs as much as water to a height of thirty-two feet; but inasmuch as the air weighs less over more elevated places, and consequently does not press equally over all points of the earth alike, it is impossible to measure exactly what is the pressure upon all parts of the world by the same process, although an approximate measure, very nearly accurate, may be taken. Thus, for instance, it may be assumed that all the places of the earth have as much pressure upon them as if there was a depth of rather more than thirty-two feet of water over them; and it is certain that this supposition is not half a foot in error.

"Now we have seen that air which is above the mountains 3000 feet high is as heavy as water to a height of twenty-nine feet. Consequently, all the air which extends from the level of the sea to the summit of the mountains weighs nearly the seventh part of the whole atmosphere.

"We gather, too, from this, that if the whole sphere of the air was compressed against the earth by a force which, driving it downward, reduced it to so small a space that it became of the density of water, it would then be only thirty-two feet high. The whole mass of air may be regarded as if it had been formerly a mass of water, thirty-two feet deep, which had become rarefied and very much dilated, and converted into the state which we call air; whereas it occupies, in truth, more space, though it preserves exactly the same weight.

"And as nothing would be simpler than to calculate what would be the weight in pounds of water surrounding the earth to a depth of thirty-two feet, we should find, by the same means, the weight of the entire mass of air.

"Curiousity led me to make this calculation, and I found that the weight of this mass of water would be about nine trillions of pounds—that is, nine followed by eighteen ciphers represents the weight, in pounds, of air surrounding the earth."

This weight is about $\frac{1}{1150000}$ part of the weight of the earth.

If all this mass of air were agglomerated into a single ball, it would weigh as much as a ball of copper with a diameter of sixty-two miles. Thus the weight of the air is far from being insignificant.

Fig. 15.—Lavoisier analyzing Atmospheric Air.

CHAPTER V.

CHEMICAL COMPONENTS OF THE AIR.

It is to the great French chemist Lavoisier that science owes the discovery of the chemical components of the air.

Let us go back to the researches of this laborious observer, and hear
from his own lips the recapitulation of his interesting studies.

Our atmosphere, he remarks, must be made up of all the substances
capable of remaining in an aëriform state at the ordinary degree of temperature and atmospheric pressure which we experience. These fluids
form a mass, almost homogeneous,* from the surface of the earth to the
highest elevation which man has ever reached, and the density of which
decreases with elevation. But it is possible that above our atmosphere
there are several strata of very different fluids.

What is the number, and what is the nature, of the elastic fluids
which compose this lower stratum that we inhabit?

After having established the fact that chemistry offers two methods
essential for the study of *bodies*—that is to say, analysis and synthesis—
Lavoisier describes as follows the celebrated experiment of the first
analysis of air :

"Taking a vessel, or long-necked tube, with a bell or globe at its extremity, containing about thirty-six cubic inches (see Fig. 16, p. 58), I bent
it (see Fig. 17, p. 58) so as to place it in the furnace while the extreme end
of the neck was under a glass cover, which was placed in a basin of mercury. Into this vessel I poured four ounces of very pure mercury ; and
then, by means of a siphon, I raised the mercury to about three-quarters
the height of the glass cover, and marked the level by gumming on a
strip of paper. I then lighted the fire in the furnace, and kept it up incessantly for twelve days, the mercury being just sufficiently heated to
boil. At the expiration of the second day, small red particles formed
upon the surface of the mercury, and increased in size and number for
the next four or five days, when they became stationary. At the end

* [*Homogeneous* must be understood to mean that the components of the atmosphere are
found mixed in the same proportion at all heights. Its usual meaning is, of course, "of uniform density."—ED.]

of the twelve days, seeing that the calcination of the mercury made no
further progress. I let out the fire and set the vessels to cool. The volume of air contained in the body and neck of the vessel before the operation was fifty cubic inches; and this was reduced by evaporation to
forty-two or forty-three. On the other hand, I found, upon carefully
collecting the red particles out of the melted mercury, that their weight
was about forty-five grains. The air which remained after this operation, and which had lost a sixth of its volume by the calcination of the
mercury, was no longer fit for respiration or combustion, as animals
placed in it died at once, and a candle was extinguished as if it had been
plunged in water. Taking the forty-five grains of red particles, and
placing them in a small glass vessel, to which was adapted an apparatus
for receiving the liquids and aëriform bodies which might become sepa-

Fig. 16.—The glass vessel.   Fig. 17.—The apparatus.

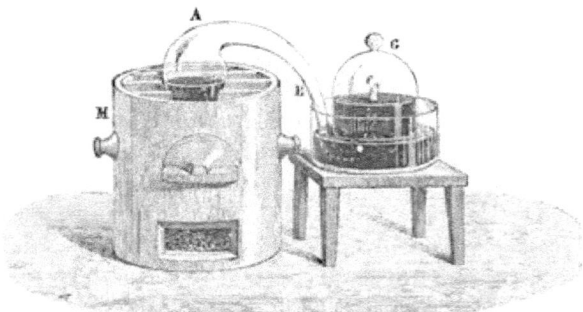

rated, and having lighted the fire in the furnace, I observed that the
more the red matter became heated, the deeper became its color. When
the vessel approached incandescence, the red matter commenced to become smaller, and in a few minutes had quite disappeared; and at the
same time forty-one and a half grains of mercury became condensed in
the small receiver, and from seven to eight cubic inches of an elastic
fluid, better adapted than the air of the atmosphere to supply the respiration of animals and combustion, passed under the glass cover. From
the consideration of this experiment, we see that the mercury, while it
is being calcined, absorbs the only portion of the air fit for respiration,
or, to speak more correctly, the base of this portion; and the rest of the
air which remains is unable to support combustion or undergo respiration. Atmospheric air is, therefore, composed of two elastic fluids of
different, and even opposite, natures."

The nature of air was thus clearly established by these experiments, which were made in 1777. Its real components were not, however, completely ascertained until the present century. The first exact analysis of air is scarcely fifty years old, and is due to Gay-Lussac and Humboldt, who analyzed it by the use of the eudiometer.

Fig. 18.—Mercury-Eudiometer, for analyzing air.

When an equal mixture of air and pure hydrogen are set fire to in the eudiometer, all the oxygen disappears in the shape of water, which becomes condensed into dew, the volume of which is insensible, and there remains a mixture formed of nitrogen and the excess of hydrogen employed. Now the hydrogen causes a volume of oxygen equal to half itself to disappear as water; whence it follows that the volume of oxygen contained in the measured air is equal to one-third of the volume that has disappeared. If the measures of the air, the hydrogen, and the gases after explosion, are made at the same pressure and the same temperature, and if, in addition, the gases were saturated with humidity before explosion, the determination would require no correction. Such is the principle of the method. Gay-Lussac and Humboldt found that there was twenty-one per cent. of oxygen, and seventy-nine per cent. of nitrogen, in the air. This analysis has since been confirmed by nearly all chemists. There is another method by means of which the relative quantities of oxygen and nitrogen contained in the air of the atmosphere can be *weighed*—a process which gives results far more accurate than the measuring of the volumes (al-

ways very small) of the gases employed in the other processes. The
apparatus used is composed—first, of a tube which brings in the air
from outside of the room where the operation is proceeding; secondly,
of a set of Liebig balls, L, containing a concentrated solution of caustic
potash; thirdly, of a tube, f, in the shape of the letter U several times
repeated, and filled with fragments of caustic potash; fourthly, of a
second set of balls, o, containing concentrated sulphuric acid; fifthly,
of a second tube, l, of the same shape as the one above mentioned,
filled with pumice-stone steeped in concentrated sulphuric acid; sixth-
ly, of a straight tube, T, of hard glass. This tube is filled with copper
filings, and laid upon a long iron furnace, so that it can be heated

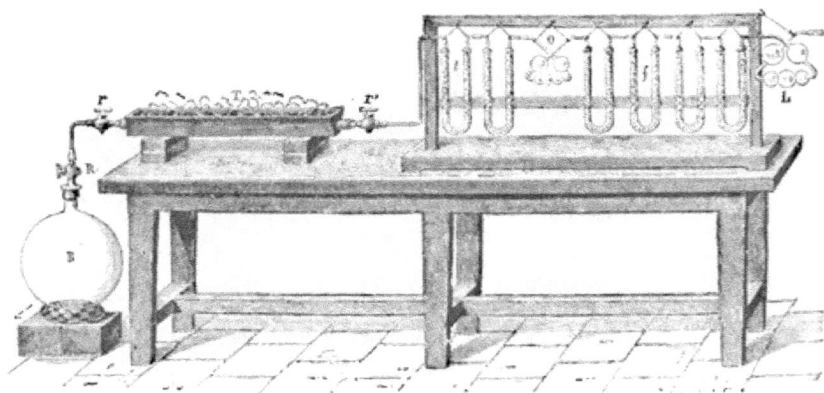

Fig. 19.—Apparatus for analyzing air by the method of weight.

throughout its whole length, and is moreover furnished at its extremities with two taps, r and r', which admit of its being emptied; seventh-
ly, of a glass globe, B, holding from two to three gallons, and the neck
of which is fitted with a tap, R.

To perform the experiment, as complete a vacuum as possible is
made in the tube T; the two taps are closed tight, and the tube, thus
emptied of air, is weighed. The glass ball B, having been emptied of
air, is also weighed. The various portions are then put together in
the order described, and the tube T is made red-hot. Then the taps
r r' of the tube T, and the tap R of the glass ball, are successively
opened. The air, entering by the suction-tube to the right, traverses
first of all the balls L and the tube f, where it parts with its carbonic
acid; then it passes into the second set of balls, o, and into the tube l,

where the sulphuric acid removes all the vapor of water it contains. Separated from these, the air makes its way into the tube T, containing the red-hot copper, which retains the oxygen, and then passes into the empty glass ball in a state of pure nitrogen. The increase of weight in the tube clearly gives the weight of the oxygen which has been deposited in the operation. The difference between the weight of the globe when empty and when full of nitrogen as clearly represents the weight of this gas. By means of this analysis, made with every conceivable precaution, MM. Dumas and Boussingault ascertained that one hundred parts of air contain—

> Oxygen, 23 in weight; 20·8 in volume.
> Nitrogen, 77   "   79·2   "

The difference between the proportion of weight and that of volume is due to the fact that oxygen is rather heavier than nitrogen.

These, therefore, are the two fundamental elements of the chemical constitution of air. But there exist other elements in far smaller quantities; such, for instance, as carbonic acid and aqueous vapor. Their quantity is determined by the apparatus described for finding the weight of the oxygen and nitrogen in the air. (See Fig. 20.) An iron vessel is filled with water, and emptied by means of a tap inserted in the lower part. The water which runs out is gradually replaced by external air, which has to

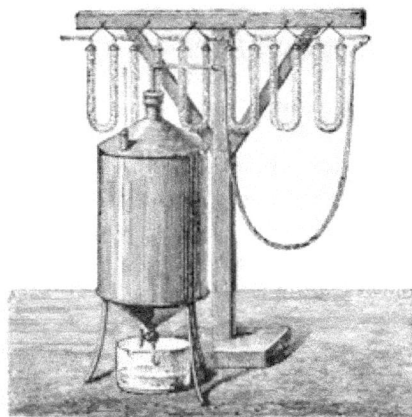

Fig 20.—Apparatus for obtaining the proportion of carbonic acid in air.

pass through the six curved tubes before it reaches the reservoir. The first two of these are filled with pumice-stone steeped in sulphuric acid, and the air, on its way through them, leaves behind the water which was mixed with it. The two middle tubes are filled with a concentrated solution of potash, which absorbs the carbonic acid. Of the last two tubes, containing pumice-stone steeped in sulphuric acid, the first is intended to extract the humidity which the potash has imparted to the air, and the other to prevent the humidity from making

its way back from the sucker into the tubes. By weighing, before
and after the experiment, the series of analyzing tubes, we obtain the
weight of the *water* and the weight of the *carbonic acid* contained in a
volume of air equal to that of the reservoir.

The atmosphere contains about $\frac{1}{10000}$ of its volume of carbonic acid.

There is also a very simple process by which the oxygen and the
nitrogen can be separated. Into a
graduated tube, containing a certain
volume of air, with its open end
placed in a vessel containing water
or mercury, is inserted a long stick
of phosphorus. (Fig. 21.) At the
expiration of six or seven hours, as
a rule, the oxygen is absorbed, and
the stick of phosphorus may be
withdrawn, and the gas which remains—that is to say, the nitrogen
—measured. The absorption is considered to be complete (the apparatus being placed in the dark) when
there ceases to be any glimmer upon

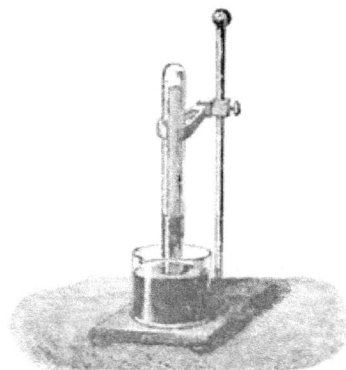

Fig. 21.—Apparatus for separating the oxygen
from the nitrogen.

the surface of the phosphorus. The rapid absorption of the oxygen
by the phosphorus may be shown by heating the gas in a bell-glass
into which a fragment of phosphorus has been introduced; the phosphorus is heated by an alcohol-lamp, and a portion of it volatilized;
and when the flame has reached all the space occupied by the gas, the
experiment is complete. Time is left for it to get cool; the volume
of nitrogen is transferred into a graduated tube and measured, the difference from the original weight giving the quantity of oxygen.

Oxygen and nitrogen are two *permanent gases*—that is to say, it has
been found impossible hitherto, either by compression or cold, to destroy their gaseous form.

The first, oxygen, is the ordinary agent of combustion, whether of
the kind which takes place in our fire-places or in our organisms.
The second, nitrogen, exercises a moderating influence over the first.

Carbonic acid, which exists in quantities varying according to time
and place, but always very small in amount, has been liquefied under
a strong pressure conjoined to intense cold; it has even been solidified.
In that state it has the appearance of light and very compressible snow,

the contact of which with the skin produces a burning sensation, this excessive cold acting upon the epidermis in the same way as great heat.* In the small quantities in which it is found, carbonic acid produces no ill effects; in larger quantities it is hurtful to the breathing, and finally produces asphyxia.

Emanations from the earth, the abundant sources of carbonic acid, are often met with in volcanic districts. When M. Boussingault explored the craters at the equator, he was shown a locality where no animals could remain; this was at Tunguravilla, not far from the volcano of Tunguragua. He thus describes his visit of 1851: "Our horses soon gave us indications that we were approaching it; they refused to obey the spur, and threw up their heads in a most disagreeable fashion. The ground was strewn with dead birds, among which was a magnificent black-cock, that our guides at once picked up. Among the victims were also several reptiles and a multitude of butterflies. The sport was good, and the game did not seem too high. An old Indian, Quichua, who accompanied us, declared that, to procure a good sleep, there was nothing like making one's bed upon the Tunguravilla."

This deleterious emanation made itself manifest by the sterility of the ground for a circle of some hundred yards; it was especially great at a point where there were many large trees lying dried up and half buried in the vegetable earth, which implies that these trees had flourished upon the spot where they have been lying since the eruption of the carbonic acid. This gas, like that which is also met with in similar circumstances in various regions of the globe, is carbonic acid more or less mixed with air, according to its distance above the soil.

Carbonic acid exercises a directly deleterious effect upon the nerves and brain. Hence the anæsthetic effects which it may produce, and which all visitors to Pouzzoles, near Naples, may have seen at a grotto which has become famous from this cause.

The keeper has a dog whose legs he ties together, to prevent his running away; he then places him in the middle of the grotto. The animal displays evident fear, struggles to escape, and soon appears to be dying. His master then takes him out into the open air, where he gradually recovers himself. One of these dogs has been used for this purpose more than three years. It is all but proved now that the con-

* [The snow-like flakes can be handled with impunity; it is only when forcibly pressed against the skin that a blister is produced.—Ed.]

vulsions of the pythonesses charged with expounding the decrees of the gods were produced by the priests with carbonic gas.

This grotto is situated upon the slope of a very fertile hill, opposite, and not far from, Lake Agnano. The entrance is closed by a gate of which the keeper retains the key. It has the appearance and shape of a small cell the walls and vault of which have been rudely cut in the rock. It is about one yard wide, three deep, and one and a half high, and it is difficult to judge from its aspect whether it is the work of man or of nature. The ground in this cavern is very earthy, damp, black, and at times heated. It is, as it were, steeped in a whitish mist, in which can be distinguished small bubbles. This mist is composed of carbonic acid gas, which is colored by a small quantity of aqueous vapor. The stratum of gas is from ten to twenty-five inches high. It represents, therefore, an inclined plane the highest part of which corresponds to the deepest portion of the grotto, and this is a physical consequence of the formation of the ground. The grotto being about on the same level as the opening leading into it, the gas finds its way out at the door, and flows like a rivulet along the hill-path. The stream may be traced for a long distance, and a candle dipped into it at a distance of more than six or seven feet from the grotto is extinguished at once. A dog dies in the grotto in three minutes, a cat in four, a rabbit in seventy-five seconds. A man could not live more than ten minutes if he were to lie down upon this fatal ground. It is said that the Emperor Tiberius had two slaves chained up there, and that they perished at once; and that Peter of Toledo, Viceroy of Naples, shut up in the grotto two men condemned to death, whose end was as rapid.

Two analyses of the air in this grotto, which had been collected at different times (see Ch. Ste. Cl. Deville and F. Le Blanc), gave in volume—

Carbonic acid	67·1	73·6
Oxygen	6·5	5·3
Nitrogen	26·4	21·1
	100·0	100·0

It is not necessary to travel so far for this predominance of carbonic acid. At Montrouge, near Paris, and in the neighborhood, there are large quarries, and even cellars, which are filled from time to time with this mephitic gas.

Upon the borders of Lake Laacher, near the Rhine, and at Aigueperse, in Auvergne, there are two sources of carbonic acid so abundant that they give rise to accidents in the open country. The gas rises out of small hollows in the ground, where the vegetation is very rich; the insects and small animals, attracted by the richness of the verdure, seek shelter there, and are at once asphyxiated. Their bodies attract the birds, which also perish.

In former times the accidents caused by this gas in caves, mines, and even in wells, gave rise to the most extravagant stories. Such localities were said to be haunted by demons, gnomes, or genii, the guardians of subterranean treasures, whose glance alone caused death, as no trace of lesion or bruise was to be found on the unfortunate persons so suddenly struck down.

In addition to the oxygen, nitrogen, and carbonic acid, the air contains a certain number of other substances, in smaller and very varying quantities.

The most important is aqueous vapor, of which I have spoken above in describing the method of analysis for determining its presence. The air always contains a certain proportion of aqueous vapor in a state of solution, and invisible. When this water passes into the state termed *vesicular*, it constitutes clouds or mists. The quantity of aqueous vapor varies with the seasons, the temperature, the altitude, the geographical position, etc. At the same temperature and under the same pressure the maximum quantity capable of being mixed with the air is invariable. The hygrometrical state of the air, for a given temperature, is but the relation between the quantity of moisture really existing in the air and the quantity which would exist if the air were saturated at the same temperature. The millions of cubic feet of *vapor of water* which, mixing with the air, form the clouds and the rain constitute the most important element of the atmosphere in respect to the circulation of life. Therefore *water* will be in a subsequent chapter the object of special study. The quantity of heat necessary for the evaporation of the water from the earth's surface has been ascertained. The volume annually evaporated may be represented by the volume of water which falls from the atmosphere in that space of time; and, in comparing the results of observations taken at different latitudes and in both hemispheres, we are led to estimate this volume as corresponding to a depth of fifty-four and a quarter inches over the whole earth. The amount of heat necessary to evaporate such a volume of water would suffice, according to Daubrée, to liquefy a thickness of ice of nearly thirty-three feet in depth

enveloping the whole globe. From the calculations of Dalton, the atmosphere contains about the 0·0142th part of its weight in water: the
upper strata are nearly free from water.

What other substances are there to be found in the atmosphere? It
unquestionably contains small quantities of ammonia, partially in a
state of carbonate of ammonia; perhaps, too, partially in a state of nitrate, or even nitrite, of ammonia. The origin of this substance must
evidently be attributed principally to the decomposition of vegetable
and animal matter; and its presence in the air is of peculiar importance
in regard to the phenomena of vegetation and the chemical statics of
plants. Several chemists have attempted to determine its exact proportion, which does not seem to exceed a few millionths of the volume of
the air.

The quantity of ammonia found in different waters is (in weight):

Rain-water	0·0000003
Fresh-water	0·0000002
Spring-water	0·0000001

From one to two grains of ammonia per cubic foot have been found
in sea-water. This is, no doubt, a very trifling quantity; but when we
reflect that the ocean covers more than three-quarters of the globe, and
when we consider also its enormous mass, it may be fairly looked upon
as a vast reservoir of ammoniacal salts, whence the atmosphere can
make good the losses which it is continually undergoing.

The streams, too, carry to the sea prodigious quantities of ammoniacal
matter. I will give one instance. According to M. Desfontaines, the
engineer, the Rhine at Lauterburg has, on the average, a flow of 39,000
cubic feet of water a second; and from a careful examination of the
amount of ammonia contained in the water, it results that the Rhine, in
its passage by Lauterburg, carries down with it every twenty-four
hours at least 22,500 lbs. of ammonia—that is, 13,000,000 lbs. a year.
The atmosphere, incessantly undergoing change (although its constitution remains unaltered) by the immense labor of human beings who,
like so many chemical pairs of bellows, are in continual motion on the
bed of the aërial ocean, is the theatre of accidental chemical modifications which play their part in the general organization. We see rising
from the ground aqueous vapor, effluvia of carbonic acid gas, nearly
always unmixed with nitrogen, sulphureted hydrogen gas, sulphurous
vapors; less frequently we notice vapors of sulphuric or hydrochloric
acid; and, lastly, carbureted hydrogen gas, which has for thousands of

years been in use among different nations for the purposes of producing warmth and light.

Of all these gaseous emanations the most numerous and abundant are those of carbonic acid. In former ages, the greater heat of the globe and the large number of crevices that the igneous rocks had not yet covered contributed considerably to these emissions. Large quantities of hot vapor and of this gas became mixed with the aërial fluid, and produced that exuberant vegetation of pit-coal and lignites which is nearly an inexhaustible source of physical strength for a nation. The enormous quantity of carbonic acid the combination of which with lime has produced the chalky rocks then rose out of the bosom of the earth under the predominant influence of volcanic forces. What the alkaline soils could not absorb spread itself into the air, whence the vegetable matter of the Old World drew continuous sustenance. Then, too, abundant emissions of sulphuric acid in vapor have led to the destruction of mollusks and fish, and to the formation of beds of gypsum. Humboldt adds, that the introduction of carbonate of ammonia into the air is probably anterior to the appearance of organic life upon the globe's surface. Besides the ammoniacal vapors, the atmosphere also contains many traces of nitrogen, and even nitric acid. Several observers have also demonstrated, especially in large towns, the presence of a small quantity of *hydrogen* in some form, probably carbureted. M. Boussingault was the first to prove, by precise experiments, the presence of a hydrogenous gas or vapor equal, at the most, to a $\frac{1}{10000}$ part of the air in volume.

Analysis has also brought to light a certain quantity of iodine. The entire, or nearly entire, absence of iodine in the air or water of certain mountainous countries has, according to M. Chatin, a close connection with the existence of goître among the inhabitants of these countries. His conclusions have been received, as a rule, with incredulity by chemists. Yet, when we consider that rain-water collected in a pluviometer contains various kinds of salts, which arise from the washing of the dust suspended in the atmosphere, and that chemists have often found evidence of the presence of iodine in rain-water, there can be no difficulty in admitting that the presence in the air of iodine, free or in combination, may be, if not a normal, at least an occasional occurrence. We now arrive at the last element ascertained by special investigations to be existent in the atmosphere, viz., *ozone*.

Van Marum, about the year 1780, by means of powerful electric machines, excited a large number of sparks in a tube full of oxygen,

about six or seven inches long. After passing about five hundred
sparks into the tube, he found that the gas had acquired a very strong
smell, which, to use his own words, "seemed clearly the smell of electric matter." Every one, indeed, is aware that if lightning strikes any
object it leaves behind it what is commonly called a sulphurous smell.
Van Marum also found that the gas acquired, after the experiment, the
property of oxidizing mercury without heat. Nearly sixty years later.
in 1839, M. Schœnbein, professor at Basle, informed the Academy of
Sciences at Munich that, having decomposed some water, he had been
struck by the smell of gas emitted. After a few researches he drew
the conclusion that a new body was brought to light by his experiment, which he called ozone, from ὄζω (to emit an odor). A large
number of contributions have been subsequently made to the subject
by various *savants*.

Ozone is interesting in a chemical point of view, both in its nature
and its energetic affinities, for it oxidizes directly both silver and mercury, at least when these metals are moist. It also liberates iodine
from potassic iodide, and forms, with the metal, an oxide which, doubt-
less, contains far more oxygen than the potash. The hydracids impart
to it their hydrogen. The salts of magnesium become decomposed by
its contact with the formation of peroxide. Chlorine, bromine, and
iodine, pass, when moist, under the influence of ozone, into chloric,
bromic, and iodic acid.

This agent has an exciting effect upon the lungs, provokes coughing
and suffocation, and presents all the characteristics of a poisonous substance.

Notwithstanding all the researches that have been made in reference
to ozone, the knowledge of it is, from a physical and chemical point of
view, very imperfect; a fact easy to understand when I state that it
is impossible, even with the most perfect methods, to transform more
than $\frac{1}{1300}$ of a mass of oxygen into pure ozone. This maximum
reached, action ceases. How can it be easy to study a body which is
spread over at least 1300 times its own volume of another gas?[*]

It has occurred to several experimentalists, such as Schœnbein,
Bérigny, Pouriau, Bœckel, Houzeau, and Scoutetten, to join to the
ordinary meteorological observations ozonometrical observations also.

---

[*] [By a continuous electrical discharge, maintained for many hours, Andrews and Tait
were enabled to transform into ozone one-twelfth of the volume of oxygen operated on.—
Phil. Trans., 1860.—ED.]

M. Schœnbein, in his experiments, boiled one part of potassic iodide, ten parts of starch, and two hundred of water, a preparation of "Joseph's paper" being afterward steeped in it. The latter is dried in a close room, and then cut up into small strips. This paper becomes blue by contact with the ozone, for the iodine is set at liberty and reacts upon the starch. The deepness of the tint, however, depends upon the quantity of oxygen which has been turned into ozone. A small strip is exposed each day for twelve hours, sheltered both from the sun's rays and the rain, and its tint is then compared with a scale of ten colors, varying from white to indigo.

In 1851, MM. Marignac and De la Rive undertook several experimental researches as to ozone; and their conclusion was, that this substance must be simply oxygen in a particular condition of chemical activity, determined by electricity. Berzelius and Faraday gave their adhesion to this opinion of the Geneva *savants;* and MM. Frémy and Becquerel demonstrated, by fresh experiments in 1852, its legitimacy. The works of Thomas Andrews, published in 1855, leave no doubt upon this head. Ozone, no matter from what source it is derived, is a unique and separate body, with identical properties and the same constitution; it is not a composite body, but an allotropic condition of oxygen. This allotropic condition is due to the action of electricity upon the oxygen. This opinion, based upon the best experiments, has now been universally accepted, and this constitution of ozone appears incontestable.

Let us further add to all these divers substances the presence of *oxygenated water*, as indicated by M. Struve, director of the Pulkowa Observatory. While engaged in a chemical analysis of the water in the River Kusa, M. Struve was struck with the presence of a certain quantity of nitrite of ammonia, which was only to be found after a fall of snow or of rain. Soon after the downfall had ceased, all trace of this substance had again disappeared; M. Struve therefore supposed that the nitrite of ammonia existed in the air, and that it had been brought away by the snow or the rain. He entered upon researches on the subject, and in the course of them made the interesting discovery of the presence of oxygenated water in the atmosphere. From these researches may be drawn the following conclusions: 1st. Oxygenated water is formed in the atmosphere like ozone and nitrite of ammonia, and becomes separated from the air through the atmospheric deposits. 2d. Ozone, oxygenated water, and nitrite of ammonia, are always intimately connected. 3d. The alterations which the atmospheric air brings about in the starch-iodine papers are caused by the ozone and oxygenated water.

One word more. In absorbing into our lungs the quantity of air due to us, we often unwittingly inhale whole hosts of microscopical animals which are in suspension in the atmospheric fluid, and even portions of antediluvian animals, mummies, and skeletons of past ages!

Paris is nearly entirely built with chalky microscopical skeletons and tortoise-shells. The shells of the *foraminifera*, for instance, in a fossil state, by themselves form entire chains of lofty hills and immense beds of building-stone. The rough chalk in the neighborhood of Paris is in some places so full of these remains that a cubic inch in the Gentilly quarries contains at least 100,000 of them. When we pass close by a house that is being pulled down, or one in course of construction, and find ourselves enveloped in a cloud of dust that penetrates down our throats, we often, beyond a doubt, inhale hundreds of these tiny atoms.

Each day and each hour we inhale and take into our chest legions of animal and vegetable life. There are the living microzoa, several species of which are the fish of our blood; there are the vibriones, which attach themselves to our teeth like oyster-banks to rocks. Then, again, there is the dust of microscopical animalcules, so small that it takes 75,000,000 to make a grain; and, besides these, there are the grains of pollen which, germinating in our lungs, further the spread of parasite life, which is out of all comparison more developed than the normal life visible to our eyes.

The winds and storms, by their violent agitation of the atmosphere; the ascending currents due to the inequalities of temperature; the volcanoes, by their incessant emission of gas, vapors, and ashes, so finely divided that they often fall at a prodigious distance, carry up and maintain in the higher regions corpuscles drawn away from the surface of the ground, or forced out of the internal and, perhaps, still incandescent portion of the globe. In the phenomena connected with the organism of plants and animals, these substances, so slight and of such different origins, the vehicle of communication for which is the air, very probably exercise a far more pronounced action than is generally believed. Their permanence is, too, placed beyond doubt by the mere evidence of the senses, when a ray of sun penetrates a darkened room. As M. Boussingault remarks, "The imagination may conceive very readily, though not without a certain disgust, what is contained in these morsels of dust which we are incessantly inhaling, and which have been aptly denominated *the refuse of the atmosphere.* They establish, in a certain sense, a contact between individuals far removed from each other; and though their proportion, their nature, and, consequently, their effects, are so varied, it is not too much to attribute to them a part of the insalubrity which generally manifests itself in all great agglomerations of human beings."

Rain carries away these morsels of dust, while it dissolves their soluble matter, among which are found ammoniacal salts, as they also dissolve the vapor of carbonate of ammonia and the carbonic acid gas diffused in the air. There must, therefore, exist in a fall of rain, at its commencement, more soluble substances than at its close; and if the rain continues uninterruptedly in calm weather, after a certain interval there can only be very insignificant indications of the existence of the substances.

Miasmas, the propagators of epidemics, are superinduced by the aërial currents; the cholera, the small-pox, the yellow fever, and the diseases which periodically attack a district, seem to have their principal source of propagation in the atmosphere—the factory of death as it is of life. The rate of mortality, which was so heavy in Paris during the early part of 1870, in consequence of small-pox, pleurisy, and inflammation of the lungs, was especially severe in the northern districts of the city, over which the southerly wind spread the miasmas of the whole town, and where there was scarcely any ozone. A knowledge of the conditions of public health will be furnished in part by a study of the relations of meteorology to the variations in the rate of mortality, which is as continually oscillating under the slight breath of the wind as under the trifling alterations in barometrical pressure.

The air which Gay-Lussac brought down with him from his aeronautical voyage, and which was collected at a height of 23,000 feet, had the same composition as that which floats upon the earth's surface. The experiments of M. Boussingault in America, and those of M. Brunner in the Alps, lead to the same conclusions. This similarity in results arises from the fact that currents of air and continual variations in density are unceasingly mixing up together the atmospheric strata.

Is it the same at a greater height? It is scarcely probable, for the nitrogen and oxygen being in a state of mixture, and not chemically combined, the gases must be ranged according to their density, allowing, of course, for the law of expansion; that is to say, there are, as it

were, two distinct atmospheres, the least dense of which does not extend so far as the other, so that the proportion of nitrogen, the density
of which is 0·972, that of the air being 1, must increase the higher one
rises in the atmosphere; while the oxygen, the density of which is
1·057 (and which is the denser of the two), must be in a greater proportion near the surface. According to this hypothesis, the latter gas,
at 23,000 feet, would constitute only $\frac{12}{105}$ of the volume of air; but at
present experiment has failed to note so great a difference, because
this calculation supposes the air to be in a state of tranquillity, whereas
at these heights it is, as a matter of fact, in a continuous state of agitation.

The composition of the air varies very little: when it rains, the condensed water dissolves more oxygen than nitrogen; in frost, the water
leaves these two gases alone; the water which evaporates returns then
to the atmosphere.

We may now ask ourselves, in terminating this study of the chemical composition of the air, if this constitution is variable over the terrestrial globe. By virtue of one of the great natural harmonies which
unite the animal and the vegetable kingdoms, while the animals act as
combustion-machines, taking the oxygen from the air and throwing it
back into the atmosphere in the state of carbonic acid, the vegetables
play the reverse part, acting as reducing-machines. Under the influence of the solar rays, the green portions of the plants react upon the
carbonic acid, decompose it, concentrate the carbon, and restore the
oxygen to the air. The atmosphere, vitiated by the animals, is puri-
fied by the action of the vegetables. The chemical equilibrium of the
air's components has thus a tendency to self-preservation by virtue of
this inverse action brought to bear upon its constituent elements.

Certain phenomena due to the decomposition of rocks through oxida-
tion seemed, at first sight, calculated to modify in the long run the composition of the air; but a series of inverse actions of reduction tends to
restore, in the shape of carbonic acid, the oxygen that has disappeared.
As Ebelmen has pointed out, in his memoir upon changes in rocks, the
process of reactions in the mineral matter upon the globe's surface seems
also calculated to establish a compensation which maintains the chemical composition of the atmosphere.

The question is whether this compensation is complete. Supposing
it does not take place—as, indeed, is possible—does the quantity of oxygen diminish? As Thenard has remarked, "This is a very important

question, the solution of which can only be arrived at in the course of several centuries, because of the enormous volume of air by which our planet is surrounded."

In their remarkable memoir upon the true constitution of the atmospheric air, MM. Dumas and Boussingault thus expressed themselves in 1841:

"Some calculations, which, though not of absolute precision, nevertheless are based upon sufficiently certain grounds, tend to prove how far an analysis should extend to reach the limit at which the variations in oxygen would be sensibly manifest. The atmosphere is unceasingly agitated; the currents, stirred up by heat, by winds, by electric phenomena, are continually being mixed up and confusing together the various strata. The whole mass would, therefore, have to be changed in order to admit of an analysis indicating the difference between one epoch and another. But this mass is enormous. If we could place the whole atmosphere into a balloon, and suspend it in one side of a pair of scales, it would be necessary to put on the other side 138,000 cubes of copper (each a mile in length, breadth, and thickness) to balance it. Let us now suppose that each man consumes a little more than two pounds of oxygen a day, that there are a thousand millions of men upon the earth, and that, through the respiration of animals and the putrefaction of organic matter, this consumption attributed to man be quadrupled. Let us further suppose that the oxygen disengaged from plants is only the compensating agent of the causes of absorption omitted in our calculation, which would assuredly be putting the chances of alteration of the air in the strongest light. Well, even on this overdrawn hypothesis, at the end of a century the whole human race, and three times its equivalent, would only have absorbed a quantity of oxygen equal to fourteen or fifteen of the cubic miles of copper.

"Thus, to assert that, with their utmost efforts, the animals which people the face of the earth could in a century render the air they breathe impure, to the extent of depriving it of the $\frac{1}{9000}$ part of the oxygen that nature has placed there, is to make a supposition far beyond the reality."

In habitations badly ventilated, the effects of the breathing of men or animals, and the phenomena of the combustion of coal or of combustible matters, may cause a sensible alteration in the state of the air. Thus, in barracks, hospital rooms, theatres, wells, mines, etc., chemical analysis, when it is accurate enough, indicates a different composition from that of the open air. Furthermore, in habitations even out of the influence of the presence of sick persons, the animal emanations which escape with the aqueous vapor in respiration and perspiration may exercise an incontestable physiological influence, often more injurious than that caused by the production of carbonic acid or the disappearance of the oxygen in small quantities.

It is especially when the air arrives at a state of saturation from the causes cited above that there is reason to consider it deleterious. There is an unanimity of opinion in the present day that, to avoid a disastrous influence upon the organic economy, dwelling-houses, and especially hospitals, should be so constructed as to give more than 20,000 cubic feet of air per day to each individual.

CHAPTER VI.

SOUND AND THE VOICE.

AMONG the works of the atmosphere in terrestrial life, one of the most important is unquestionably that of serving as a vehicle for human thought, and enveloping the world in a sphere of harmony and activity which could not exist without it.

What is *sound?*

It is a movement produced in the air, and transmitted therein by successive undulations. To be perceived by the ear, this vibratory movement must be neither too slow nor too rapid. When the air, agitated by sound, vibrates at the rate of sixty undulations a second, it emits the dullest sound which can reach the ear. When the vibrations are 40,000 per second, they convey the sharpest sound which the auditory nerves can perceive.

To appreciate the nature of the sonorous movement, let us suppose that between the chaps in a vise, A (see Fig. 22), is fixed one of the extremities, C, of an elastic blade, C D; that the upper end, D, is pulled back to D', and then let go. By virtue of its elasticity, the blade will return to its primitive position; but in consequence of the speed it has acquired, it will pass it and go on to D'', executing on both sides of C D a series of oscillations the amplitude of which will gradually decrease, and in a more or less short space of time altogether cease.

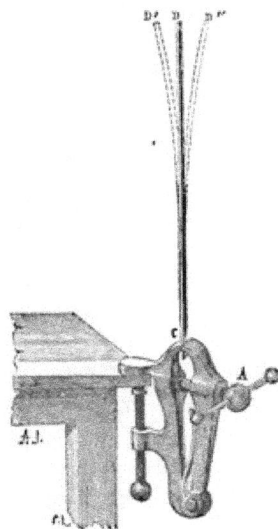

Fig. 22.—Vibrations of a blade.

The longer the elastic blade is, the slower will be the vibrations; while, in proportion as the blade is shortened, the vibratory movement will become more rapid, and at a certain point will be imperceptible to the eye. But when the organ of vision ceases to play a part, so to speak, that of the organ of hearing begins, and the ear can distinctly catch a sound, the nature of which depends upon the physical conditions of the vibrating body. Another instance of the production of sound is furnished by the vibration of a piece of cord fastened at its extremities, A B, and pulled in the middle.* Its vibration is rendered perceptible by the fact of the cord presenting the shape of a bobbin. By reason of the persistent impression upon the retina, and the speed of the vibratory movement, the eye sees the cord in all its positions together, as it were, the time of a vibration being less than that of a luminous impression, which is the tenth of a second. Sound, therefore, is but an impression upon the organ of hearing, caused by the vibrating movement of a given body. But the existence of a vibratory body on the one hand, and of an ear on the other, is not enough to cause an impression: a relation must be established between that body and the organ of hearing, and this is effected by a ponderable medium,

Fig. 23. — Vibration of a cord.

liquid or gaseous, constituted of more or less elastic matter. If we imagine a body vibrating in a complete vacuum, or in the centre of a space entirely devoid of elasticity, the ear, at a certain distance off, would catch no sound. Sound, in the proper sense of the word, does not exist in such a case.

We may in fact form, from what is mentioned above, the following definition of sound:

Sound is an impression produced by the vibrations of a body transmitted to the organ of hearing by the intervention of a ponderable and elastic medium.

At what rate is sound propagated?

The first exact measurements were made in 1738 by a commission of the Academy of Sciences, of which Lacaille and Cassini de Thury were members. Several pieces of ordnance were placed upon the heights of Montmartre (then outside the walls of Paris) and at Montlhéry (an elevated position in the department of the Seine-et-Oise, distant about 16 miles from Paris), and it was arranged that from a given hour a gun should be fired at equal stated intervals. The persons engaged in the experiment counted the time that elapsed between the flash and the arrival of the report; and this was found to be, on an

---

* [This is, of course, the principle of all stringed instruments—the harp, violin, etc. It is difficult to hold the cord sufficiently tight by the hand to produce a note.—ED.]

average, 1 minute 24 seconds for a distance of about 95,000 feet, which is at the rate of 1037 feet per second.

These experiments were repeated in 1822 by the Bureau des Longitudes—a section of the Academy of Sciences—the persons taking part in them being Arago, Gay-Lussac, Humboldt, Prony, Bouvard, and Mathieu. Villejuif and Montlhéry, distant from each other 61,000 feet, were the places selected; and it was found that at a temperature of 61° the velocity of transmission was 1047 feet a second.

A great number of similar experiments have been made in different countries. Very recently, M. Regnault investigated this subject, employing all the resources of modern physics, and especially telegraphic signals, for registering instantaneously the discharge and the arrival of the sound.

The velocity of sound varies with the density and the elasticity of the air, and therefore with its temperature. According to the most accurate measurements, the following table may be given in reference thereto:

Temperature (Fahr.)	Velocity per Second.	Temperature (Fahr.)	Velocity per Second.
5°	1056 feet.	68°	1122 feet.
14°	1070 "	77°	1132 "
23°	1079 "	86°	1142 "
32°	1089 "	95°	1152 "
41°	1096 "	104°	1161 "
50°	1102 "	113°	1171 "
59°	1112 "	122°	1181 "

Sound is propagated in the air by successive undulations, which may be roughly compared to the circular waves which are produced on the surface of water around a point disturbed by the fall of a stone. But they are, in reality, very different phenomena. In the liquid waves, the molecules are alternately raised and lowered in regard to the general level, but undergo no change of density; while this change is, on the contrary, a characteristic of the waves of sound. There is, however, one circumstance common to both these phenomena which is worth pointing out—and that is, that the wave causes no real progressive movement. Thus, when waves of water follow each other, if we notice any small floating object, it is seen to alternately rise and fall, but it remains in the same place upon the surface of the water. Similarly, in the waves of sound, the molecules of the air execute oscillatory movements in regard to the propagation of sound, but the centre of these movements remains unchanged.

Scientific education should teach us to behold in nature the invisible as well as the visible—to depict to the eyes of the intellect what escapes the eyes of the body. We may, with a little application, form a true idea of a sound-wave; we may mentally see the molecules of air first pressed the one against the other; then, immediately after, this condensation brought away again by an opposite effect of dilatation or rarefaction. We thus learn that a wave of sound is composed of two parts: in one the air is condensed; while in the other, on the contrary, it is rarefied. A condensation and a dilatation are then the essential constituents of a sound-wave. But, if the air is necessary to the propagation of sound, what happens when a sounding body, such as the bell of a clock, is placed in a space destitute of air? The result is that no sound proceeds from the empty space; the hammer strikes the bell, but silently. Hawksbee demonstrated this fact in a memorable experiment in 1705, before the Royal Society of London. He placed a clock under the receiver of an air-pump, in such a way that the striking of the clapper would continue after the air had been exhausted. While the receiver was full of air, the sound was quite audible; but it was no longer so (or at least in a very slight degree) when a vacuum had been created. The appended illustration is that of a contrivance which enables us to repeat Hawksbee's experiment in an improved manner. Under the receiver, B, placed firmly on the plate of an air-pump, will be seen the works of a striking clock, A. The hammer is kept back by a spring and ratchet, c. As much as possible of the air is exhausted; then, by means of a stem, g, which passes out through the top of the receiver, without letting in the exterior air, the trigger d, which holds back the hammer b, is pulled. The bell, a, vibrates *silently.* But if we let the air into the recipient, we at once hear a sound, very feeble at first, but growing louder as the air becomes denser. At great heights in the atmosphere, the intensity of sound is notably less. According to the calculations of Saussure, the detonation of a pistol upon the summit of Mont Blanc is about equal in intensity to that of a common cracker at the level of the sea.

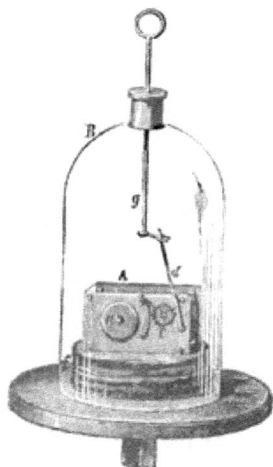

Fig. 24.

Since it is proved that there is no sound in a vacuum, fearful catastrophes might take place in the planetary regions without the slightest audible notice of them reaching the surface of the earth.

The vibratory movement of the air has been represented as being a circular wave, which spreads out in all directions with equal velocity, and diminishes in intensity as it advances. Where does it cease? where is it extinguished? We must regard this as taking place at the point in space where it is no longer sensible to the most delicate ear; and we all know how much this limit varies with the organization and habits of different individuals. At the same time, there can be no doubt that the aërial wave continues to spread out after the most practiced ear has ceased to be sensible of it. In the places where there is a numerous population, the incessant noise kept up in the air by so many thousands of people creates a characteristic difference between day and night; the noises become confounded together, and are propagated in a confused mass. During the night there is nothing to lessen the intensity of sound, and the ear perceives in all their force the howling of the tempest, the blast of the winds, the roaring of the waves, the shrill cry of the bird of prey or the wild beast; and it is then that pusillanimous fears and superstitious terror take possession of the timid. Traveling in a balloon over the plains of Charente, the stream of a river seemed to make as much noise as that of a great cascade, and the croaking of the frogs were audible at the height of 3000 feet. Above two miles all noise ceases. I never encountered a silence more complete and solemn than in the heights of the atmosphere—in those chilling solitudes to which no terrestrial sound reaches.

"Two conditions determine essentially," says Tyndall, " the velocity of the sound-wave, viz., the elasticity and density of the medium which it passes through." The elasticity of the air is measured by the pressure which it supports, and to which it forms an equilibrium. We have seen that, at the level of the sea, this pressure is equal to that of a column of quicksilver 29·92 inches high. Upon the summit of Mont Blanc the barometrical column scarcely exceeds half this height, and, therefore, at the highest point of this mountain, the elasticity of the air is only half what it is upon the sea-coast.

If we could increase the elasticity of the air without at the same time augmenting its density, we should increase the velocity of sound. We should also effect that object if we could diminish the density without making any change in the elasticity. The air heated in a closed vessel.

in which it can not become dilated, has its elasticity increased by the warmth, while its density remains the same. Sound will, therefore, be propagated more rapidly through air thus heated than through the air at its normal temperature. In like manner, air which is free to dilate has its density diminished by heat,* while its elasticity remains the same, and consequently it will propagate sound more rapidly than cold air—this, indeed, takes place when our atmosphere is heated by the sun; the air becomes dilated and much lighter, volume for volume, while its pressure, or, in other words, its elasticity, remains the same. This is the explanation of the statement that the velocity of sound in air is 1090 feet a second at the temperature of melting ice. At a lower temperature the velocity is less, and at higher temperatures greater, with an average difference of about one foot for one degree (Fahr.). Under the same pressure—that is to say, with the same elasticity—the density of hydrogen is much less than that of the air, and, in consequence, the velocity of sound through hydrogen gas considerably exceeds its velocity through air. The reverse is the case with carbonic gas, which is denser than air; for under the same pressure sound travels less rapidly through this gas than through air.

The fact that air, even when very rarefied, can transmit intense sounds, is proved by the explosion of meteors at a great height above the earth, though it is true that, for this to be the case, the initial cause of the atmospheric disturbance must be very violent.

The movement of sound, like all others, is less in amount when it communicates from a light body to one more dense. The action of hydrogen on the voice is a phenomenon of this kind. The voice is formed by the injection of air from the lungs into the larynx; in its passage through this organ the air is set vibrating by the vocal chords, which thus give rise to sound; and if one wishes to speak when the lungs are full of hydrogen, the vocal chords still impress their movement on the hydrogen, which transmits it to the air outside. But this transmission of a light gas to one much denser causes a considerable diminution in the intensity of the sound. The effect of this is very remarkable. Tyndall demonstrated it to the Royal Institution in London. Having, by a great effort of inhalation, filled his lungs with hydrogen, he began to speak, and his voice, generally powerful, was hoarse and hollow; there was no ring in it; it seemed to issue from the depths of the grave.

* [The air must be contained in a vessel so constructed that the elasticity (pressure) is *kept* the same (for instance, in a cylinder in which fits a piston of constant weight).—ED.]

The intensity of sound mainly depends upon the density of the air from which it proceeds, not on that of the air in which it is heard.

The wave of sound, propagated in all directions from the point where the sound has been produced, diffuses itself in the mass of air in which the motion takes place, and consequently lessens the amount of movement at any point. Let us imagine around the centre of disturbance a spherical layer of air, with a radius of a yard; another layer of the same thickness, with a radius of two yards, contains four times as much air; one with a radius of three yards contains nine times as much; one with a radius of four yards, sixteen times as much; and so on. The quantity of matter set in motion increases, therefore, as the square of the distance from the centre of disturbance; the *intensity* of the sound diminishes in the same degree. This law is expressed by the statement that the intensity of sound varies inversely as the square of the distance from the point of initial disturbance. The decrease in the sound in inverse ratio to the square of the distance would not occur if the sound-wave spread in such a way as to prevent its being diffused laterally. By producing a sound in a tube the interior surface of which is perfectly smooth these conditions may be realized, and the wave thus confined reaches a great distance, with but a slight loss of intensity. In this way Biot, noting the transmission of sound through the conduit pipes that supply Paris with water, found that he could carry on a conversation in a low tone at a distance of 3300 feet; the faintest murmur of the voice was heard at this distance, and the firing of a pistol at one end of the pipes extinguished a candle placed at the other end.

Echoes depend, in a great measure, upon the compressibility and elasticity of the air. The sound-wave, as has been stated, spreads indefinitely, and is finally lost in space; but if it encounters a body capable of opposing it, it undergoes a reflection like that of light when it falls upon a smooth surface. For an echo to be distinctly produced, there must be a distance of fifty-five feet at least—the tenth of a second in time—between the person speaking and the reflecting surface. When the former is nearer, the echo is replaced by a confused resonance, which, in some buildings, renders it impossible for a speaker to make himself heard.

Whether acute or grave, sounds have the same velocity*—that of

---

* [This is proved by the fact that, if a band of music be heard at a distance, the sounds are not confused, the distinctness of the tune being unaffected by the distance, though the loudness is of course diminished.—Ed.]

1115 feet a second in air of 61° (Fahr.). At half this distance the echo gives back four syllables rapidly pronounced; at a greater distance it will distinctly reflect a larger number of syllables and whole phrases. The echo in Woodstock Park repeats seventeen syllables in the day-time and sixty at night. Pliny tells us that a portico was built at Olympia which repeated sounds twenty times. The echo at the Château de Simonetti was said to repeat the same word forty times. The theory is the same for the multiplied echoes; they result from the re-flecting surfaces against which the aërial wave is thrown back several times from the one to the other, like a ray of light between two parallel glass plates. Perceptible sounds are included between the limits of 30,000 and 40,000 simple vibrations a second, except in the case of ears which are exceptionally sharp. The undulations of the ether, which produce light, are far more rapid.* Visible colors are the result of vibrations so rapid that between 400 and 800 billions take place in a second.

Of perceptible sounds, the extreme limits of the human voice are the lowest, *fa*, of 87, and the highest, *ut*, of 4200 vibrations.

Sound has four fundamental properties—duration, height, intensity, and *timbre* or quality. The first three are defined by the words used to express them. As to the *timbre*, it is the resonance peculiar to each instrument and to each voice which enables us to clearly distinguish the sounds of a violin from those of a clarionet or a flute, and to recog-nize a person by hearing him speak or sing.†

The *timbre* of sounds has long been an insoluble enigma to natural philosophers and physiologists. It is only within the last few years that the excellent experiments of Helmholtz have proved that it de-pends upon the number of harmonic sounds which are produced simul-taneously with the fundamental tone, and upon their relative intensity.

The intensity of sounds generated upon the surface of the earth spreads upward far more readily than in any other direction, and is transmitted to great heights in the atmosphere. Citing some few in-stances from my aeronautical travels, I will, in the first place, mention

* [It must be borne in mind that there is only a general analogy between light and sound. In the latter, the vibrations consist of condensations and rarefactions in the air (or other gas), which are longitudinal—*i. e.*, take place in the direction in which the sound is proceeding; while, in light, the vibrations are transversal (*i. e.*, perpendicular to the direction of the ray), and take place in an *ether* which is supposed to pervade all space.—En.]

† It is, of course, more difficult to recognize a person by his song than by his speech.

that a noise, immense, colossal, and indescribable, is ever to be heard at 1000 to 1500 feet above Paris. In rising from a relatively quiet garden—as from the Observatory—we are astonished to hear a chaos of sound and a thousand various noises. The following details will, however, illustrate more strikingly this ascent of sound:

The whistle of a steam-engine may be heard at 10,000 feet; the noise of a train at 8200;* the barking of a dog at 6000; the report of a gun attains the same height; the shouts of people sometimes are audible at 5000 feet, as also the crowing of a cock or the tolling of a bell. At 4500 feet the beating of a drum and the sound of a band are audible; at 3900 feet the rumble of vehicles upon the pavement; and at 3300 feet the shout of a single individual. At this last height, during the silence of the night, the current of a stream at all rapid produces the same effect as the rush of a cascade; and at 2950 feet the croaking of frogs is plaintively distinct. At 2620 feet the slight noises made by the cricket are heard very plainly.

This does not hold good of sound when descending. While we hear distinctly the voice of a person speaking from 1600 feet underneath us, it is impossible to catch what is said at a height of more than 300 feet above us.

The occasion upon which I was most struck by this astonishing transmission of sounds vertically upward was in an ascent that took place on June 23, 1867. Having been in the midst of the clouds for several minutes, we were surrounded by a white and opaque veil that concealed both the sky and the earth, when I noticed with surprise a singular increase of light taking place around us, and all at once the sounds of a band reached our ears. We could follow the piece of music as distinctly as if the band had been in the clouds, a few yards distant from us. We were then just above Antony, a village near Paris. Having mentioned the fact in a newspaper, I was glad to receive, a few days afterward, a letter from the President of the Philharmonic Society in that place, informing me that his society had seen the balloon above them, and had purposely played a very soft piece, in the hope that they might be of service to us in our researches.

In this case the balloon was about 2950 feet above the place. At 3280, 3940, and even 4590 feet, the parts were still distinctly audible. Far from being an obstacle to the transmission of sound, the clouds increased its intensity, and made the band seem close to us.

* [On June 26, 1863, I heard a rail-way train when at the height of 22,000 feet.—Ed.]

When sound has ceased, there still continues in the air a movement which may cause to vibrate membranes placed to receive and to interpret these impressions. M. Regnault has measured these *silent waves;* he has determined the distance traversed both by the sonorous wave and the silent wave which continues after the former has ceased. In a gas-pipe, twelve inches in diameter, a pistol, with a charge of fifteen grains of gunpowder, was heard at the other extremity, 6250 feet off; and when the pipe was closed with an iron plate, the echo of the report was perceptible to any one listening attentively. The limit of the sonorous wave was therefore, in this instance, 12,600 feet; that of *silent waves* is much greater.

Air, the vehicle of sound, is at the same time the vehicle of smells and of all the emanations that are exhaled from the terrestrial surface. But smells are due not only to the vibratory movement, like sound and light. Fourcroy was the first to establish the fact that they are in part caused by the volatilization of vegetables or other matter; that smells are caused by actual molecules suspended in the air—material particles, very slender and volatilized in the atmosphere. But the matter seems to become almost intangible.

Nothing can give a more faithful idea of the divisibility of matter than the diffusion of smells. Three-quarters of a grain of musk placed in a room develop a very strong smell in it for a considerable time, without the musk perceptibly losing weight, and the box containing the musk will retain the perfume almost indefinitely. Haller states that papers perfumed with a grain of ambergris were quite odoriferous at the expiration of forty years. I remember purchasing upon the quay in Paris, some twelve years ago, a pamphlet which had a pronounced odor of musk about it. It had, no doubt, been there many months, exposed to the sun, the wind, and the rain. Since that time it has remained upon a library shelf, where the air has full access to it, and having just opened its pages, I find it as fully scented as ever.

Smells are transported by the air to great distances. A dog can recognize his master's approach from a distance; and it is asserted that at twenty-five miles from the coast of Ceylon the delicious perfume of its balmy forests is still borne upon the wind. These sweet perfumes, like the harmony and the activity of the terrestrial surface, we owe to the atmosphere.

CHAPTER VII.

AERONAUTICAL ASCENTS.

THE air being a fluid possessing weight, analogous to water in regard to the principles of pressure,[*] but, as we have seen, very much lighter, an instant's reflection will suffice to show that, if a body lighter than air be placed in the atmosphere, it will rise just as a body lighter than water—such as wood or cork—will, if placed at the bottom, at once ascend to the surface, because of its less specific gravity.

If the atmosphere formed a homogeneous ocean above the surface of the globe, equally dense throughout, and terminated, like the sea, by a defined surface, every body the density of which was less than the density of this aërial ocean would rise, when left to itself, by the ascensional force of a pressure dependent on the difference of densities, and would remain floating upon the upper surface of this atmosphere. This was the notion of several of the predecessors of Montgolfier; among others, of the worthy Father Galien, in his fantastic scheme for aërial navigation, published in 1755. His famous ship was to contain "fifty-four times as much weight as Noah's ark," its dimensions were to be equal to those of the town of Avignon; for the hypothesis of this excellent ecclesiastic was that this vast iron vessel would float in the atmosphere in virtue of the same principle as that by which a ship floats upon the ocean. But as the density of the atmospheric strata diminishes with elevation, all objects lighter than the lower strata mount merely to the region the density of which is such that the weight of the body is equal to the weight of the volume of fluid displaced.

Archimedes established for liquids a principle which we can apply with precision to the atmospheric fluid, enunciating it as follows: All bodies situated in the atmosphere lose a portion of their absolute weight, equal to the weight of the air which they displace.

This actual loss of weight in the air is proved by means of a pair of scales specially constructed for the purpose, as the name indicates.

---

[*] [It must be borne in mind that water is very slightly compressible indeed; while air is an elastic fluid, capable of almost indefinite compression or expansion.—ED.]

of *seeing* the weight—the baroscope. One extremity of the beam has attached to it a hollow copper sphere; the other end carries a small

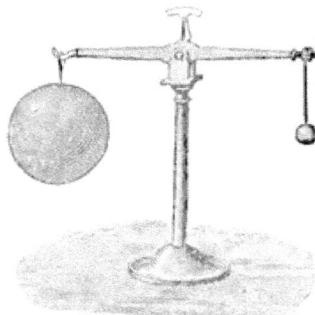

Fig. 25.—The Baroscope.

piece of lead, balancing in the air the copper sphere. If this apparatus is placed under the glass-receiver of an air-pump, as soon as a vacuum has been created the balance inclines to the side of the sphere, showing that in reality it weighs more than the mass of lead which was in equilibrium with it when in the air; or, in other words, that it has lost in the air a portion of its weight, because its volume was larger than that of the piece of lead. To verify, by means of the same apparatus, that this loss is just equal to the weight of the air displaced, the volume of the sphere must be measured, and if it holds say about a pint, or 34·6 cubic inches, the weight of this volume of air being 11·3 grains, the corresponding weight must be attached to the piece of lead, and the equilibrium will be re-established in the vacuum, but will be destroyed upon the re-introduction of air.

Let us note, *en passant*, in reference to this subject, that when any object is weighed in scales it is never its exact weight which is obtained, but its apparent weight. To get at the actual weight, the object must be weighed in vacuum. This is a source of continual error which is rarely taken into consideration. But, on the other hand, it may be asked, what is the real weight of any particular body? and the reply must be, there is no such thing. It is a purely relative matter, resulting from the volume and density of the planet which we inhabit. A pound weight does not constitute an absolute quantity, notwithstanding appearances to the contrary. The proof of this is, that if a pound weight were transported to the surface of the sun it would weigh nearly twenty-eight pounds;[*] whereas it would weigh two pounds and three quarters, nearly, upon the surface of Jupiter, and only one-sixth of a pound at the moon! And even without going so far as this, if we imagine our atmosphere gradually becoming denser and denser, we

[*] [The weighing must, of course, be made by means of a spring-balance, or other balance of the same kind. If a certain object balances a pound weight on the earth in a pair of scales, it would do so also anywhere else—on the sun, moon, etc.—ED.]

should, in that case, become lighter; or, again, if the earth revolved seventeen times faster than it does, the inhabitants of tropical countries would have no weight at all, and only weigh a few grains in the latitude of London or Paris. This may serve to confirm the doctrine of those English philosophers who, with Berkeley at their head, argued that the only real fact is, that there is nothing real in the world.

But let us return to the weight of the air. A balloon is, in fact, merely a body lighter than the weight of the air which it displaces, and which consequently rises in search of its equilibrium into higher regions of less density, where it will only displace a volume of air equal to its own weight. It is clear that, far from being in opposition to the laws of gravity, the ascent of balloons is, on the contrary, a special confirmation of them.

Whatever may be the substance which is used for filling a globe of silk or other material, if the whole apparatus—the gas which fills the envelope, the car, the net to which it is attached, the aeronauts, etc.— weighs less than the air which it displaces, it constitutes by that very fact an aerostatical machine, and rises in the atmosphere.

When Montgolfier launched, for the first time, a balloon into the air, his balloon was simply inflated with hot air. The density of air heated up to 122° (Fahr.) is 0·84, that of air at 32° being represented by 1. The density at 212°, the temperature of boiling-water, is 0·72, giving scarcely a difference of one-third for the ascensional force. The density of pure hydrogen is only 0·07; that is, one-fourteenth of that of air. The density of carbureted hydrogen is about 0·55; that is, about one-half the density of air. The latter of these two gases is generally used for filling balloons.

By a happy coincidence not rare in the history of science, hydrogen gas was discovered almost simultaneously with the invention of balloons. In 1782, Cavallo exhibited before audiences, at his London lectures, soap-bubbles formed of hydrogen, which rose by their less specific gravity up to the ceiling of the hall. In the following year (June 5, 1783) Montgolfier launched the first aerostat. With a little study and energy, Cavallo might have deprived the Annonay manufacturer of the immortality of his invention.

A balloon inflated with hot air is still often called a Montgolfier balloon, after its inventor. A balloon inflated with gas is denominated a gas-balloon, and often, popularly, an air-balloon. Gas has been adopted almost exclusively since its first trial, which was made at

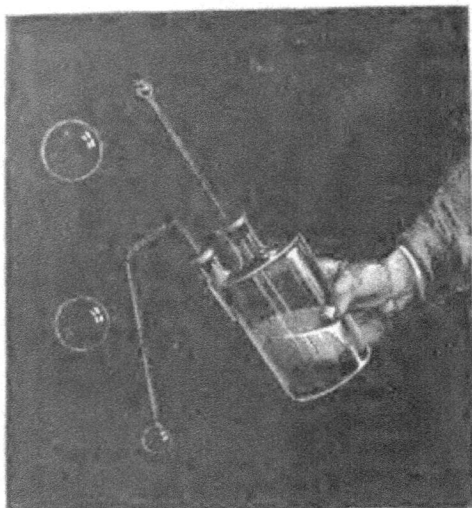

Fig. 26.—Soap-bubbles inflated with hydrogen.

Paris, on the 27th of August, 1783, by M. Charles, Member of the
Academy of Sciences, and the Brothers Robert.

The first time that a car was suspended to a balloon was on the 19th
of September, 1783, in presence of Louis XVI. and Marie Antoinette,
at Versailles; and the earliest passengers were a sheep, a cock, and a
duck. The first real aërial voyage was accomplished on the 21st of
October following, by Pilâtre des Rosiers and the Marquis d'Arlandes,
who rose, by means of a fire-balloon, from the Château de la Muette
(the Bois de Boulogne), and made their descent at Montrouge (on the
south side of Paris), after having crossed the capital.

To say that one *feels* one's self being carried up by a balloon perhaps
scarcely gives a correct idea of the situation. It is better to say, *sees*
one's self carried up, for the voyager feels no kind of movement, and
the earth seems to him to be descending.

As personal impressions are unquestionably those the recital of
which comes nearest to the reality, I will take the liberty of citing
some. My first ascent took place on Ascension-Day (May 25) in 1867.
Eugène Godard, the aeronaut, having verified the perfect equilibrium
of the balloon, orders the four assistants to let slip through their hands,
without losing hold of them, the ropes which secured the car, and thus
we find ourselves a few yards above the ground. The sky is clear, the

wind light, and the balloon, filled with hydrogen gas, becomes impatient and endeavors to rise. Then, taking a sack of ballast in his hand, Godard gives the word to " let go," throwing over a few pounds of sand, and the aerostat rises with majestic ease.

The balloon rises in an oblique curve, caused by two component forces—its ascensional power on the one hand, and the velocity of the wind on the other. If, as is proper from all points of view, we take care to let the balloon have only a slight ascensional force, the most magnificent of panoramas is slowly developed before the charmed gaze. If we wish only to ascend to a height of 3000 to 4000 feet, the balloon is allowed to move horizontally as soon as it reaches an atmospheric stratum of this elevation, whose density is then equal to that of the balloon. For higher ascents, the balloon is lightened by throwing out ballast.

The aeronaut, the meteorologist, or the astronomer who thus hovers in the air, is in a most enviable position for studying the atmosphere. Penetrating into the very midst of the clouds, traversing them to determine the light and heat which influence them, following the storm in its mysterious formation, studying the production of rain, snow, and the hail, transporting himself, in fact, into the very regions where these phenomena are occurring, it is there alone that the observer is really master of the globe. The *savant* may in vain spend years by his fireside in forming hypotheses by the aid of books and apparatus; but in this, as in most other things, the surest method of ascertaining what is going on, is "to go and see for one's self," as the old proverb has it. And, assuredly, no attempt can yield more fruitful results.

I do not intend to revert to a subject which was largely and completely dealt with in 1870 in a work specially devoted thereto. The purpose of this chapter is not to record my travels in the air; the scientific results flowing from them will be found embodied in the various explanations which compose the present book. It was merely necessary to lay down the general theory of the ascent of a balloon in its relations to the study of the atmosphere, and to give some idea of the effects of the higher regions.

If aërial travels may be profitably applied to the study of the forces at work in the atmosphere, and of the laws which preside over its multiform movements, they are also a special subject of interest for the observer, and open for him an exclusive vista of vast and useful contemplation. Borne into the fields of the sky by the invisible breath of the winds, the solitary balloon rises above the earth, and the traveler views its surface as a map stretched out on a boundless plain seen with all the characteristics of its local topography. Capitals situated on the banks of rivers, the central cities of provinces, innumerable villages disseminated over the country, and succeeding each other in hundreds like the little châteaux one used to see dotted down in old-fashioned maps, hillsides brown with the vine, furrows golden with grain, verdant meadows, cragged mountains whose tops are covered with sombre forests, sparkling streams and sinuous rivers running to the distant ocean—all the charms, soft or stern, of landscape and perspective are slowly revealed to the delighted gaze of the aeronaut who, without feeling the slightest movement, hovers as in a dream until he again sets foot upon the earth that he has been contemplating from on high. A less powerful impression, but of a similar kind, is derived from a mountain ascent.

The purity of the upper air, and the variation in atmospheric pressure, are physical elements which must be taken into account in order to explain the benefit of a sojourn at a moderate altitude. The peculiar action which may be exercised upon impressionable organizations by the contemplation of mountains, where nature has bestowed so liberally that mixture of the gracious and the terrible which tends to make up the picturesque, is undeniable. J. J. Rousseau says: "Every one must feel, though he may not observe it, that in the purer and more subtle air of the mountains he has a greater facility of breathing, more nimbleness in the body, more serenity of mind; the pleasures are less ardent there, as the passions are more subdued. Meditation assumes a certain tranquil voluptuousness, which is not in the least sensuous or bitter. It seems that, as we rise above the abode of man, we leave all terrestrial and base sentiments behind, and as we approach the ethereal regions, the soul gains something of their inalterable purity. We become grave without being melancholy, placid without indolence, content to live and to think. I doubt whether any violent agitation, any hysterical affection, could hold out against a lengthened sojourn there; and I am astonished that a bath of the healthy mountain air is not one of the greatest medical remedies."

It is, however, proper to state that, beyond moderate altitudes, the human organism is susceptible of a deleterious influence, owing to the change in atmospheric pressure, the dryness of the air, and the cold.

The physiological uneasiness and disturbances which are felt at great heights have long been ascertained facts. As early as the fifteenth century they were observed and described by Da Costa, under the name of *mal de montagne*. Later, all mountain explorers in the Alps, the Andes, and the Himalayas, as well as aeronauts, have noted these singular perturbations of organism, and have published theories more or less plausible in explanation of them. The principal cause assigned since De Saussure has been merely the rarefaction of the air; but by what series of actions and reactions does this rarefaction affect the human body? That was the point which needed elucidation.

In 1804, Gay-Lussac and Biot rose as high as 13,000 feet in a balloon. Gay-Lussac's pulse went up from 62 to 80 a minute; that of Biot from 79 to 111. In the memorable ascent of July 17, 1862, Messrs. Glaisher and Coxwell attained the enormous elevation of 37,000 feet. Previous to the start, Glaisher's pulse stood at 76 beats a minute, Mr. Coxwell's at 74. At 17,000 feet the pulse of the former was at 84—of the latter, at 100; at 19,000 feet Glaisher's hands and lips were quite blue, but not his face; at 21,000 feet he heard his heart beating, and his breathing was becoming oppressed; at 29,000 feet he became senseless, and only returned to himself when the balloon had come down again to the same level; at 37,000 feet the aeronaut could no longer use his hands, and was obliged to pull the string of the valve with his teeth. A few minutes later he would have swooned away, and probably lost his life. The temperature of the air was at this time 12° below zero. In aerostats, however, the explorer remains motionless, expending little or none of his strength, and he can therefore reach a greater elevation before feeling the disturbance which brings to a halt at a far lower level the traveler who ascends by the sole strength of his muscles the steep sides of a mountain.

De Saussure, in his ascent of Mont Blanc on the 2d of August, 1787, has given an account of the uneasiness which his companions and himself began to experience when a long distance from the summit. Thus, at 13,000 feet, upon the Petit-Plateau, where he passed the night, the hardy guides who accompanied him, to whom the few hours' previous marching was absolutely child's play, had only removed five or six spadefuls of snow in order to pitch the tent, when they were obliged to give in and take a rest, while several felt so indisposed that they were compelled to lie upon the snow to prevent themselves from fainting. "The next day," De Saussure tells us, "in mounting the last ridge which leads to the summit, I was obliged to halt for breath at every fifteen or sixteen paces, generally remaining upright and leaning on my stock; but on more than one occasion I had to lie down, as I felt an absolute need of repose. If I attempted to surmount the feeling, my legs refused to perform their functions; I had an initiatory feeling of faintness, and was dazzled in a way quite independent of the action of the light, for the double crape over my face entirely sheltered the eyes. As I saw with regret the time which I had intended for experiments upon the summit slipping away, I made several attempts to shorten these intervals of rest. I tried, for instance, a momentary stoppage every four or five paces, instead of going to the limit of my strength, but to no purpose, as at the end of the fifteen or sixteen paces I was obliged to rest again for as long a time as if I had done them at a stretch; indeed, the uneasy feeling was strongest about eight or ten seconds after a stoppage. The only thing which refreshed me and augmented my strength was the fresh wind from the north. When, in mounting, I had this in my face, and could swallow it down in gulps, I could take twenty-five or twenty-six paces without stopping."

Bravais, Martins, and Le Pileur, in their celebrated expedition to Mont Blanc in 1844, experienced and investigated the same phenomena upon the Grand Plateau. In clearing the tent, which was half filled with snow, the guides had continually to stop for breath. An internal uneasiness, according to Martins, made itself apparent in many different ways. The appetite was gone. The strongest, biggest, and most hardy of the guides fell upon the snow, and was nearly in a fit when the doctor, Le Pileur, felt his pulse. On nearing the summit, Bravais was anxious to see how far he could go without a rest; at the thirty-second step he was obliged to stop short.

All the indispositions felt by the *savans* of whom we have been speaking, and by many other travelers, at great elevations, have been classed in the following list:

Breathing.—The breathing is accelerated, impeded, laborious; and there is a feeling of extreme dyspnœa at the least movement.

Circulation.—The great majority of travelers have noticed palpitations, quickening of the pulse, beating of the carotids, a sensation of plenitude in the vessels, and sometimes the imminent approach of suffocation and various kinds of hemorrhage.

Innervation.—Very painful headache, a sometimes irresistible desire to sleep, dullness of the senses, loss of memory, and moral prostration.

Digestion.—Thirst, strong desire for cooling drinks, dryness of the tongue, distaste for solid food, nausea, and eructations.

Functions of Locomotion.—Pains more or less severe in the knees and legs; walking causing great fatigue and exhausting all strength.

These disturbances are not regular, they do not all come on at once, and evidently depend a good deal upon the strength, the age, the habits, and the previous actions of the individual. They seem to have a greater effect upon Alpine climbers than in other mountainous regions. Thus, at the Great St. Bernard, the monastery of which has an altitude of only 8117 feet, most of the monks become asthmatic. They are compelled to descend frequently into the valley of the Rhône to regain their health, and at the end of ten or twelve years' service to quit the monastery for good, under penalty of becoming quite infirm; and yet, in the Andes and Thibet, there are whole cities where people can enjoy as good health as anywhere else. Boussingault says, that "when one has seen the activity which goes on in towns like Bogota, Micuipampa, Potosi, etc., which have a height of from 8500 feet to 13,000 feet; has witnessed the strength and agility of the toreadors in a bull-fight at Quito (which is 9541 feet); when one has seen young and delicate women dance for the whole night long in localities almost as lofty as Mont Blanc, where De Saussure had scarcely the strength to read his instruments, and where the vigorous mountaineers fainted; when one remembers that a celebrated combat, that of Pichincha, took place at a height as great as that of Monte Rosa (15,000 feet), it will be admitted that man can become habituated to the rarefied air of the highest mountains."

The same writer is also of opinion that in the vast fields of snow, the discomfort is increased by an emission of vitiated air under the action of the solar rays, and he bases this impression upon an experiment of De Saussure, who found the air near the surface of snow to contain less oxygen than that of the surrounding atmosphere. In certain hollows and inclosed valleys of the higher part of Mont Blanc—in the *Corridor*, for instance—people generally feel so unwell in traversing it, that the guides long thought that this part of the mountain was impregnated with some mephitic exhalation. Thus, even now, whenever the weather permits, people ascend by the *Bosses* ridge, where a purer air prevents the physiological disturbances from being so intense.

Notwithstanding that one may become gradually accustomed to the attenuated air of high elevations, certain animals can not live there. Thus cats, taken up to the altitude of 13,000 feet, invariably succumb,

after having been subject to singular attacks of tetanus, of gradually
increasing intensity; and, after making tremendous leaps, succumb
from fatigue, and die in convulsions.

We will conclude these remarks by mentioning that the highest inhabited spot in the world is the Buddhist cloister of Hanle (Thibet),
where twenty priests live at the enormous height of 16,500 feet.
There are other cloisters built at a nearly equal height in the province
of Guari Khorsum, upon the banks of the lakes Monsaraour and Ba-
kous, and they are inhabited all the year round. In these equatorial
regions one can live very easily for ten or twelve days at an altitude
of 18,000 feet, but not for a longer time. The Brothers Schlagintweit,
when they explored the glaciers of the Ibi-Gamin in Thibet, encamped
and passed the night, with eight men of their expedition, from the 13th
to the 23d of August, 1855, at these exceptional elevations, which are
rarely visited by a human being. For ten days their encampments
varied from 18,000 to 21,000 feet; that is to say, the greatest altitude
at which a European ever passed the night. These three brothers
succeeded, on August 19, 1856, in mounting to an elevation of 24,339
feet—farther than man has ever yet reached. At first they suffered a
good deal when they got to 17,000 feet; but, after a few days, they felt
nothing but a passing uneasiness even at 19,000 feet. It is, however,
probable that a prolonged stay at this altitude would have produced
ill effects.

Three or four years ago, Professor Tyndall, in order to take scientific
observations, passed the whole night upon the summit of Mont Blanc,
sheltered only by a small tent. The guides who accompanied him were
so unwell that the next morning they were obliged to make their way
downward as quickly as possible.

A year or two ago, M. Lortet, who had several times ascended to
14,000 feet upon Mont Blanc without discomfort, and who doubted
whether another 1600 feet could superinduce the symptoms asserted,
went to the summit to judge for himself. He writes: "I am now convinced, and am compelled to admit, *de visu* and rather at my expense,
that there really do exist causes of disturbance at this height which affect a person who ascends so far, especially if he is *in motion*, in this
rarefied air. This is also the result of my personal observations; and
I have satisfied myself that it is much less hurtful to the organic functions to rise to great heights when sitting still in a car than by climbing
over the snows."

To complete our atmospheric panorama, it is interesting to see what are the highest points of the mountainous peaks upon which man is living, and what are the highest points of the mountain chains which raise into the rarefied atmosphere their silent and icy peaks. The highest spots of the earth which are inhabited are:

The Buddhist cloister of Hanle (Thibet)	16,532 feet.
Cloisters on the sides of the Himalaya	14,764 to 16,404 "
The post-house of Apo (Peru)	14,377 "
The post-house of Ancomarca (do.)	14,206 "
The village of Tacora (do.)	13,491 "
The town of Calamarca (Bolivia)	13,651 "
The vineyard of Antisana (Republic of Ecuador)	13,455 "
The town of Potosi (Bolivia), ancient pop.: 100,000	13,323 "
The town of Pano (Peru)	12,871 "
The town of Oruro (Bolivia)	12,155 "
The town of La Paz (do.)	12,225 "

Quito, capital of the Ecuador Republic, is situated at an altitude of 9541 feet; La Plata, capital of Bolivia, at 9331 feet; Santa Fé de Bogota, at 8730 feet. The highest inhabited spot in Europe is the Mónastery of Mount St. Bernard, which is 8117 feet high.

The highest passes of the Alps are—the pass of Mount Cervin, 11,188 feet; the Great St. Bernard, 8110 feet; the Col de Seigne, 8074 feet; and the Furka, 8002 feet. The highest passes in the Pyrenees are— the Port d'Oo, 9843 feet; the Port Viel d'Estaube, 8402 feet; and the Port de Pinede, 8202 feet.

The highest mountains in the world are:

Asia : The Gaurisankar, or Mount Everest (Himalaya)			29,003 feet.
The Kanchinjinga (Sikkim, Himalaya)			28,156 "
The Dhaulagiri (Nepaul,	do.	)	26,825 "
The Juwahir (Kemaon,	do.	)	25,670 "
Choomalari (Thibet,	do.	)	23,945 "
America: The Aconcagua (Chili)			22,422 "
The Sahama (Peru)			22,349 "
The Chimborazo (Republic of Ecuador)			21,424 "
The Sorota (Bolivia)			21,283 "
Africa: The Kilimanjaro			20,001 "
Mount Woso (Ethiopia)			16,601 "
Oceania : The Mowna-Roa, volcano (Sandwich Isles)			15,874 "
Europe: Mont Blanc			15,797 "
Monte Rosa			15,211 "

The birds, of course, represent the population of the very highest altitudes. In the Andes the condor, in the Alps the eagle and the vulture, hover above the topmost peaks. Fitted for the longest journeys, they are the greatest sailors in the atmospheric ocean, just as the petrels and the gigantic sea-swallows are the great sailors over the Atlantic. The choucas (a kind of jackdaw), with its intensely black plumage and yellow beak and red legs, does not rise so high into the atmosphere, but it is especially the bird of the highest peaks, of the region of snows and barren cones. It is met with at the summit of Monte Rosa and at the Col du Géant, at over 11,500 feet.

There are also birds more graceful in form which live in the region of hoar-frost, and lend a little animation to those bleak and unchanging landscapes. The snow-chaffinch has so great a preference for this cold region that he rarely descends to the zone of the woods. The *accenteur* of the Alps also follows him to great elevations, preferring the stony and barren region which separates the zone of vegetation from that of perpetual snow, and both of these birds sometimes soar as high as 11,000 to 15,000 feet in pursuit of insects.

The engraving (see Fig. 27) represents the principal kinds of birds according to the maximum height to which they fly. The earth has its birds, like the air. Certain kinds never use their wings but for a few moments, when it is impossible for them to move along the ground: for instance, all the gallinaceous kinds. The region of snow has its own kind, just as it has its characteristic sparrows. The ptarmigan, or snow-hen, is met with in Iceland as in Switzerland. It soars far above the everlasting hoar-frosts, and is so fond of the snow that at the approach of summer it mounts farther in search of it, plunging into it with evident delight. A few lichens, grains brought up there by the air, suffice for its food. It looks for insects, with which it nourishes its young.

The insects are, indeed, the only animals which are abundant in these bleak regions—a fresh analogy with the polar countries. It is also the class of coleoptera which predominate in the higher Alpine regions. They attain to 9800 feet on the southern slope, and to 7900 feet on the opposite side. Their wings are so short that they scarcely seem to have any; one would imagine that nature had intended to protect them from the strong currents of air which would undoubtedly carry them away if their wings had not been, so to speak, reefed. One does every now and then encounter other insects, neuroptera and butterflies, which the winds have taken up to these heights, and which are afterward lost amidst the snows. The seas of ice are covered with victims that have

Fig. 27.—Distribution of kinds of Birds according to height of flight.

. Condor (has been seen as high as 9000 metres, or 29,560 feet) ; 2. Griffon ; 3. Vulture ; 4. Sarcoramphus ; 5. Eagle ; 6. Urubu ; 7. Kite ; 8. Falco ; 9. Sparrow-hawk ; 10. Fly-bird ; 11. Pigeon ; 12. Buzzard ; 13. Swallow ; 14. Heron ; 15. Crane ; 16. Duck and Swan (found in lakes at an altitude of 1600 metres, or 5509 feet) ; 17. Crow ; 18. Lark ; 19. Quail ; 20. Parrot ; 21. Partridges and Pheasants ; 22. Penguin.

perished in this way. Nevertheless, there are certain kinds which appear to travel freely as high as 13,000 or 16,800 feet. In my aërial voyages, I have met with butterflies at heights to which the birds of our latitudes do not ascend, and at more than 9800 feet above the ground. Dr. J. D. Hooker noticed some at Mount Momay, at an altitude of more than 17,700 feet. Such is the scale of animal life in these Alpine zones, where the fauna gradually becomes scarcer, finally giving way to solitude and desolation. Beyond the last stage of vegetation, beyond the extreme region attained by the insect and mammifers, all becomes silent and uninhabited; yet the air is still full of microscopic animalcules, which the wind raises up like dust, and which are disseminated to an unknown height.

BOOK SECOND.

LIGHT AND THE OPTICAL PHENOMENA OF THE AIR.

CHAPTER I.

THE DAY.

As the atmosphere is the organizer of life; as all beings, animal and vegetable, are so constituted as to be able to breathe in its midst and construct, by means of its fluid molecules, the solid tissue of their organisms, we must now turn our attention with admiration to the atmosphere, as being still further the ornament of nature, and we shall see that we owe to it not only the picture, but also the frame.

Whether the sky be clear or cloudy, it always seems to us to have the shape of an elliptic arch; far from having the form of a circular arch, it always seems flattened and depressed above our heads, and gradually to become farther removed toward the horizon. Our ancestors imagined that this blue vault was really what the eye would lead them to believe it to be; but, as Voltaire remarks, this is about as reasonable as if a silk-worm took his web for the limits of the universe. The Greek astronomers represented it as formed of a solid crystal substance; and so recently as Copernicus, a large number of astronomers thought it was as solid as plate-glass. The Latin poets placed the divinities of Olympus and the stately mythological court upon this vault, above the planets and the fixed stars. Previous to the knowledge that the earth was moving in space, and that space is everywhere, theologians had installed the Trinity in the empyrean, the angelic hierarchy, the saints, and all the heavenly host. . . . A missionary of the Middle Ages even tells us that, in one of his voyages in search of the terrestrial paradise, he reached the horizon where the earth and the heavens met, and that he discovered a certain point where they were not joined together, and where, by stooping, he passed under the roof of the heavens. . . . And yet this vault has, in fact, no real existence! I have myself risen higher in a balloon than the Greek Olympus was supposed to be situated, without being able to reach this limit, which, of course, recedes in proportion as one travels in pursuit of it—like the apples of Tantalus.

What, then, is this blue, which certainly does exist, and which veils from us the stars during the day?

The vault which we behold is formed by the atmospheric strata which, in reflecting the light that emanates from the sun, interpose between space and ourselves a sort of fluid veil, which varies in intensity and height with the density of the aërial zones. The illusion referred to above took a long time to dispel, and it was also a work of time to make it known that the shape and dimensions of the celestial vault change with the constitution of the atmosphere, with its state of transparency and its degree of illumination. One part of the celestial rays sent from the sun to our planet is absorbed by the air, the other part is reflected; the air, nevertheless, does not act equally on all the colored rays of which white light is composed, but acts like a glass, allowing the rays toward the red end of the solar spectrum to pass more readily than those in the neighborhood of the blue end. This difference is only perceptible when the light passes through a great thickness of air. De Saussure pointed out that the blue color of the sky was due to the reflection of light, and not to a hue peculiar to aërial particles. "If the air were blue," he says, "the distant mountains, which are covered with snow, would appear blue also, which is not the case." An experiment made by Hassenfratz also proves that the blue ray is more reflected; in fact, the thicker the atmospheric stratum is which a ray traverses, the more do the blue rays disappear to give place to the red; and as, when the sun is near to the horizon, the ray has to traverse a greater thickness of air, the sun therefore appears red, purple, or yellow. The blue rays are also frequently absent in rainbows which make their appearance just before sunset.

We shall see further on that it is the vapor of water accumulated in the air which plays the principal part in this reflection of the light, to which we owe the azure of the sky and the brightness of day.

Very recently, Professor Tyndall reproduced the blue of the sky and the tint of the clouds in an experiment at the Royal Institution. Vapor of different substances, of nitrite of butylene, of benzone, and of carbonic sulphide, is introduced into a glass tube; a succession of electric sparks is then passed through it, and the condensation and rarefaction of the vapor augmented *ad libitum*. As soon as the vapors employed, no matter what their nature is, are sufficiently attenuated, the reflection of the light first manifests itself by the formation of a blue like that of the sky. There is, I will suppose, in the tube a half atmosphere of air mixed with vapor, and another half atmosphere of air that has passed through hydrochloric acid. The proportion and density of the gas can, of course, be varied.

The vaporish cloud, after having first assumed the blue tint, becomes more condensed and white, and as it thickens, becomes exactly like real cloud, presenting, as regards polarization, the same variation of phenomena.

The atmospheric air is one of the most transparent bodies known. When it is not charged with mist or obscured by other bodies, we can see objects at an immense distance, and mountains only disappear from our view when they are below the horizon; but, in spite of its slight power of absorption, the air is not completely transparent; its molecules absorb a portion of the light which they receive, permit the passage of another part, and reflect the third; and hence it is that they give rise to what appears a vault, that they light up terrestrial objects which the sun does not reach directly, and effect an imperceptible transition between day and night.

It is easy to convince one's self of the decrease in the intensity of the solar light during its passage through the atmosphere by daily observations. If an object situated near the horizon is watched for several days together, it will be seen that it is more visible at one time than at another. The distance at which its details fade out of sight is at one moment less than at another, as may be proved by direct measurement; the transparency of the air can be even expressed numerically, as has been done by De Saussure through the instrumentality of the *diaphanometer.* The distance at which objects disappear does not depend upon the angle of vision alone, but also upon the manner of their illumination, and the contrast which their color offers to surrounding objects. This explains why the stars, despite their small diameter, are so visible in the vault of heaven. It is the same with some terrestrial objects. It is difficult to distinguish a man, as he stands out in the fields, as against dark surfaces; but he is very easily seen if he is placed upon an elevation so as to stand out against the clear sky. Hence the optical illusions so common in mountainous countries.

While the chain of the Alps, seen from the plain at a great distance, is visible in its minute details, the spectator standing upon one of its peaks can distinguish hardly any thing in the plain. From the Faulhorn, for instance, it is easy to make out very distinctly the chain of the high Alps; but every thing in the valley below is dim and confused. The summits of the Pilate, the Black Forest, and the Vosges are clearly defined at a great distance, whereas nothing can be distinguished in the plain between the Alps and the Jura. Any one who

has passed a few months amidst the lakes and mountains of Switzerland must have noticed the same variations in the visibility of objects.

To measure the intensity of the blue color, De Saussure invented the
cyanometer, which is composed simply of a strip of paper divided into
thirty rectangles, the first of which is of the deepest cobalt blue, while
the last is nearly white, the intermediate colors offering every conceivable shade between dark blue and white. If it be found that the blue
of one of these rectangles is identical with that of the sky, this identi-
ty is then represented by a number corresponding to one of the rectangles, and all that remains to be done is to arrange the scale of the
instrument.

Humboldt perfected this apparatus, and rendered it capable of giving
very precise measurements of the blue tint.

The mere contemplation of the heavens tells us that their color is not
the same at every altitude, being generally deeper at the zenith, and
gradually becoming lighter toward the horizon, where it is often nearly
white. The contrast is rendered the more striking by the use of the
cyanometer. Thus it will be found that sometimes the color corresponds to the number twenty-three in the neighborhood of the zenith,
and to the number four near the horizon. But the color of the same
part of the sky also changes pretty regularly during the day, as it becomes darker from morning until noon, and lighter again from noon
until evening. In our climates the deepest blue is when, after several
days of rain, the wind drives away the clouds.

The color of the sky is modified by the combination of three tints—
the blue, which is reflected by the aërial particles; the black of infinite
space; and the white of the vesicles of mists and snow-flakes which
float at the high elevations. If we rise sufficiently high in the atmosphere, we leave a part of the vesicles of vapor below us. Thus the
white rays reach the eye in a lesser proportion, and, the sky being covered with fewer particles which reflect the light, its color becomes of a
deeper blue.

The nature of the ground also plays an important part in these effects
of reflection and atmospheric transparency.

In the regions where there are vast surfaces devoid of vegetation, as
in a great part of Africa, the air is very dry, and loses part of its transparency, especially in consequence of the dust borne by the winds and
the absence of heavy rain to cleanse the air. In the other parts of the
intertropical zone, upon the Atlantic, on the American continent, in the

South Sea Islands, and in certain regions of India, aqueous vapor, in a state of transparent gas, is abundantly mixed with the air; and in place of the grayish hue which it possesses in our climates and in sandy deserts, the sky presents a strongly-marked tint of azure blue, which specially characterizes the regions about the zenith, and sometimes even the sky near the horizon.

The limiting surface of the atmosphere being parallel to that of the earth, and the visible portion being that only which is above the plane of the horizon, it is clear that rays of light reaching the eye in different directions have traversed different thicknesses of air. If the sun were at the zenith, its rays would pass through the thinnest stratum of air; the nearer the sun approaches the horizon, the thicker becomes the mass of air which its rays have to pierce, and consequently the weaker its rays become. The light of the sun at its meridian passage is dazzling, whereas we can look at it with the naked eye when near the horizon; and for the same reason the regions situated near the horizon seem always to be without stars.

The color of the sky is thus explained by the reflection of light upon the molecules of the vapor of water which invisibly pervades the air.

How are we now to explain the very perceptible shape of an elliptical *vault* which the sky presents, whether cloudy or entirely clear?

This may be explained as a simple effect of perspective.

I will suppose we have before us an avenue of poplars, all of the same height. Every one knows that this height will apparently decrease with distance, and that the top of the trees at the extreme end of the avenue will appear to be at the height of our eyes.

The trees' roots are upon a horizontal surface, because we ourselves are upon the ground. It is by the *top line* that the inclination toward the ground operates. If we were in the upper branches of the nearest tree, then it would be *from below* that the perspective inclination would operate. The same train of reasoning may be applied to the clouds. Starting from those which are vertically above our heads, they successively decline in height according to their distances above the horizon.

When we are above the clouds in a balloon, they no longer seem to sink toward the earth like a vault, but to extend like the plane surface of an immense ocean of snow. When but a few miles above them, they describe a curve in the contrary direction.*

* [Having been led theoretically to expect such a phenomenon, I always, when some miles above the clouds, attentively looked for its appearance, and invariably without success. It

With a clear sky, the surface of the earth, seen from a great height, is *hollow* underneath the car of a balloon, and gradually rises around up to the circular horizon. Far from appearing convex, as might be expected if one imagined that at a great height in the atmosphere the spherical shape of the globe would be recognized, the surface of the ground is hollowed out underneath us, rising till it reaches the horizon, which seems always to be on a level with the eye.

This aspect of the earth, hollowed out like a basin, surprised me very much the first time I saw it from a balloon, for at the height which I had attained I had expected to see it convex.

Thus the sinking of the apparent vault of the sky above our heads is due to an effect of perspective, as we can not estimate vertical heights in the same way as horizontal lengths. A tree forty-five feet high seems much longer on the ground than when standing. A tower three hundred feet high would appear far more if laid along the ground than when vertical. Being in the habit of walking along the ground, and not of soaring into the air, we appreciate lengths at their true estimate, whereas heights are beyond our powers of direct judgment.

It results from the apparent shape of the celestial vault that the constellations seem to us much larger toward the horizon than at the zenith (as, for instance, the Great Bear when it skirts the horizon, and Orion when he rises), and that the sun and the moon appear to have larger disks at their rising and setting than at their culminating points. It further results that we are constantly in error in estimating the height of stars above the horizon. A star which is at 45° of altitude— that is, just half-way between the horizon and the zenith—seems to us much higher; and when we point out a star as being at 45°, it may happen that it is only at 30°.*

Modern treatises on physics and meteorology have not gone into this

is true that, the dip of the horizon being very small, objects on the horizon practically appear to be on the same level as the eye, while the ground underneath of course seems far below, so that, in this sense, the appearance of the earth is cup-shaped. But, in point of fact, if the day be clear, the distance of the horizon is so much greater than is that of the ground below, that the effect is no more noticeable than it is from the top of a hill. If the air be not clear, all traces of the appearance are of course absent.—Ed.]

* [Most people imagine they are looking at the zenith when they are looking at a point 10° or 20° below it, and on this account their estimates of heights are too great. As regards the shape of the celestial sphere, it may be remarked that the distance to the horizon would appear greater than to the zenith, if it were only because of the intervening objects which occur in the former case; while, looking upward, there is nothing to aid the eye in its estimation.—Ed.]

curious question of the aspect of the sky. I find it discussed in certain works of the seventeenth and eighteenth centuries, but rather from a philosophical point of view than in its purely geometrical aspect. After a long dispute between Mallebranche and Régis upon this point, Robert Smith examined it in his "Optics" (1728), and concluded that the horizontal diameter of the celestial vault must seem to us six times as long as the vertical diameter. He is of opinion that this is due to the fact that "our view does not extend distinctly to the point at which the objects form an angle of the 8000th part of an inch in our eye, so that all objects seem to us to sink under the horizon at a distance of 25,000 yards."

The mathematician Euler, in his "Letters to a German Princess" (1762), devotes several chapters to an explanation of it, which may be stated in a few words. First, the light of the stars which are near the horizon is much weakened, because their rays have a greater distance to travel through our lower atmosphere than those which are at a greater height; secondly, being less luminous, we deem them to be farther off, because we always take the objects which are most clear to be nearest to us (for instance, a conflagration at night seems much closer to us than it really is); thirdly, this apparent distance of the celestial objects which are near the horizon gives rise to the imaginary elliptic vault of the heavens.

The logical arrangement of these last two points seems the inverse of the theory explained above, yet it may be seen that these two facts do not follow the one from the other, but are simultaneous in our observation. Perspective is due to the distance and to the diminution in brightness, and it gives a clear explanation of the apparent shape presented by the atmospheric strata, and the variation in size according to the elevation above the horizon. There is, so to speak, a double effect of geometrical and luminous perspective.

We do not appreciate the beauty or the practical importance of the diffusion of light by the air, because it is always present to us. A sojourn of a few hours in our neighbor the moon would suffice to show us the enormous difference there is between an atmospheric day and one without air.

As Biot remarked, in a very correct simile, the air is around the earth a sort of brilliant veil, which multiplies and disperses the sunlight by an infinity of *repercussions*. It is to it that we owe the light which we enjoy when the sun is below the horizon. After the latter has risen

there is no spot so secluded, provided the air has access to it, which
does not receive some light, although the sun's rays may not reach it
directly. If the atmosphere did not exist, each point of the terrestrial
surface would only receive the light reaching it directly from the sun.

The strange effect of the absence of the atmosphere would be far
more complete and striking if we had the power of transporting ourselves into our satellite. Let us compare the cheerful spectacle that
the earth presents, partly covered with its humid and wavy mantle,
and decked with flowers, to the aspect of the moon, with its stony or
metallic surface, abounding with crevasses and vast mountainous deserts, with its extinct volcanoes and peaks that seem like gigantic tombs,
with its sky invariably black and shapeless, in which reign, day and
night, stars without scintillation, the sun and the earth. There day-
time is, so to speak, nothing but night lighted up by a rayless sun.
No dawn in the morning, no twilight in the evening. The nights are
pitch-dark. Those parts of the lunar hemisphere which are toward us
are lighted by an earth-light, the first quarter of which coincides with
sunset, the *full earth* with midnight, and the *new earth* with sunrise.*
In day-time the solar rays are lost against the jagged ridges, the sharp
points of the rocks, or the steep sides of their abysses, designing here
and there grotesque shapes against the angular contours, and only striking the surfaces exposed to their action to become at once reflected and
lose themselves in space—fantastic shadows standing out in the midst
of a sepulchral world. Fig. 28 represents a landscape taken in the
moon, in the centre of the mountainous region of Aristarchus. There
is nothing but white and black. The rocks reflect passively the light
of the sun; the craters remain partially wrapped in shade; fantastic
steeples seem to stand out like phantoms in this glacial cemetery; the
absence of the atmosphere leaves the black space of the starry heaven
perpetually hanging over this dismal region, to which, fortunately, the
earth can offer no sort of analogy.

* [The moon always turns the same face to the earth; so that there is one-half of the
moon's surface that has never been seen from the earth. The words *one-half* must not be
taken quite literally, as, owing to a slight oscillatory motion of the moon, called libration,
we sometimes see a little more round the corner, as it were, than at other times. Speaking
generally, therefore, an inhabitant of the moon, if he saw the earth at all (i. e., was on the
hemisphere turned toward us), would always see it in the same position in the sky (and in
size about four times as large as the moon appears to us). The statement in the text is only
true for a spectator placed at the middle point of the visible hemisphere of the moon; the
lunar *day* is of course about four weeks.—ED.]

Fig. 28.—Lunar Day.

CHAPTER II.

EVENING.

LIGHT, that imponderable agent which enables us to see objects, and which by its qualities illuminates the magnificent atmospheric world in which we live, gives rise to an ever-changing series of effects. The atmosphere not only bathes the landscape with light by reflection, but also decomposes it by refraction, and gives additional variety to the beauties of the earth and sky.

When a ray of light passes from one transparent medium to another, it undergoes a deviation caused by the difference of density of the two media.* In passing from air to water the ray is bent toward the vertical, because water is denser than air. It is the same with a ray which passes from a higher to a lower stratum of air, for, as we have seen, the lower strata are denser than those above.

If a ray of common light be admitted through a small hole in a darkened room, and, after passing through a glass prism, be received on a screen, it will be seen that the ray of white light has been decomposed by refraction through the prism into seven colors — violet, indigo, brown, green, yellow, orange, red— which occupy different positions, in the above order, on the screen. The red rays, being the least bent from the direction of the original ray, are said to be least refrangible, and the violet rays, which form the other end of the spectrum, are said to be most refrangible.

In refracting light the air produces two distinct effects. On the one hand, it causes a ray of light which has its origin beyond the earth's at-

---

* [M. Flammarion here adds the sentence, "A stick plunged into water appears bent at the surface of the liquid, and the immersed portion appears more nearly vertical." As this illustration of the effect of refraction is given in many popular works, I think it worth while to point out its inaccuracy. A ray of light entering a denser fluid (the surface of which is horizontal) is bent nearer to the vertical; but a stick is not a ray of light, and in no way resembles one. The immersed portion of the stick is seen by rays that have been refracted at the surface of the water; and it easily follows, from the principles of optics, that the part under water appears bent *from* (not *toward*) the vertical. This any one can verify for himself experimentally. The sentence quoted above is therefore not only erroneous in theory, but also incorrect in fact. The apparent bending of the stick is only indirectly due to refraction.—ED.]

8

mosphere to become bent as it approaches the earth, so that we see the sun, moon, planets, comets, and the stars, as if they were higher in the heavens than they really are. On the other hand, it causes a more or less considerable separation between the various rays that constitute white light, according to its state of transparency and density.

The first effect mainly produces twilight; the second gives that soft, undulating beauty which is seen in the serenity of the evening.

Refraction is greater or less, in proportion as the luminous ray traverses the atmosphere in a direction more or less inclined to the vertical, being greatest for horizontal and vanishing for vertical rays. Astronomical observations would all be false with regard to the positions of objects if they were not corrected for the effect of refraction. Thus,

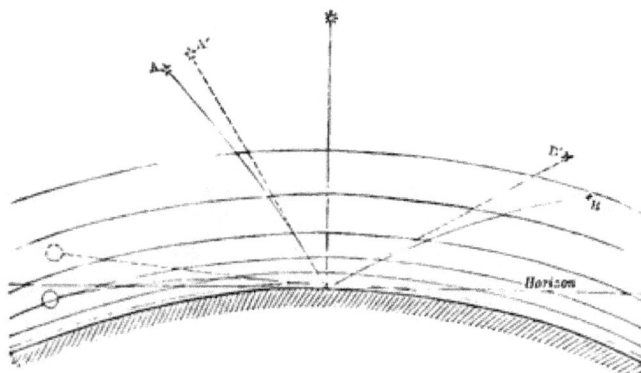

Fig. 29.—Atmospheric refraction.

for instance, the star A is seen at A'; the star B at B'; at the zenith alone stars are where they appear to be, there being no alteration in the direction of the ray of light due to refraction. To make these necessary corrections, tables have been constructed giving refractions, based upon the hypothesis of a uniform disposition of the different strata of air lying one above the other. The refracting power of the air is determined on the hypothesis that it contains only oxygen and nitrogen: but we have seen that it further contains from four to six parts in 10,000 of carbonic acid, and an ever-varying quantity of the vapor of water. The refracting power of the vapor of water differs so little from that of air properly so called, that the correction depending on it need not, as a rule, be taken into the calculation.

To calculate the amount of correction to be applied to any observation, it is only necessary to note at the time the temperature of the air and the pressure of the atmosphere at the place of observation.

To illustrate the effect of refraction, I have selected from a table of refractions a few numbers, at different zenith distances. They show to what extent objects are apparently raised by its influence:

TABLE OF REFRACTIONS.

Distances from the Zenith.	Refractions.	Distances from the Zenith.	Refractions.
90 deg.	33 min. 17 sec.	74 deg.	3 min. 20 sec.
89 "	24 " 22 "	72 "	2 " 57 "
88 "	18 " 23 "	70 "	2 " 38 "
87 "	14 " 28 "	65 "	2 " 4 "
86 "	11 " 48 "	60 "	1 " 40 "
85 "	9 " 54 "	55 "	1 " 23 "
84 "	8 " 30 "	50 "	1 " 9 "
83 "	7 " 25 "	45 "	0 " 58 "
82 "	6 " 34 "	40 "	0 " 48 "
81 "	5 " 53 "	30 "	0 " 33 "
80 "	5 " 20 "	20 "	0 " 21 "
78 "	4 " 28 "	10 "	0 " 10 "
76 "	3 " 50 "	0 "	0 " 0 "

From this table we see that an object situated just upon the horizon is raised by more than 33', or about $\frac{1}{160}$ of the distance from the horizon to the zenith. Neither the sun nor the moon is 33' in diameter. When, therefore, they appear to have just risen, they are still entirely below the horizon. In the same way, the sun does not appear to begin to set until after sunset has actually taken place.

It follows from these considerations that the sun may be seen in the west and the moon in the east at the time of full moon, and even an eclipse of the moon may be visible while the sun is still above the horizon, although the earth is then exactly between the two luminaries, and the latter are both, astronomically speaking, below the horizon. This is due to refraction. This curious circumstance was noted during eclipses of the moon on June 16, 1666, and May 26, 1668.

Owing to the same cause, the sun and the moon seem to be flattened both at their rising and setting, the rays proceeding from the lower edge of the luminary being more refracted than those proceeding from the upper, so that the apparent vertical diameter is diminished, while the horizontal diameter remains, of course, unaltered. The length of the day is thus increased, and that of the night decreased. It is for this reason that at Paris the longest day of the year is sixteen hours seven minutes, and the shortest eight hours eleven minutes, instead of being fifteen hours fifty-eight minutes, and eight hours two minutes. We see that the length of the day at Paris at the time of the solstices is thus prolonged by nine minutes, and by seven minutes at the equinoxes. At the North Pole the sun seems to be in the horizon, not when it arrives at the spring equinox, nor when its angular distance from the North Pole is 90°, but when it is 90° 33′; it then remains visible until, having passed to the autumnal equinox, its polar distance has again become equal to 90° 33′. Care must always be taken to keep account of refraction in calculating the hours of sunrise and sunset.

Twilight is that light which remains after the sun has set or which is seen before sunrise. The duration of twilight is, in many respects, a useful element to be acquainted with. It depends chiefly upon the angle to which the sun has descended below the horizon; but it is modified by several circumstances, the chief of which is the degree of clearness of the atmosphere. The direct light of the sun at the time of sunset reaches to the west; as the sun sinks, its boundary-line rises, and some little time afterward crosses the zenith, when civil twilight ends; the planets and large stars then become visible to the naked eye. The eastern half of the sky being thus first deprived of direct solar light, night begins there. Afterward, the boundary-line (the crepuscular curve) itself disappears in the west; then the astronomical twilight ceases and night has fully set in. Twilight begins or ends when the sun is at a certain distance below the horizon; this distance is variable, depending upon the state of the atmosphere. It may be taken that civil twilight ends when the sun is about 8° below the horizon, and that astronomical twilight ends when the sun is about 18° below the horizon. The phenomena of twilight are hardly known in tropical climates; as soon as the sun has descended below the horizon, darkness sets in suddenly. This was remarked by Bruce at Senegal, where, however, the air is so transparent that Venus may sometimes be distinguished at midday, and in the interior of Africa night succeeds day almost immediately after sunset. At Cumana, Humboldt tells us, twilight lasts but a very few minutes, although the atmosphere is higher under the tropics than in other regions.

The following tables give the length of the civil and astronomical twilight in France for the various seasons and for the fifteenth day of each month. By adding the duration of twilight to the hour of sunset, the time at which each of the twilights terminates is readily obtained, and subtracting it from the hour of sunrise, the times of their commencement are found. France, from the Pyrenees to Dunkirk, is within the 41st and 42d degrees of latitude. It will be seen that, even within these trifling limits, there is a perceptible difference. The shortest civil twilights take place on the 29th of September and the 15th of March, the longest on the 21st of June; the shortest astronomical twilights fall upon the 7th of October and the 6th of March, the longest on the 21st of June. North of 50° latitude, the astronomical twilight continues all night for some time both before and after the summer solstice.

TABLE OF THE LENGTHS OF THE LONGEST AND SHORTEST DAYS.

Latitude.	Length of the Day.	
	The longest: June 21.	The shortest: December 21.
42 degrees.	15 hrs. 13 min.	9 hrs.
44 "	15 " 28 "	8 " 47 min.
46 "	15 " 44 "	8 " 30 "
48 "	16 " 2 "	8 " 14 "
50 "	16 " 24 "	7 " 55 "

TABLE OF THE DURATION OF CIVIL TWILIGHT.

Month.	Latitude.				
	42 deg.	44 deg.	46 deg.	48 deg.	50 deg.
January	34 min.	35 min.	36 min.	38 min.	40 min.
February	32 "	33 "	34 "	35 "	37 "
March	31 "	32 "	33 "	34 "	35 "
April	32 "	33 "	34 "	36 "	36 "
May	35 "	36 "	38 "	40 "	42 "
June	37 "	39 "	41 "	44 "	46 "
July	36 "	38 "	39 "	42 "	44 "
August	33 "	34 "	36 "	37 "	39 "
September	31 "	32 "	33 "	34 "	36 "
October	31 "	32 "	33 "	35 "	36 "
November	33 "	34 "	35 "	37 "	39 "
December	34 "	36 "	37 "	39 "	41 "

TABLE OF THE DURATION OF ASTRONOMICAL TWILIGHT.

Month.	Latitude.									
	42 deg.		44 deg.		46 deg.		48 deg.		50 deg.	
	H.	M.	H.	M.	H.	M.	H.	M.	H.	M.
January	1	31	1	33	1	36	1	40	1	45
February	1	24	1	26	1	29	1	32	1	36
March	1	24	1	26	1	29	1	33	1	37
April	1	33	1	35	1	39	1	44	1	50
May	1	46	1	52	2	1	2	11	2	26
June	1	56	2	5	2	19	2	36	3	13
July	1	48	1	54	2	4	2	14	2	31
August	1	32	1	37	1	42	1	47	1	54
September	1	24	1	26	1	30	1	34	1	38
October	1	23	1	25	1	29	1	33	1	36
November	1	30	1	32	1	35	1	39	1	43
December	1	34	1	36	1	40	1	45	1	50

In warm countries, the presence of humidity in the air not only gives
to the sky its dark azure tint, but also has the effect of modifying the
vital power of the solar rays. At the equator it adds to the thousand
other wonders of nature an incomparably beautiful display of light both
at sunrise and sunset. The sunset, in particular, affords a spectacle
indescribably magnificent—a superiority over sunrise attributable to
the presence of moisture in the air. This is more abundant in the
evening, after the heat of the day, than in the morning, when it is partially condensed into dew by the effect of the cooler temperature of
night.

It is not in our climate that the finest sunsets are seen. The celestial
blue of distant mountains, the rose or violet tints which in turn tinge
the nearer hills, and the warm tones of the soil, harmonize in a marvelous manner, when the sun disappears below the horizon, with the
gleaming gold of the west, the red or roseate tints that crown it in the
sky, the dark azure of the zenith, and the more sombre and often, in
contrast to the others, greenish hue which prevails in the east. In the
equinoctial regions, these soft and delicate tints, joined to the varied aspect of the earth's configuration and the richness of vegetation, produce
more striking effects than with us. Sometimes light and roseate clouds,
fringed with a coppery red, produce peculiar effects, similar to certain
sunsets in our regions ; but whenever the sky is clear the shades differ
entirely from those of the temperate zone, and present a special character. Sometimes, too, the indentations of mountains situated below the
horizon, or invisible clouds intercepting a part of the solar rays which,
after sunset, still reach the elevated regions of the atmosphere, give rise
to the curious phenomenon of crepuscular rays. Then may be seen,
starting from the point where the sun has disappeared, a series of rays,
or rather of diverging "glories," which sometimes extend as far as 90°,
and even in some instances are prolonged as far as the point opposite
to the sun. "Upon the ocean," as M. Liais remarks, "when the sky,
near the equator, is free from cloud in the visible part, and when the
diverging rays mingle with the crepuscular arcs, the play of light assumes a form and brilliance which defy all description or pictorial illustration. How, indeed, is it possible to depict completely the rosy tints
of the arc fringed by the crepuscular rays that border the segment
which is still strongly lighted up from the west, the segment itself being
tinged with a bright gold hue? How, above all, is it possible to describe the tint of an inimitable blue, different from that of noonday, and

occupying that portion of the sky which is included between the ordinary azure and the crepuscular arc?

"To all this splendor of the western sky must be added the description of its fires as reflected upon the surface of the waters agitated by the trade-wind, the dark blue color of the sea to the east, the white foam of the wave, which sharply defines upon this gloomy background the pale roseate arc of the eastern sky, and the sombre and greenish segment of the horizon."

What spectacle can be more sublime than a sunset at sea? We have attempted in the illustration to recall this beautiful spectacle. The colored clouds which float in this western sky are *cirro-cumuli*, which will be described in the chapter upon the Clouds.

The setting sun is nearly always accompanied by these cirro-cumuli clouds, which serve to display those aspects of the sky which are of so remarkable a beauty in the west. In consequence of the curvature of the earth, sea-clouds which are sometimes seen from Paris are more than two miles above the ocean, and are formed of ice and snow, even in the month of July. These are nearly the highest clouds, and produce the varied forms of mountains, fishes, animals, and other fantastic shapes, which one may discern of an evening upon a bright and rich ground of every tint that light can give.

To the preceding remarks may be added one of a more general and curious nature, in reference to the influence of the evening light in the construction of cities. Towns grow in a westward direction. Paris, the cradle of which was the *Ile de la Cité*, has, in its successive aggrandizements, constantly extended toward the west. Two thousand years ago, Paris was situated on the north-east slope of Mount St. Geneviève, where the arenas have recently been discovered. Under the Merovingians it commenced its descent toward the west, and has unceasingly progressed in that direction ever since. The wealthy classes have a pronounced tendency to emigrate westward, leaving the eastern districts for the laboring populations. This remark applies not only to Paris, but to most great cities—London, Vienna, Berlin, St. Petersburg, Turin, Liége, Toulouse, Montpellier, Caen, and even Pompeii.

Whence arises this tendency? A fact so universal can not be due to accident. Is it the stream of the Seine which has taken Paris westward in its wake? Not so, for the Thames flows in a contrary direction, while London has none the less extended to the west like Paris. Twelve years ago, Doctor Junod (*Comptes-Rendus* of the Academy of Sciences in 1858) offered, as an explanation of this fact, the statement that the east wind is that which raises in the greatest degree the barometrical column, while the west wind lowers it the most, and therefore inundates the eastern part of a town with deleterious gases, so that the latter has to put up not only with its own smoke and miasmas, but also with those coming from the western portion. It may, in fact, be admitted that people prefer going where fresh air is to be found, and in the direction from which the wind blows most frequently.

But the wind is not the same in all countries. For my own part, I am more inclined to see in this fact an evidence of the attraction of light. And the suggestion is an extremely simple one. It may be remarked that people, as a rule, take their promenade of an evening, and not of a morning, and always, or nearly always, in the direction of sunset. This disposition has led to the formation of gardens, country houses, and places of public resort, and, little by little, the wealthy population of a large city extends in this direction.

CHAPTER III.

THE RAINBOW.

THE general action of light in nature is always evident to our eyes; its effects in the atmosphere are of very different kinds, and produce a thousand optical phenomena, always curious, often fantastic, but all capable of explanation in these days by physical laws. We shall devote the following chapters to the examination of the phenomena that are due to this agent, at once so powerful and so delicate.

The most common of these phenomena is the rainbow, and the explanation of it will aid us in understanding the others.

There are few persons who have not remarked in water falling from a fountain or cascade the production of a miniature rainbow analogous to the majestic arch which crosses the sky. Whenever these small rainbows are seen, three circumstances will be observed in connection with them: first, that drops of water must be present; secondly, that the sun must be shining; and, thirdly, that the observer must be between the sun and the water.

These three conditions in regard to the production of the rainbow will explain the phenomenon in which the Jewish religion saw the presence of Jehovah, and the Greek mythology the auspicious influence of the goddess Iris. In order to see a rainbow as a result of the action of light, whether on artificial rain or on the drops of rain falling from the clouds in the atmosphere, the spectator's back must be to the sun. In this position, the solar rays which shine upon the drops of water are reflected and refracted as follows: Let us suppose a drop of water, A I I', in the atmosphere. A solar ray reaches this drop at I, and passes into it, being deflected from a straight line by refraction,

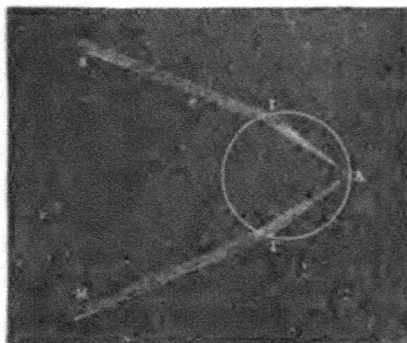

Fig. 30.—Simple reflection of rays in a drop of rain.

inasmuch as the ray passes into a medium of different density. Arriving at A on the surface of the small sphere of liquid which constitutes the drop, it is reflected and returns in the direction of A I', being refracted on emergence into the direction I' M.

This ray, so decomposed by refraction, presents all the colors arranged in regular order, as each color possesses a different degree of refrangibility. The inclination increases from red to violet; that is to say, that if the red ray from a particular drop reaches the eye, the other rays proceeding from the same drop can not reach it too; but a drop at a less elevation in the air can send a violet ray which will be visible at the same time. Thus, the observer sees, in the direction of these drops, a red hue above and a violet hue below. The intermediate drops similarly emit rays which, when seen by the eye, are of the colors included between red and violet, forming a solar spectrum, the colors of which, starting from the lowest are, are *violet, indigo, blue, green, yellow, orange, red.*

Let us now imagine a conical surface passing through the drop, and having for axis the straight line drawn from the eye of the observer to the sun. Every drop of water which is upon this surface of the cone produces the same effect, so that there is a mass of spectra forming a circular band, in which the simple colors succeed each other in the order indicated, the violet, *a* (see Fig. 33, p. 124), being inside, and the red. *b*, outside.

The phenomenon continues as long as the drops of water go on falling in the same region of space, the luminous appearance being incessantly renewed by the falling of the drops, so that the arch appears permanent while the rain lasts.

Calculation has shown that the angle of the cone of the red rays is 42° 20′, and that of the violet rays 40° 30′. This is, therefore, the distance from the arc to the centre or the point of the sky on which the shadow of the head of the spectator, P (see Fig. 33), would be cast. The diameter, H H′ (see Fig. 33), of the whole arc subtends an angle of about 84°, the width of the arc being 2°, or nearly four times the apparent diameter of the sun.

The rainbow, therefore, demonstrates the existence of small spheres of liquid water, falling as rain in the midst of the atmosphere. The arch is more brilliant as their size increases. They must be much larger than those which form the clouds for the eye to be able to distinguish the colors, and that is the reason why mists and clouds do not

produce any rainbow. Knowing that the rainbow is caused by the refraction of the sun's rays through drops of rain as they fall, we may deduce therefrom not only the size of this arch, but also the conditions
without which it could not exist. If the sun were on the horizon, the
shadow of the spectator's head would be cast there also, and as the axis
of the cone would be horizontal, it follows that we should see a semi-
circle of an apparent radius of 41°. As the sun rises, the axis of the
cone is inclined, and the arch becomes smaller; and finally, when the
sun reaches a height of 41°, the axis of the cone forms the same angle
with the plane of the horizon, and the top of the arch just touches this

Fig. 31.—Formation of the rainbow.

latter plane. If the sun were still higher, the arch would be projected
upon the ground. The phenomenon is rarely visible under this last condition. The secondary rainbow, of which I am about to speak, disappears when the sun reaches an altitude of 52°, for which reason a rainbow can not be seen at noon in summer. The observer standing upon
the earth can, therefore, never see more than half a circumference (viz.,
when the sun is on the horizon); and, as a rule, the arch is only 100°
to 150° in length. When the earth does not stand in the way of the
production of the lower part, more than a semi-circumference, and even
a whole circumference, may be seen. This occurred to me once in a

balloon; and by a curious coincidence (the upper part being concealed),
I saw *a rainbow upside down*, in
which the violet color was inside.

A second arch, in which the colors appear in an inverse order to those in the rainbow described above, is frequently remarked. This second arch is explained by a double reflection, S I A B I′ M (see Fig. 32) and s′a′o, s′b′o (see Fig.

Fig. 32.—Double reflection of rays in a drop of rain.

33). In this case, the deviations of the rays after they emerge from the liquid sphere are 51° for the red rays, and 54° for the violet rays. This secondary arch is always paler than the first.

The zone comprised between the principal and the secondary arch is generally darker than the rest of the sky, and appears to me, after numerous observations, to be a region of absorption for the luminous rays.

It is ascertained that a larger number of reflections may be produced, and that other arches, more and more pale in hue, may exist. But the diffused light prevents them from being seen. However, a third has been seen, at 40° from the sun. By causing the solar rays to fall upon a jet of water in a dark place, as many as seventeen rainbows have been counted.

It may happen that the sun is reflected toward a cloud by the surface of a piece of still water; and then this reflection will also

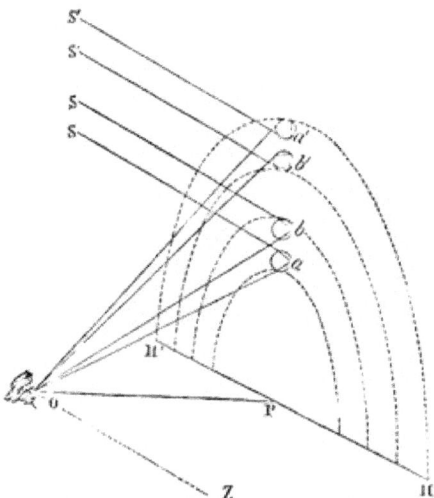

Fig. 33.—Theory of the two arches of a rainbow.

give rise to a rainbow. It has been found that in this case the rainbow must cut the arch formed by the direct rays at a height dependent upon that of the sun. If the two phenomena produce a secondary arch, the four curves intertwined form a very beautiful spectacle. A case in which they were quite complete and perfectly distinct is cited by Monge. Halley observed three arches, one of which was formed by the rays reflected upon a river. This arch first intersected the exterior arch so as to divide it into three equal parts. When the sun sunk toward the horizon, the points of meeting were drawn close together. There soon was seen but one single arch, and as the colors were in inverse order, pure white was formed by the superposition of the two series. The sun, too, may produce, after being reflected upon a piece of water, a complete circle, the upper part of which being sometimes invis-

Fig. 34.—Triple rainbow.

ible, gives rise to the singular phenomenon of a rainbow upside down. The Academicians dispatched to the polar regions to measure an arc of the meridian, observed upon the Ketima Mountain, on July 17, 1736, a *triple* rainbow analogous to that of which Halley speaks. In the lower bow the violet was underneath, the red outside as usual: this was the principal arch. The second, which was parallel to it, was the secondary arch. In this the red was underneath and the violet at the top. The third arch, starting from the extremities of the first, crossed the second, and had, like the principal one, the violet inside and the red outside. This is the phenomenon drawn in Fig. 34.

Seeing, then, that the rainbow is due to the refraction and reflection of the solar rays upon little drops of water falling in the air, it is easy to conceive that moonlight may cause an analogous appearance, though less intense; and this indeed is the case, though a lunar rainbow is not very common. The illustration represents a lunar rainbow which I had an opportunity of remarking one spring evening at Compiègne.

Many observers have remarked and described this nocturnal rainbow. I gather from the writings of Americ Vespuce (1501) that he had several times observed "the iris at night." He considers that the red of the arch is due to fire, the green to the earth, the white to the air, and the blue to the water; and, he adds, "this sign will cease to appear when the elements are used up, forty years before the end of the world."

I notice in an ancient treatise on meteorology (that of P. Cotte) that, in addition to the ordinary rainbow, the secondary rainbow, the reflected arches, and the lunar rainbow, there has been mentioned yet another optical effect, called the " marine rainbow," formed upon the surface of the sea, and composed of a large number of zones. It sometimes appears upon wet meadows lying opposite to the sun. This fifth aspect is a kind of anthelion, which I will allude to in the next chapter. The name of " white rainbow " has also been given to the anthelical circle, which will also be considered in the same chapter.

Lastly, there are sometimes seen colored bands below the violet of the ordinary rainbow, which appear to belong to an arch lying over the first. This arch then takes the name of *supernumerary* arch, and is due to very complex effects of interference of light, explainable on the undulatory theory.

The first person who attempted to explain the phenomenon of the rainbow by the reflection of light upon the interior of the drops of rain was a German monk of the name of Theodoric; the second an archbishop, A. De Dominis (1611). But the true theory was first given by Descartes, with the exception of the separation of colors, which was only determined by the discovery of Newton as to the unequal refrangibility of the rays of the solar spectrum.

A. Marie pinx^t   Eug. Cicéri chromolith^t

LUNAR RAINBOW SEEN AT COMPIÈGNE

CHAPTER IV.

ANTHELIA: SPECTRE-SHADOWS UPON MOUNTAINS—THE ULLOA CIRCLE—CIRCLE SEEN FROM A BALLOON.

TREATISES on meteorology have not, up to the present day, classified with sufficient regularity the diverse optical phenomena of the air. Some of these phenomena have, however, been seen but rarely, and have not been sufficiently studied to admit of their classification. We have examined the common phenomenon of the rainbow, and we have seen that it is due to the refraction and reflection of light on drops of water, and that it is seen upon the opposite side of the sky to the sun in day-time or the moon at night. We are now about to consider an order of phenomena which are of rarer occurrence, but which have this property in common with the rainbow, viz., that they take place also upon the side of the sky opposite to the sun. These different optical effects are classed together under the name of *anthelia* (from ἀνθί, opposite to, and ἥλιος, the sun). The optical phenomena which occur on the same side as, or around the sun, such as halos, parhelia, etc., will form the subject of the next chapter.

Before coming to the anthelia, properly so called, or to the colored rings which appear around a shadow, it is as well first to note the effects produced on the clouds and mists that are facing the sun when it rises or sets.

Upon high mountains, the shadow of the mountain is often seen thrown either upon the surface of the lower mists or upon the neighboring mountains, and projected opposite to the sun almost horizontally. I once saw the shadow of the Righi very distinctly traced upon Mount Pilate, which is situated to the west of the Righi, on the other side of the Lake of Lucerne. This phenomenon occurs a few minutes after sunrise, and the triangular form of Righi is delineated in a shape very easy to recognize.

The shadow of Mont Blanc is discerned more easily at sunset. MM. Bravais and Martins, in one of their scientific ascents, noticed it under specially favorable circumstances, the shadow being thrown upon the snow-covered mountains, and gradually rising in the atmosphere until it

Fig. 35.—The Spectre of the Brocken.

spectre. He then called another person to him, and placing themselves in the very spot where the apparition was first seen, the pair kept their eyes fixed on the Achtermannshohe, but saw nothing. After a short interval, however, two colossal figures appeared, which repeated the gestures made by them, and then disappeared.

Some few years ago, in the summer of 1862, a French artist, M. Stroobant, witnessed and carefully sketched this phenomenon, which is drawn in Fig. 35. He had slept at the inn of the Brocken, and rising at two in the morning, he repaired to the plateau upon the summit in the company of a guide. They reached the highest point just as the first glimmer of the rising sun enabled them to distinguish clearly objects at a great distance. To use M. Stroobant's own words, "My guide, who had for some time appeared to be walking in search of something, suddenly led me to an elevation whence I had the singular privilege of contemplating for a few instants the magnificent effect of mirage, which is termed the Spectre of the Brocken. The appearance is most striking. A thick mist, which seemed to emerge from the clouds like an immense curtain, suddenly rose to the west of the mountain, a rainbow was formed, then certain indistinct shapes were delineated. First, the large tower of the inn was reproduced upon a gigantic scale; after that we saw our two selves in a more vague and less exact shape, and these shadows were in each instance surrounded by the colors of the rainbow, which served as a frame to this fairy picture. Some tourists who were staying at the inn had seen the sun rise from their windows, but no one had witnessed the magnificent spectacle which had taken place on the other side of the mountain."

Sometimes these spectres are surrounded by colored concentric arcs. Since the beginning of the present century, treatises on meteorology designate, under the name of the *Ulloa circle*, the pale external arch which surrounds the phenomenon, and this same circle has sometimes been called the "white rainbow." But it is not formed at the same angular distance as the rainbow, and, although pale, it often envelops a series of interior colored arcs.

Ulloa, being in company with six fellow-travelers upon the Pambamarca at day-break one morning, observed that the summit of the mountain was entirely covered with thick clouds, and that the sun, when it rose, dissipated them, leaving only in their stead light vapors, which it was almost impossible to distinguish. Suddenly, in the opposite direction to where the sun was rising, "each of the travelers beheld, at about seventy feet from where he was standing, his own image reflected in the air as in a mirror. The image was in the centre of three rainbows of different colors, and surrounded at a certain distance by a fourth bow with only one color. The inside color of each bow was carnation or red, the next shade was violet, the third yellow, the fourth straw color, the last green. All these bows were perpendicular to the horizon; they moved in the direction of, and followed, the image of the person whom they enveloped as with a glory." The most remarkable point was that, although the seven spectators were standing in a group, each

Fig. 36.—The Ulloa circle.

person only saw the phenomenon in regard to his own person, and was disposed to disbelieve that it was repeated in respect to his companions. The extent of the bows increased continually and in proportion to the height of the sun; at the same time their colors faded away, the spectres became paler and more indistinct, and finally the phenomenon disappeared altogether. At the first appearance the shape of the bows was oval, but toward the end they became quite circular. The same apparition was observed in the polar regions by Scoresby, and described by him. He states that the phenomenon appears whenever there is mist and at the same time shining sun. In the polar seas, whenever a rather thick mist rises over the ocean, an observer, placed on the mast, sees one or several circles upon the mist.

These circles are concentric, and their common centre is in the straight line joining the eye of the observer to the sun, and extended from the sun toward the mist. The number of circles varies from one to five; they are particularly numerous and well colored when the sun is very brilliant and the mist thick and low. On July 23, 1821, Scoresby saw four concentric circles around his head. The colors of the first and of the second were very well defined; those of the third, only visible at intervals, were very faint, and the fourth only showed a slight greenish tint.

The meteorologist Kaemtz has often observed the same fact in the Alps. Whenever his shadow was projected upon a cloud, his head appeared surrounded by a luminous aureola.

To what action of light is this phenomenon due? Bouguer is of opinion that it must be attributed to the passage of light through icy particles. Such, also, is the opinion of De Saussure, Scoresby, and other meteorologists.

In regard to the mountains, as we can not assure ourselves directly of the fact by entering into the clouds, we are reduced to conjecture. The aerostat traversing the clouds completely, and passing by the very point where the apparition is seen, affords one an opportunity of ascertaining the state of the cloud. · This observation I have been able to make, and so to offer an explanation of the phenomenon.*

As the balloon sails on, borne forward by the wind, its shadow travels either on the ground or on the clouds. This shadow is, as a rule, black, like all others; but it frequently happens that it appears alone on the surface of the ground, and thus appears luminous. Examining this shadow by the aid of a telescope, I have noticed that it is often composed of a dark nucleus and a penumbra of the shape of an aureola. This aureola, frequently very large in proportion to the diameter of the central nucleus, eclipses it to the naked eye, so that the whole shadow appears like a nebulous circle projected in yellow upon the green ground of the woods and meadows. I have noticed, too, that this luminous shadow is generally all the more strongly marked in proportion to the greater humidity of the surface of the ground.

Seen upon the clouds, this shadow sometimes presents a curious as-

* [The explanation of the phenomenon offered by M. Flammarion (viz., that it is due to diffraction) was generally recognized long previous to M. Flammarion's ascents,—ED.]

pect. I have often, when the balloon emerged from the clouds into the clear sky, suddenly perceived, at twenty or thirty yards' distance, a second balloon distinctly delineated, and apparently of a grayish color, against the white ground of the clouds. This phenomenon manifests itself at the moment when the sun re-appears. The smallest details of the car can be made out clearly, and our gestures are strikingly reproduced by the shadow.

On April 15, 1868, at about half-past three in the afternoon, we emerged from a stratum of clouds, when the shadow of the balloon was seen by us, surrounded by colored concentric circles, of which the car formed the centre. It was very plainly visible upon a yellowish white ground. A first circle of pale blue encompassed this ground and the car in a kind of ring. Around this ring was a second of a deeper yellow, then a grayish red zone, and lastly, as the exterior circumference, a fourth circle, violet in hue, and imperceptibly toning down into the gray tint of the clouds. The slightest details were clearly discernible— net, ropes, and instruments. Every one of our gestures was instantaneously reproduced by the aërial spectres. The anthelion remained upon the clouds sufficiently distinct, and for a sufficiently long time, to permit of my taking a sketch in my journal and studying the physical condition of the clouds upon which it was produced.* I was able to determine directly the circumstances of its production. Indeed, as this brilliant phenomenon occurred in the midst of the very clouds which I was traversing, it was easy for me to ascertain that these clouds were not formed of frozen particles. The thermometer marked 2° above zero. The hygrometer marked a maximum of humidity experienced, namely, 77 at 3770 feet, and the balloon was then at 4600 feet, where the humidity was only 73. It is therefore certain that this is a phenomenon of the diffraction of light simply produced by the vesicles of the mist.

The name of diffraction is given to all the modifications which the luminous rays undergo when they come in contact with the surface of bodies. Light, under these circumstances, is subject to a sort of deviation, at the same time becoming decomposed, whence result those curious appearances in the shadows of objects which were observed for the first time by Grimaldi and Newton.

The most interesting phenomena of diffraction are those presented by

* A colored illustration of this remarkable phenomenon is given in the *Voyages Aériens*, which was published by MM. Glaisher, De Fonvielle, and G. Tissandier, in conjunction with myself, part 2, p. 292.

gratings, as are technically denominated the systems of linear and very narrow openings situated parallel to one another and at very small intervals. A system of this kind may be realized by tracing with a diamond, for instance, on a pane of glass equidistant lines very close together. As the light would be able to pass in the interstices between the strokes, whereas it would be stopped in the points corresponding to those where the glass was not smooth, there is, in reality, an effect produced as if there were a series of openings very near to each other. A hundred strokes, about $\frac{1}{25}$ of an inch in length, may thus be drawn without difficulty. The light is then decomposed in spectra, each overlapping the other. It is a phenomenon of this kind which is seen when we, look into the light with the eye half closed; the eyelashes, in this case, acting as a net-work or grating. These net-works may also be produced by reflection, and it is to this circumstance that are due the brilliant colors observed when a pencil of luminous rays is reflected on a metallic surface regularly striated.

To the phenomena of gratings must be attributed, too, the colors, often so brilliant, to be seen in mother-of-pearl. This substance is of a laminated structure; so much so, that in carving it the different folds are often cut in such a way as to form a regular net-work upon the surface. It is, again, to a phenomenon of this sort that are due the rainbow hues seen in the feathers of certain birds, and sometimes in spiders' webs. The latter, although very fine, are not simple, for they are composed of a large number of pieces joined together by a viscous substance, and thus constitute a kind of net-work.

If the sun is near the horizon, and the shadow of the observer falls upon the grass, upon a field of corn, or other surface covered with dew, there is visible an aureola, the light of which is especially bright about the head, but which diminishes from below the middle of the body. This light is due to the reflection of light by the moist stubble and the drops of due. It is brighter about the head, because the blades that are near where the shadow of the head falls expose to it all that part of them which is lighted up, whereas those farther off expose not only the part which is lighted up, but other parts which are not, and this diminishes the brightness in proportion as their distance from the head increases. The phenomenon is seen whenever there is simultaneously mist and sun. This fact is easily verified upon a mountain. As soon as the shadow of the mountaineer is projected upon a mist, his head gives rise to a shadow surrounded by a luminous aureola.

The *Illustrated London News* of July 8, 1871, illustrates one of these apparitions, "The Fog Bow, seen from the Matterhorn," observed by E. Whymper in this celebrated region of the Alps. The observation was taken just after the catastrophe of July 14, 1865; and by a curious co-incidence, two immense white aërial crosses projected into the interior of the external arc. These two crosses were no doubt formed by the intersection of circles, the remaining parts of which were invisible. The apparition was of a grand and solemn character, further increased by the silence of the fathomless abyss into which the four ill-fated tourists had just been precipitated.

Other optical appearances of an analogous kind are manifested under different conditions. Thus, for instance, if any one, turning his back to the sun, looks into water, he will perceive the shadow of his head, but always very much deformed. At the same time he will see starting from this shadow what seem to be luminous bodies, which dart their rays in all directions with inconceivable rapidity, and to a great dis-tance. These luminous appearances—these aureola rays—have, in ad-dition to the darting movement, a rapid rotatory movement around the head.

CHAPTER V.

HALOS: PARHELIA—PARASELENES—CIRCLES SURROUNDING AND TRAV-
ERSING THE SUN—CORONAS—COLUMNS—VARIOUS PHENOMENA.

THE description of optical phenomena now brings us to one of the
most singular and complicated effects of the reflection of light in the atmosphere. Under the name of *halo* (ἅλως, area) is designated a brilliant circle which, under certain atmospheric conditions, surrounds the
sun at a distance of 22° or 46°; while, under the name of *parhelia*, or
mock suns (παρά, near, and ἥλιος, sun), are designated luminous circular spaces, generally of a red, yellow, or greenish color, which appear
both to the right and to the left of the sun, at the same distance (viz.,
about 22°), bearing a sort of rough resemblance to the sun itself. The
same appearances may be seen about the moon; and it is, indeed, easier
to observe them, as the diminished brilliancy of the moon's light renders an examination of the area around it less difficult. These luminous spaces are called *paraselenes* (παρὰ, near, and σελήνη, moon), or
mock moons. The two cases only differ as to the intensity of the luminary from which they are derived—a difference similar to that which
may be observed between ordinary solar and lunar rainbows.

In addition to the halo and the two parhelia, a number of other circles, arches, bands, or luminous spots, are sometimes seen upon the sky.
These are more or less bright, and accompany the halo.

It is well known that, when a triangular prism of glass is submitted
to the action of the sun's rays, part of the light falling on it is reflected
from the surface of the prism as upon a mirror, and another part penetrates into the glass and leaves it in a direction different from that by
which it entered, producing an image formed of different colors. It is
upon this fact that Mariotte based the explanation of the phenomenon
which we are about to consider. The origin of halos, in his opinion, is
to be discovered in the crystals of ice in the shape of equilateral triangular prisms in the air. These prisms may be situated at all possible
angles, and in all directions in the atmosphere, some among them being
in such positions as to produce the absolute minimum of deviation of
the rays of light which, entering by one of the three lateral surfaces of

the prisms, traverse one of the other two on their way out of it. Mariotte has shown that, at an angular distance from the sun equal to that of minimum deviation, which is 22°, a brilliant circle must be formed, and this is the ordinary halo. If from some cause or other all the prisms become vertical, the halo is replaced by two parhelia. The tangent arcs seen near the ordinary halo, the halo with a radius of 46° and the parhelion circle, have been explained by Young upon the hypothesis that, in certain cases, the prisms may be situated in such a way that their axes are all horizontal.

Twenty years ago, Bravais devoted to the analysis of these phenomena a work which will be useful to us as a guide. The theory of these phenomena is somewhat complex, and demands a certain amount of attention in order to be intelligible. Voltaire confessed that he was obliged to read the same things twice over in order to comprehend them thoroughly; and perhaps those of us who do not consider ourselves more acute than the sage of Ferney will do well to imitate him in this instance.

When a halo appears upon the sky, light cirri clouds (of which we shall speak presently) are generally seen, and it is upon them that the phenomenon appears to be delineated. Sometimes, too, these cirri are collected into one single mass, so that the eye can not seize their shapes: a white vapor predominates in that part of the sky near to the sun; and the blue tint of the atmosphere is replaced by a kind of light mist, the brilliancy of which is sometimes unbearable to the eye. But these light clouds of snow, placed high in the air, are so distant that it is difficult to decide upon their real nature. Hence we see how easily the mode in which the phenomenon is produced might for a long period have remained unknown; and this is unquestionably one of the reasons why halos and parhelia were in early ages deemed marvelous phenomena, signs of celestial ire, presages of the death of princes, etc., etc.

It is not enough for the clouds of the higher strata of the atmosphere to be formed of snowy particles for the phenomenon of the halo to become visible; the two following conditions are further necessary. The cloud must be of a certain degree of thickness; for, if too thin, the halo would not occur; if too dense, the light would be intercepted. The crystallization of the water must also proceed slowly and not be disturbed by wind, as with a rapid, and therefore irregular, crystallization the points lose their transparency, the angles of the facets their consistency, and the surfaces by which the rays enter and leave, their smoothness.

The appearance of halos is less rare than might be supposed. It is calculated that in our latitudes the number of days on which this phenomenon occurs, in the rudimentary state at least, are fifty a year, and in the north of Europe many more.

The most simple form of crystals of ice, snow, or hoar-frost—viz., that seen in the earliest process of crystallization—is a right prism, having for its section a regular hexagon, and terminated by two bases perpendicular to the lateral surfaces, which are rectangular.

These simple forms are, however, rarely seen in a fall of snow, because, before reaching the ground, lateral crystallization, due to the condensation of vapor in the lower strata, makes an addition to the primitive nucleus.

The hexagonal prism gives rise to all the spots or curves, the appearance of which has been placed beyond doubt by numerous observations.

The halo, with all its aspects, is explained on the hypothesis of snow or ice-crystals falling slowly in a calm atmosphere.

It is therefore due simply to the refraction of the solar rays upon crystals of ice. The different positions of the prisms of ice are the cause of the diversity of the appearances. The situation of these sharp-pointed needles of ice in the atmosphere may be divided into three classes: 1st, prisms placed at any angle; 2d, prisms axes of which are vertical; 3d, prisms placed horizontally.

In order to comprehend the production of the phenomena, let us, as in explaining the rainbow, take the first case and examine its effects. If a prism is turned round, the ray which emerges from it is seen to make a variable angle with that which enters it. But there is a certain position in which the entering and departing rays make the smallest angle possible with each other; the prism then relative to the incident ray is said to be in its position of minimum deviation. Now, in this position, the prism may be turned a little one way or a little the other without causing any perceptible change in the direction of the refracted ray.

If a prism of this kind turns upon its own axis in the atmosphere, rays are incessantly emanating from it, which reach the eye and disappear immediately afterward; but, as has just been remarked, it is clear that the ray will catch the eye for the greatest length of time when its deviation is a minimum. If the number of these prisms is very great, we receive at the same time the rays refracted by a prism at the moment at which the others disappear, so that the impression upon our eye is persistent, although the rays are not transmitted to it by the same crystals. A solar ray enters a triangular prism by the surface A (see Fig. 37), and undergoes a deviation. This ray is, of course, decomposed. Let us suppose the violet portion, after emergence from the surface B, reaches the eye of the spectator placed at O. Another prism, c, nearer to the direction o s of the sun, will send red rays which have deviated the least, so that in fact the cone passing through A will be violet, the cone passing through c red, and the intermediate one colored with various intermediate colors of the spectrum.

Fig. 37.—Theory of the halo.

Refraction of the solar rays will thus produce all round the sun, and at the same distance, a series of luminous impressions. The deviation is about 22°, but is not the same for all colors. Calculation, coinciding with observation, gives 21° 37′ for the red, which is the least refrangible color, 21° 48′ for the yellow, 21° 57′ for the green, 22° 10′ for the blue, and 20° 40′ for the violet. This circle of 22° radius which is thus formed around the sun and the moon is the ordinary halo which is seen most frequently. The red is inside; then we have orange, yellow, green; but these colors gradually become weaker, because they are influenced by the prisms, which are not in the position of minimum deviation.

The red remains most visible. The sun, however, is not, as we have assumed, a mere luminous point, but each part of its disk contributes to the production of this phenomenon; and this circumstance tends to blend still further the various colors, which are, in consequence, never very clearly defined, and the halo generally appears as a bright ring with a reddish tint on the inside, 2° to 3° in width, and inclosing a circular area of which the sun occupies the centre.

By a well-known optical effect, a spectator not previously instructed upon the point would be inclined to attribute an elliptic shape to the halo, considering it an oval with a longer vertical axis; but this illusion, which also takes place when an entire rainbow is seen, disappears before angular measurement. From a similar cause, the halo appears to get smaller as the sun rises, just as the moon loses, at a certain elevation, the gigantic proportions that its disk presented soon after rising. In addition to the halo of 22° radius, a second is also frequently seen, the diameter of which is about twice as large as that of the preceding one.

The latter is produced by the refraction of light across the dihedral angles of 90° that the sides of the prisms make with the bases, just as the angles of 60° produce the ordinary halo. Like the latter, it is composed of a succession of rings, the first of which (viz., the one nearest to the sun) is red. But, by a superposition of colors similar to that which occurs in the halo of 22°, there is scarcely discernible more than a ring, reddish upon its inside and yellowish in the middle, whereas the external part seems of a whitish hue, and gradually becomes lost in the general light of the atmosphere. The total width of this halo is rather large, embracing about the 3° between 45° and 48° distance from the sun, the white light that borders it included.

These two circles are, therefore, formed by the reflection of light upon the prisms of ice placed at all angles in the air. Let us now consider what effects may be produced by prisms placed vertically. When the light is reflected across the dihedral angles of 60°, which the six sides of the prisms of ice falling vertically form between them, there are two *parhelia* produced, one to the right, the other to the left of the sun, and both situated at the same height as the latter. To rightly understand the reason of this phenomenon, the principle must first be enunciated that the light given by a group of prisms, all of which have their axes vertical, but which are situated in every conceivable position as to the direction of their sides, is similar to that which would be transmitted

by a single prism turning rapidly on its own axis. It follows, in
fact, that the prism, in the movement indicated above, passes in succession through all the positions compatible with the verticality of its
axis.

When the sun is on the horizon, the distance at which these appearances are formed is exactly the angle of minimum deviation, or, in other words, the radius of the halo. If the halo and the parhelia are seen
together, the latter appear to be situated just upon the circumference of
the prism, and occupy in height a distance equal to the diameter of the
sun. The various tints are clearer than in the halo; the yellow is very
distinct, and so is the green, but the blue is pale, and scarcely visible;
while the violet, overlapped by the other colors, is too indistinct to be

Fig. 38.—Halo seen in Norway.

seen. The phenomenon is completed by a tail of white light, sometimes very indistinct, but occasionally attaining a length of from 10° to
20° in the opposite direction to the sun, and parallel to the horizon.
This light is due to those prisms, the positions of which are somewhat
out of the line that corresponds to the minimum deviation.

When the sun rises about the horizon, the luminous rays traverse the
prisms, moving in oblique planes, and the smallest of the deviations
produced during the rotation is greater than the absolute corresponding
minimum, when the sun is at the horizon. This shows that the parhelia must emerge slowly from the circumference of the halo, in proportion as the latter rises in height; but on the other hand, as the halo is
nearly 2° in width (including the white light that borders it), the par-

helia only become completely separated from it when the sun is at an elevation of 20° or 30°.

Optical considerations show that the formation of parhelia becomes impossible when the sun has reached an elevation of 60°.

Parhelia are sometimes very brilliant, and their brightness may then be in a certain measure compared to that of the sun itself, in which case it is quite conceivable that each parhelion may become in its turn the origin of two others, which are then the parhelia of parhelia, or *secondary parhelia*. The effect caused by the refraction of light across angles of 90°, which produce the large halo, is still more remarkable. The solar rays enter obliquely at the upper base of the prisms, and, passing through it, emerge by one of the vertical surfaces.

If we imagine, as we have already done in the case of the parhelia, that the prism on the upper base of which the rays are falling, turns rapidly upon its own axis, it may be proved by optics that the light emerging from it will be scattered in the form of a bright curve with its axis vertical, whence it is easy to conclude that the corresponding optical appearance upon the celestial sphere will be a luminous arc parallel to the horizon and situated at a great distance above the sun.

The arc thus produced, which may be termed the *upper tangent arc of the halo of* 46°, or, more briefly, the *circumzenithal arc*, deserves special notice, for it is unquestionably the most remarkable of all the appearances which may accompany the halo. The brightness of the tints, the distinctness of the colors, the precision with which its edges, as well as its extreme limits, are shown upon the sky, give it the characteristic of a real rainbow. Of the respective rings composing it, the red is nearest to the sun, the violet fringes the concave part of the arc, and is on the opposite side; the width of the various rings is about the same as in the rainbow, though rather less, owing to the illusion caused by the proximity of the zenith. When the halo of 46° is visible, the circumzenithal arc generally appears to touch it at its highest point, the red of the arc being then in contact with the red of the halo, the orange with the orange, and so on with the other colors; but very often the circumzenithal arc is seen without the halo of 46°, just as the parhelia may appear without the halo of 22°, although they owe their existence to the same kind of dihedral angles.

From the observations that have been made of this arc, it appears that it never is to be seen when the height of the sun is less than 12° or more than 31°.

It follows also from optical consideration that prisms, falling and turning upon their sides, can reflect the sun, forming upon the celestial sphere a luminous horizontal band, extending right round the horizon and passing through the exact centre of the sun. As reflection does not separate the colors which compose white light, this circle will appear to be quite white, and its apparent width will be equal to the diameter of the sun. Such is the origin of the white circle called the *parhelical ring*. It is upon its circumference that the ordinary parhelia always appear, as also the secondary parhelia situated at about 45° from the sun; hence the name.

Sometimes the solar rays experience two successive reflections upon the vertical surfaces of one of the prisms. There is then visible, at 120° from the sun, a white image more or less diffuse, which has received the name of *paranthelion*. The horizontal bases of the ice-crystals reflect also the solar light, but in an upward direction, which prevents the spectator from perceiving it, unless he be upon the summit of a steep mountain, or in the car of a balloon, above the cloud containing the icy particles. It will be readily admitted that these conditions can be rarely fulfilled; but MM. Barral and Bixio were, fortunately, able to realize them on July 27th, 1850. The image of the sun thus reflected appeared almost as luminous as the sun itself. Bravais suggested for this phenomenon, at once so remarkable and so rare, the name of *pseudohelion*.

Finally, the prisms of ice which are *horizontal* in the atmosphere give rise, by reflections and refractions analogous to the above, to tangent arcs which often appear on each side of the halo.

The most complete halo that has yet been seen is that which Lowitz observed at St. Petersburg, on June 29, 1790, from 7 hours 30 minutes A.M. to 12 hours 30 minutes P.M. Since that time there have, of course, been a great number of halos observed; but this is, perhaps, the most complete that has been recorded. MM. Bravais and Martins observed one at Pitéo, in Sweden, on October 4, 1839, which was also very remarkable, but less complete than that seen by Lowitz.

The examination which we have made of the general phenomenon of halos leads us to speak of other optical effects, the explanation of which is more or less akin to that of the above.

The columns of white light, the crosses, and the different luminous aspects sometimes visible at sunrise and sunset, are due to the reflection of light upon the surfaces of crystals of ice situated high in the atmosphere. It is well known that if we look at the reflection of any light (such as the sun, the moon, or a street-lamp) on the surface of rather troubled water, the reflection extends vertically; the motion of the water gives rise to a multitude of small surface-planes which are oscillating unceasingly about the horizontal, in all possible directions. This is the exact reproduction of what is going on in the region of the ice-cloud: the small coruscating bases of the prisms, to which I have attributed above the reflection of the sun as seen from a balloon, are perpetually shifting their position. The reflection produced will therefore also be very elongated, and its upper part may, at sunrise or sunset, rise several degrees above the horizon.

Such is the origin of those columns of white light which are sometimes seen to form at the moment of sunset, and to increase in size as the sun gradually sinks lower. It is scarcely necessary to add that, when the sun has descended below the horizon, the reflection of the light takes place at the lower and not at the upper surfaces of the prisms.

Previous to sunset, on April 22, 1847, four luminous columns, each about 15° in extent, were seen from Paris, presenting the appearance of a cross with the sun in the centre. After sunset one of these four columns (the uppermost of the four, of course) still remained visible for some little time.

When the sun is near the horizon, part of a vertical circle may rise above that luminary in the shape of a column. On June 8, 1824, appearances of this kind were seen in several parts of Germany. At Dohna, near Dresden, at eight in the evening, just as the sun was about to disappear behind the mountains, Lohrmann perceived a luminous band, perpendicular to the crepuscular arc, and similar to the tail of a comet. This column was 30° high and 1° in width. It is more unusual to see a band below the sun or the moon, and more unusual still to see also a horizontal arc passing the sun in such a way that it is situated in the middle of a cross. Roth saw very distinctly a phenomenon of this kind at Cassel, on January 2, 1586. Before the sun appeared, a luminous vertical column, with a diameter equal to that of the sun, was visible at the spot where the sun was about to rise, resembling a brilliant flame, except that its brightness was of uniform intensity throughout. Soon after there appeared a reflection of the sun, so brilliant that it was taken for the sun itself; and this parhelion had scarcely risen above the horizon when the sun rose immediately under it, followed by a column resembling that which had appeared above it.

This latter, with its three suns, remained continuously vertical. The three suns were each exactly similar in appearance, but the true sun was the most brilliant. The phenomenon lasted about an hour.

If the sun, instead of being on the horizon, is some few degrees above it, the luminous column which rises from the pseudohelion then situated below the horizon, and consequently invisible, may reach to the centre of the sun, but can not extend perceptibly beyond it. We then have the appearance of a luminous ascending column, which seems to support the solar disk. Instances of this are afforded by the observations taken by Parry at Melville Island on March 8, 1820; by Sturm on December 9, 1689; and by many others.

The vertical gleams which, passing through the centre of the sun, extend symmetrically above and below it, without having their base at the horizon, and which accompany the sun in his apparent course from east to west, seem due to the same cause. It is easy to see that they are caused by the rays twice reflected upon the horizontal bases of the vertical prisms, or, at all events, by some even number of successive reflections. They are never seen but at heights less than 25°; and are far more frequently seen about the moon than about the sun—a fact which is no doubt due to the greater brightness of the latter, which thus eclipses all the gleams near to it. The reverse is the case with the columns which are seen at sunset, because the sun then being below the horizon, the phenomenon is projected upon a partially lighted ground, and may thus be seen in all its brilliancy.

The combination of the parheliacal circle with the vertical column passing through the centre of the sun, produces the phenomenon of the solar or lunar crosses which are often seen when the halo of 22° is not visible. Sometimes the arms of the cross may be nearly equal in length, and sometimes the horizontal are larger than the vertical.

The vertical columns, and the lunar and solar crosses, are mostly seen in northern countries during the long winters which envelop those regions in snow and ice.

To these optical effects must be added, finally, the *coronas* (see Fig. 39) which appear around the sun and the moon when the air is not clear, and when small drops of vesicular vapor, or light clouds, are passing before their bodies.

These colored rings, which are frequently seen round the moon, owe their origin not to refraction, but to diffraction; they have the red outside, and the violet inside, like the primary rainbow, and their colors

are the converse of those of the two halos concentric with the sun and
moon. The diameters of coronas of the same color are in the proportion of the natural numbers, 1, 2, 3, 4, etc., but the diameter of the first
ring seems enlarged. This diameter, varying from 1° to 4°, depends
upon that of the vesicles of water interposed between the luminary and
the observer. Generally, the color of it is blue mixed with white for a
certain distance round the luminary; then follows a red circle, and then
other colored circles, as in Newton's rings. For the phenomenon to

Fig. 39.—Corona formed around the moon by diffraction.

take place there must be a certain number of globules of the same character, and, indeed, a far greater number of this diameter than of any
other. If the diameters of the spherules of cloud were all different, the
corona would not be produced. An exactly similar effect is observable
when a luminous object is examined through a piece of glass that has
been sprinkled with lycopodium powder, or, in a less marked degree, if
the glass has merely been breathed upon before use.

To these different effects, due to the refraction and reflection of light
in the atmospheric strata, must also be added the deformation of the

sun at the horizon, which occasionally gives rise to most singular appearances, in consequence of the want of homogeneousness in the lower strata, and the curious action of atmospheric refraction.

With the progress of astronomy and physics, the decadence of astrology, and the expansion of inquiry, these optical phenomena lose their supernatural attributes. For the last century they have undergone a calm and impartial study and analysis; while we see in this chapter that they may be explained upon theory, and savans merely recognize them as so many physical facts belonging to the vast domain of meteorology. The historian Josephus relates that at the beginning of the siege of Jerusalem by the Romans, A.D. 70, the Jews foresaw their disaster "in armies marching upon red clouds." Nearly analogous apparitions were visible at the commencement of the siege of Paris in September, 1870, to say nothing of the aurora borealis on the 24th of October; but we now know that the physical effects are purely natural, and are produced merely by the action of light in the atmosphere.

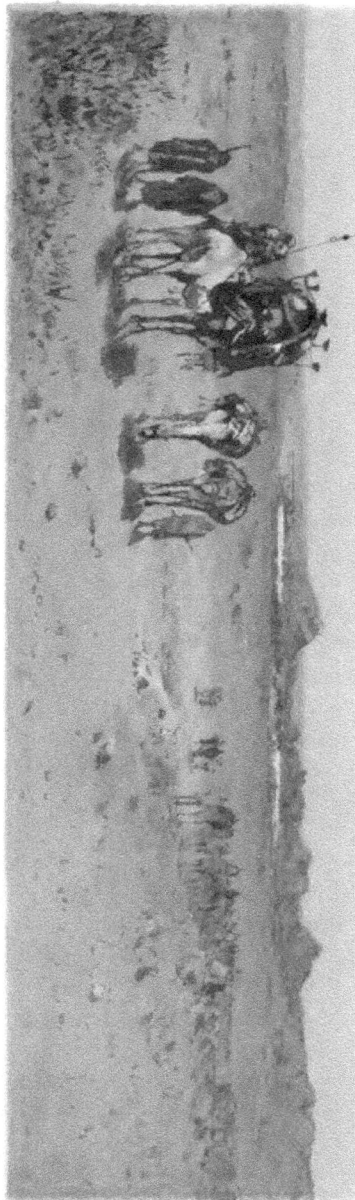

CHAPTER VI.

THE MIRAGE.

NOT only does the atmosphere produce remarkable phenomena in the aërial heights, but it gives play to its fancy even in the lower regions where we move, and the very surface of the ground and of the water is occasionally the field of strange metamorphoses due to the rays of light in the air.

Under the name of *mirage* we designate those optical apparitions caused by a peculiar state of the *densities* of the atmospheric strata—a state which produces variations in the ordinary refractions which we considered in the previous chapter.

In consequence of these variations distant objects appear either deformed, transported to a certain distance, or inverted and reflected, according to the deviation which the abnormal density of the air causes in the luminous rays.

The mirage is no new phenomenon. In Diodorus Siculus we read: "An extraordinary phenomenon occurs in Africa at certain periods, especially in calm weather; the air becomes filled with images of all sorts of animals, some motionless, others floating in the air: now they seem running away, now pursuing; they are all of enormous proportions, and this spectacle fills with terror and awe those who are not accustomed to it. When these figures overtake the traveler whom they seem to be pursuing, they surround him with a cold and shivering feeling. Strangers not used to this extraordinary phenomenon are seized with fear; but the inhabitants, who are in the habit of seeing it, take no particular notice of it.

" Certain physical philosophers attempt to explain the true causes of this phenomenon, which seems extraordinary and fabulous. They say that there is no wind, or scarcely any, in this country. The masses of condensed air produce in Libya what the clouds sometimes produce with us on rainy days, viz., images of all shapes rising on every side in the air. These strata of air, suspended by light breezes, become mixed with other strata. executing at the same time very rapid oscillatory movements; and when calm again sets in they descend toward the ground by their own weight, preserving the shapes that they had accidentally assumed. If no cause occurs to disperse them, they spontaneously attach themselves to the first animals which come near. Their movements do not appear to be the effect of volition, for it is impossible for an inanimate being to progress or go backward. But it is the animated beings who, unwittingly, produce these voluntary movements, for, as they advance, they cause a violent recoil in the images which seem to fly before them. Similarly, those which recoil seem, by producing a void and a relaxation in the strata of the air, to be pursued by the aërial spectres. The persons running away are probably struck, when they stop or return to their former position, by the matter of these figures, which break against their bodies and produce, at the moment of the shock, the chilly sensation."

We see that, before the epoch of Diodorus, the mirage had been observed; the philosophers of the period were nevertheless far from being in possession of the true scientific explanation, although it was then attributed to a change of density in the aërial strata.

This same phenomenon (of which Quintus Curtius has also spoken) has long been remarked by the Arabs, and it is often discussed in the treatises of Oriental writers. Among other instances may be cited the Koran, which says that "the works of the incredulous are like the mirage (scrab) of the plain; the thirsty man takes it for water until he draws nigh to it, and then he discovers that it is nothing."

In about the middle of the seventeenth century the mirage began to attract the special attention of physicists. The discovery of telescopes rendered possible a great number of observations, which were beyond the power of the naked eye; and the knowledge of the laws of the refraction of light, and of the variations in the density of the air caused by changes in its temperature, prepared the way for the theoretical explanation of these singular apparitions. It is not till 1783 that we find the first really scientific work treating of the mirage. This was from the pen of Professor Busch, who observed its effects on the Elbe, near Hamburg, and on the coasts of the Northern and Baltic seas. He often made use of a telescope, and this method of observation disclosed to him many details hitherto unknown. He saw upon several occasions a *mirror of the waters* and *mock bank*, beneath which figures upside down seemed to be delineated; he saw ships suspended in the air, and bearing beneath their keels the reversed image of their masts and sails. On the 5th October, 1779, he saw, at the distance of two German miles from

Bremen, the ordinary image of that town and a second image, very distinct but upside down; between him and the town there was a large and verdant common. The principal circumstances of the phenomenon are clearly indicated in his work, without, however, the theoretical explanation of them.

It was during Bonaparte's expedition to Egypt that the true explanation of the phenomenon was first given.

The soil of Lower Egypt forms a vast and perfectly horizontal plain, the uniformity of which is only broken by gentle eminences upon which are built the villages that are thus protected from the overflowings of the Nile. At morning and evening there is no change in the aspect of the country; but when the sun has heated the surface of the soil, it seems, at a certain distance off, to be inundated; the villages look like islands in the middle of an immense lake, and below each village is to be seen its inverted reflection. To complete the illusion, the ground vanishes, and the vault of the firmament is apparently reflected in still water. It is easy to understand the cruel disappointment of the French army. Exhausted by fatigue, with a devouring thirst under the burning sky, the men fancied they had reached a great pool of still water in which they saw reflected the shadow of the villages and the palm-trees; but as they gradually approached, the limits of this seeming inundation retreated; the imaginary lake, that appeared to surround the village, drew back, and finally melted away altogether, the same illusion being repeated in the case of the next village. The savans attached to the expedition who witnessed this phenomenon were not less surprised than the rest of the army; but Monge succeeded in giving the explanation of it.

The theory of the mirage, in order to be perfectly understood, demands very special attention. The phenomenon occurs when the luminous rays, through whose agency we see objects, are made (before they reach our eye) to undergo a deviation caused by differences of density in the strata of air they pass through. We have seen that when a luminous ray penetrates from a less dense into a more dense medium, it undergoes a deviation which bends it nearer to the line perpendicular to the boundaries of the two surfaces; and when it passes from a more dense to a less dense medium, it suffers a deviation bending it from the perpendicular.

Further, the angle of refraction is greater than the angle of incidence, and at a given moment a certain ray will, after refraction, make an angle of 90° with the perpendicular to the surface. This is called the critical angle.

Beyond this angle the rays are reflected, and do not enter the medium at all; this is known in physics under the name of *total reflection*.

An illustration of this fact may be obtained by filling a glass with water and holding it so as to see the surface of the liquid from underneath; this surface acts like a mirror, and appears very bright. A spoon dipped into it is reflected. Another instance: a prism of glass properly placed at the opening of a dark room is capable of intercepting entirely the passage of light by this very fact of total reflection. In fact, when a luminous ray tends to emerge from a more reflecting medium into one that is less so, at an angle greater than the critical angle, the ray is entirely reflected.

This being taken for granted, we may now affirm that the mirage is a phenomenon of total reflection.

By the action of the solar rays, when the atmosphere is calm, the strata of air which are in contact with the soil become very much heated, and it may happen that for a short distance up their density may increase as they are farther from the ground. This is a purely accidental fact, which depends upon various circumstances peculiar to the place where it occurs; it does not extend very far, and consequently in nowise affects the general law of the decrease of density in proportion to the

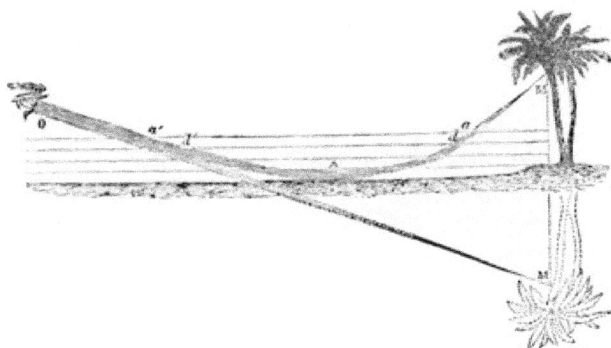

Fig. 40.—Explanation of the ordinary mirage.

elevation. In the event of these physical conditions happening, the following may be the result: a luminous ray, starting from the point M (see Fig. 40). is successively refracted in *a' d'*, as it is bent from the normal; at a given moment the direction will coincide with that of the stratum of air A, and this latter will serve as a mirror: the ray will follow, therefore, in an opposite direction a path, A d' a', similar to that which it has already taken, and will reach the eye of the spectator, who will see in the lower direction, O M, the reflection of the palm-tree M, at the same time that he will see the object directly. It is, therefore, the stratum of air which, at a given moment, becomes a mirror, and consequently acts in the same way as a piece of reflecting water, which gives rise to the phenomenon. Such is the ordinary or inferior mirage.

This lower and reflected deviation of the luminous rays does not always attract attention so much as might be fancied. Many people will pass by it without remarking it, and, even when their attention is called to the fact, will declare that they perceive nothing extraordinary or worthy of notice. To clearly discern the mirage, a person must not only possess long and very accurate eyesight, but must also know how to observe details, and be accustomed to the view. To travelers, sailors, and meteorologists, this is a practice that has become familiar; but very frequently non-scientific eyes fail to distinguish these details.

Yet, in some cases, and especially in certain regions of the globe, the mirage is so plainly evident that it arrests the most inattentive gaze. Such is at times the mirage upon the coasts of the Gulf of Messina; and such, it appears, is that seen upon the sandy plains of Arabia and Egypt.

The mirage is sometimes visible upon the surface of the sea, and of lakes and large streams; sometimes upon the great dry and sandy plains, or upon high-roads or the sea-shore.

Very frequently these misleading appearances, due to the action of solar rays, and to their prismatic reflection across the strata of air of unequal density, present purely imaginary shapes which one is inclined to consider as real, although their origin is as fortuitous as that of the appearances occasionally seen in the clouds. The same may be said of those unknown islands which rise up in mid-ocean before the astonished navigator, and which lead him astray toward imaginary lands. The Swedish sailors for a long time went in search of a magic island that seemed to rise between those of Aland and of Upland; it turned out to be only a mirage. The towns which seem evolved by the wand of a fairy are sometimes but the reflection of distant towns; but more frequently there is nothing to explain, if not their nature, at least their origin. M. Grellois says, "During the summer of 1847, I was proceeding one very hot day on horseback, at a walking pace, between Ghelma

and Bône, in company with a young friend who has since died. When
we had arrived within about two leagues of Bône, toward one in the
afternoon, we were suddenly brought to a halt at a turn in the road by
the appearance of a marvelous picture unfolded before our eyes. To
the east of Bône, upon a sandy stretch of ground which a few days before we had seen arid and bare, there rose at this moment, upon a gently
sloping hill running down to the sea, a vast and beautiful city, adorned
with monuments, domes, and steeples. The illusion was so complete,
that reason refused to admit that this was only a vision which held us
entranced for nearly half an hour. Whence came this apparition?
There was no resemblance to Bône, still less to La Calle or Ghelma,
both distant twenty leagues at least. Are we to suppose it was the reflected image of some large city on the Sicilian coast? That seems to
me very improbable."

The inferior mirage is sometimes affected by simple effects of refraction, by a change or magnifying of the objects observed. Thus, for instance, in May of 1837, during the Algerian expedition, which preceded
the treaty made with Abd-el-Kader, M. Bonnefont observed, among others, the curious mirage described below:

"A flock of flamingoes, birds of prey which are very common in this
province, were seen upon the south-east bank, about three miles and a
half off. These birds, as they left the ground to fly to the surface of
the lake, assumed such enormous dimensions as to give the idea of
Arab horsemen defiling one after the other. The illusion was for a
moment so complete, that General Bugeaud sent a Spahi forward as a
scout. The latter crossed the lake in a straight line; but when he had
reached a point where the undulations commenced, the horse's legs became so elongated, that both steed and rider seemed to be borne up by
a fantastic horse several yards high, and disporting itself in the midst
of the water that appeared to submerge it. All eyes were fixed on this
curious phenomenon, until a thick cloud intercepting the sun's rays
caused these optical illusions to disappear, and re-established objects in
their natural shape.

"Sometimes another effect, which became a source of amusement to
the soldiers, was produced. If, while the sun was in the east and the
wind blowing from an opposite direction, a small and buoyant object,
susceptible of being floated along by the wind, was cast into the lake,
it was curious to observe how it became larger as it got farther off, and,
as soon as the wind had made it undulate, it suddenly took the shape of

a small boat, the movement of which, above the waves, was in proportion to the shaking it experienced from the wind. The objects that an-
swered best for the experiment were thistle-heads, as they were most
easily influenced, even by the lightest breeze, and rendered the illusion
complete. At about half-past eight on the morning of June 18, with a
temperature of 26° centigrade, while a somewhat strong breeze was
blowing from the east, and a nebulous stratum was beginning to dissi-
pate the heat, a certain number of these thistle-heads were launched
upon the water, and no sooner had the wind driven them to the point
where undulation commenced, than they presented the curious spectacle
of a fleet in disorder. The vessels seemed to dash one against the other,
and then, driven by the wind to a great distance, they disappeared as
completely as if they had gone down."

We now come to a second kind of mirage which is often seen, but the
effects of which are less striking, and which has consequently been less
studied, viz., the approach of objects situated beyond the horizon, and
which are raised above it. In the ordinary mirage which we have just
described, the density of the air increases with the height, the trajectories being convex toward the earth, at least in their lower parts. In
the case under consideration, the density decreases and the trajectories
become very concave toward the ground. A luminous ray, at first horizontal, should, as it moves through the void, remain rectilinear; but
the ordinary atmospheric refraction inflects this trajectory, imparting to
it about a twelfth part of the terrestrial curvature. But if the condition
of the strata is modified, and if, by the effect of an abnormal increase in
the temperature, the density decreases with the height much more than
is usual, the refracting effect of these strata may impart to these trajectories a greater curvature, amounting to a quarter, a half, or even the
whole of the curvature of a great circle of the earth. Indeed, sometimes this action may cause it to exceed this latter limit.

In these fresh conditions, the various trajectories passing through the
eye and situated in the same vertical plane, instead of cutting each other
two and two, as in the case of an ordinary mirage, generally diverge.
Hence it results that we can not obtain two reflections of one object.
If the depression of the apparent horizon is measured, it is found to be
very much raised, sometimes to the level of the rational horizon; and
objects, usually invisible by reason of their great distance and curvature of the earth, may become visible. The accidental position of these
objects beyond the apparent contour of the visible horizon makes them

appear to be much nearer than usual, while another circumstance favors the illusion, viz., the transparency of the air during the occurrence of the phenomenon. It is clear that, as no reversal of the objects takes place, one would be less struck with this particular form of mirage than with that which corresponds to the cases previously described. Woltmann and Biot point out that when the atmosphere is in this particular condition the sea seems to be concave, at the same time the horizon is seen above the hulls of ships, distant shores take the shape of high cliffs, and very distant objects seem to rise in the air like clouds.

An optical circumstance well worthy of attention is the following: at the same time that some objects are thus raised above others by which they are ordinarily hidden from view, or when they are apparently removed to this side of the apparent horizon, they seem to the eye to be very much nearer. Heim has described a case of this kind observed in the mountains of Thuringia, where he suddenly beheld three lofty peaks appear above an intermediate chain which generally concealed them from sight; and these peaks appeared to be so clearly defined that he was able to distinguish, with an ordinary glass, tufts of grass that were distant four German miles. M. de Tessan saw a phenomenon of the same kind in the harbor of San Blas (California).

A letter from Teneriffe, published in the *Courrier des Sciences*, states that from the summit of this mountain, whence the view embraces a horizon of fifty leagues radius, a mirage rendered visible the Alleghany Mountains in North America, a thousand leagues distant. I scarcely can venture to credit this story.

Having explained the two great classes of facts relating to the phenomena of mirages, one of which is due to the depression of the objects, and the other to their elevation, we now come to the consideration of another effect scarcely less curious, viz., the *superior mirage*.

This presents three different aspects. Sometimes the reflection is seen inverted above the object, and, above the former, a second reflection, erect as the object; sometimes the first reflection alone is seen, the upper one having disappeared; and, thirdly, the upper reflection remains without any inverted reflection beneath it.

Woltmann noticed the superior mirage on three different occasions; objects appeared to be reflected in the sky; in the air was seen the reflection of the horizon of the waters, and below were suspended, upside down, the shores, houses, trees, hills, and windmills. Frequently a lay-

Fig. 41. — Mirage seen at Paris in 1869.

er of air separated the objects turned upside down from those beneath, but usually the reflection and the object were in contact.

Welterling made analogous observations upon the Svenska-Hogar, islands situated at the entrance of the harbor of Stockholm. He says: "Above each of the sand-banks a black spot rises and appears in the air; these spots then become elongated downward, and finally reach the sand-bank, which assumes the appearance of a column nine or ten times higher than it really is. Hence there results a mock horizon, to which all the objects are transported, all thus appearing in a straight line upon the same level, though their absolute height differs considerably."

Crauz saw in Greenland the shores of the Kokernen Islands, raised in the shape of high cliffs, ancient towers, and ruined edifices. Brandes several times witnessed the superior mirage; as a rule the reflection of objects were not seen very distinctly by him, for he adds that the upper or direct reflection was generally wanting, and he attributes this fact to the want of spherical shape in the homogeneous strata. He also remarks that this is a very local phenomenon, being seen often upon the houses in the eastern part of Daingast, and at the same time being invisible upon those in the west part of the town.

In December, 1869, between the hours of three and four in the morning, a mirage was seen in Paris, as represented in the opposite plate.

These objects are occasionally delineated in the sky at a considerable height above the horizon. Some move very rapidly, and others are stationary, while they are sometimes tinged with colors. In proportion as the light augments, the shape becomes more airy, and they vanish entirely when the sun is shining with full brightness. Mirage may also be produced by two strata of air separated by a vertical plane. This notably occurs in the case of large walls with a southern aspect, when they are heated by the sun, and then the ordinary mirage is formed. It is in this case termed the *lateral mirage*. The wall in this instance acts in the same way as the soil when exposed to the solar rays, and a line perpendicular to the wall replaces the vertical line in the horizontal mirage. But, as the heated strata of air are easily renewed as they rise along the wall, the disturbing influence of the densities does not extend very far. The eye must, therefore, be placed in front of the plane of the wall, and must view in a parallel direction any objects that may approach and recede. The persons who approach the doors in the wall, the images which cross in the sky the vertical parallel to that of the wall, are always seen inverted, as indicated in the theory of the ordinary mirage. Gruber seems to have been one of the earliest spectators of this phenomenon. Blackader has described a lateral mirage that he saw upon a wall at Leith. It was also observed by Gilbert.

Let us add to the above the multiplied mirage which is seen when several reflections, all inverted, are superposed upon the object. Biot and Arago saw phenomena of this kind from the mountain Desserto de las Palmas, and observed at night, with the repeating circle, an illuminated reflector in the island of Ivyza. Besides the ordinary reflection, two, three, or even four false reflections, superposed in the same vertical line, have been seen. Scoresby observed, on July 18, 1822, a brig with three reflections superposed, all inverted, and in each of them the vessel was in contact with the reflection, also inverted, of the field of ice beyond which it was situated.

The mirage does not always present such regular characteristics as we have indicated; sometimes the second reflection is seen above the original one; sometimes the two are seen beside each other; and, lastly, the reflections sometimes are not inverted.

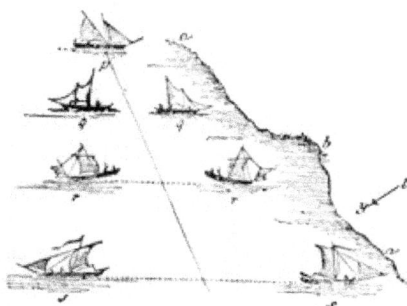

Fig. 42.—Lateral mirage seen on Lake Geneva.

Dr. Vince relates several remarkable observations. From Ramsgate, in fine weather, may be seen the tops of the four highest towers of Dover Castle. The remainder of the edifice is concealed by a hill, which is about twelve miles from Ramsgate. On the 6th of August, 1866, Dr. Vince, looking toward Dover at seven in the evening, perceived, not only the four towers as usual, but the entire castle from roof to base, as distinctly as if it had been transported to the hill near Ramsgate.

In the polar regions, the action of refraction is seen under the most capricious and extraordinary conditions. Admiral Wrangell writes: "The extreme condensation of the air in winter, and the vapor diffused in the atmosphere in summer, give great power to refraction in the frozen sea. In these circumstances the mountains of ice often assume the most grotesque shapes; sometimes, indeed, they seem to be detached from the icy surface which serves as their base, so as to appear to be suspended in the air."

Very frequently Admiral Wrangell and his companions thought they perceived mountains of a bluish color, whose shapes were clearly defined, and between which they thought they could discern valleys and even rocks. But just as they were congratulating themselves on having discovered the long-sought land, the bluish mass, carried away by the wind, extended on each side, and finally embraced the whole horizon. Scoresby, who collected so much interesting information in these Greenland regions, has also pointed out that ice assumes at the horizon the most regular shapes, and even appears, at many points, suspended in the air.

The most curious phenomenon was to see the reflection, inverted and very distinct, of a vessel below the horizon. He says: "We had already observed similar apparitions, but this one was peculiar for the distinctness of the reflection, in spite of the great distance of the vessel. Its contour was so well defined that, in looking at it with a Dolland's glass, I could distinguish the details of the masts and the hull of the ship, which I recognized as that of my father. On comparing our books, we saw that we were 34 miles from each other, that is, 19¼ miles from the horizon, and far beyond the limits of vision."

Upon the shores of the Orinoco, Humboldt and Bonpland discovered that at noon the temperature of the sand was 127°, while at six yards above the ground the temperature of the air was only 104°. The hillocks of San Juan and Ortez, the chain called the *Galera*, situated three or four leagues off, seemed suspended in the air; the palm-trees appeared to have no hold on the ground, and, in the midst of the savanna of Caraccas, these savans saw, at a distance of a mile and a half, a *herd of oxen apparently in the air.* They noticed no double reflection. Humboldt also remarked a herd of wild cattle, part of which seemed to be above the surface of the ground, while the remainder were standing upon the soil.

Mirages are not exclusively phenomena of warm climates; as we have seen, they have been observed in the very heart of the polar seas.

When, instead of occurring in plane and regular strata, refractions and reflections take place in the curved and irregular strata, a mirage is produced, the reflections of which are deformed in all directions, broken or repeated several times, and very far distant from one another.

This is the case with the fantastic aërial vision, formerly attributed to a fairy—the *Fata Morgana*—which sometimes attracts crowds of people to the sea-shore at Naples and at Reggio, upon the Sicilian coast.

The phenomenon generally occurs of a morning in very calm weather. For an extent of several leagues the sea upon the Sicilian coast assumes the appearance of a chain of sombre mountains, while the waters upon the Calabrian side remain quite unaffected. Above the latter is seen depicted a row of several thousands of pilasters, all of equal elevation, of equal distance apart, and of equal degrees of light and shade. In the twinkling of an eye these pilasters sometimes lose half their height, and appear to take the shape of arcades and vaults, like the Roman

Fig. 43.—La Fata Morgana.

aqueducts. There is often, also, noticeable a long cornice upon their summits, and there are also seen countless castles, all exactly alike. These soon fade away, and give place to towers which in turn disappear, leaving nothing but a colonnade, then windows, and lastly pine-trees and cypresses, several times repeated.

Similar fantastic apparitions were noticed with great surprise in the neighborhood of Edinburgh on the 16th and 17th of June, 1870, previous to a severe thunder-storm. These are unquestionably among the most curious kinds of mirage that exist.

CHAPTER VII.

SHOOTING-STARS— BOLIDES— AEROLITES— STONES FALLING FROM THE SKY.

NONE of my readers will have failed to have been struck with surprise, during the calm of a fine starry night, by the spectacle of a star gliding noiselessly through the celestial vault to extinction. Some, perhaps, of those who peruse these pages, may have enjoyed the rare privilege of beholding, not only a *shooting-star*, but a more brilliant and sometimes very exciting phenomenon, viz., the rapid passage through space of a flaming *bolide*, scattering a gleaming light in all directions— a globe of fire, leaving a luminous track behind it, and sometimes bursting with an explosion like that of an enormous shell, and a report like that of a cannon. Some, perhaps, also, by a still more fortunate chance, have had an opportunity of picking up a fragment of an exploded bolide—a piece that has fallen from the sky—an aerolite or stone that has come down from the heights of the atmosphere.

We here have three distinct facts, which nevertheless seem to be related to each other in their origin. The progress made during the last few years in the special study of these meteors is a reason for considering them separately, taking first the shooting-stars, then the bolides, and lastly the aerolites.

The first point to consider in the study of shooting-stars is the measurement of the height at which they are seen. Two spectators, placed at a distance of some miles from each other, notice the passage of a shooting-star among the constellations; its path is not exactly the same to both observers, owing to perspective. From the observation of these two paths the distance can be obtained. This method, as early as 1798, two German savans, Brandes and Benzemberg, had already made use of. From the latest researches upon this head made by Alexander Herschel (grandson of the famous Sir William Herschel), by Professor Newton, of New Haven, Conn., and by Father Secchi, Director of the Observatory at Rome, it has been concluded that the average height of a shooting-star is seventy-five miles when first seen, and fifty miles at the end of its visible journey.

The velocity varies from seven to forty miles a second.

Shooting-stars are not common to all nights of the year alike, for the result of observations shows that there are yearly, monthly, and daily periods of recurrence of certain sets of shooting-stars. Great showers of shooting-stars on particular nights have been remarked since the last century; Brandes relates that, on December 6, 1798, during a carriage-drive to Bremen, he counted four hundred and eighty from the coach window; and he estimates that, at that rate, there must have been at least two thousand in the course of the night.

During the night of the 11th to the 12th of November, 1799, Humboldt and Bonpland witnessed a perfect shower of shooting-stars at Cumana (South America). Bonpland states that there was no part of the sky equal in extent to three diameters of the moon that was not continuously being filled with shooting-stars. The inhabitants of Cumana were terrified by this phenomenon, and the oldest of them remembered an analogous occurrence in 1766, accompanied by an earthquake.

This shower of stars at the close of the last century had been nearly forgotten, when a fresh shower was seen in America on November 13, 1833. Professor Olmsted, of New Haven, Conn., basing his calculations upon data which had been transmitted to him, regards the number of shooting-stars that appeared in certain districts on that occasion as over two hundred thousand. Olmsted was the first to point out that the great display of November must be periodical, and would be reproduced every year at the same epoch. A very considerable increase in the number of shooting-stars at that date has, in fact, been noticed, but not to the extent of the extraordinary phenomenon in America in 1833. The astronomer Olbers, writing on the same subject in 1837, says: "We shall, perhaps, have to wait until 1867 for the recurrence of the splendid phenomenon witnessed in 1799 and 1833." This bold prediction was completely realized just a twelvemonth earlier, in 1866.

From a general discussion of the observations, it results that the number of shooting-stars which ordinarily appear over the whole extent of the visible sky in the space of an hour is, on an average, from ten to eleven.

Now, at the time of the maximum on November 12 and 13, this hourly number, which was equal to fifty in 1834, gradually fell annually, until it was reduced to thirty in 1839, to twenty in 1844, to seventeen in 1849; three or four years later the maximum had disappeared, and was replaced by a normal appearance of from ten to eleven an hour

Matters remained in this condition until 1863, when a maximum of thirty-seven an hour again occurred at the same epoch, rising to seventy-four an hour the next year, and thus acting as a precursor of the great phenomenon of 1866, when Olbers's prediction was fulfilled. Another

Fig. 44.—Shooting-stars.

maximum occurred on August 10, and was noticed by M. Quételet so long ago as 1837. The maximum hourly number of shooting-stars was, on that night, fifty-nine. There was a progressive rise in the number to seventy-nine in 1841, to eighty-five in 1845, and to one hundred and ten in 1848, from which date it gradually decreased each year, standing at thirty-eight in 1859, since which time it has alternately risen and fallen, varying between the numbers thirty-seven and sixty-seven.

Here we have a well-ascertained *annual* variation in these periodical showers. The researches of Coulvier-Gravier clearly establish the existence of a *monthly* variation, the number of shooting-stars being greater in autumn than in spring. There is, also, a *daily* variation. The hourly numbers, from six in the evening to six in the morning, are twice as great as for the corresponding hours in the day-time.

Shooting-stars are seen in all parts of the sky; but if the directions whence they seem to come are examined, it is found that the different parts of the horizon furnish different numbers. There is thus a variation in this respect which is termed the *azimuthal variation*, and which has been thoroughly studied by means of carefully registered observations. Many more shooting-stars come from the east than from the west, but nearly equal numbers from the north and the south.

At the periods of the maxima. toward the 12th and 13th of November, and toward the 9th and 10th of August, the shooting-stars, instead of appearing in all the regions of space indifferently, nearly all come from given directions. Some (those of November) start from the constellation *Leo;* the others (August) emanate from the constellation *Perseus.* What path in space is then taken by these periodical showers, the existence of which is ascertained?

It has been observed that the speed of the meteors is equal to that of comets descending toward the earth from the depths of space, and their orbit has been also assimilated to the orbits of the comets. Signor Schiaparelli, Director of the Milan Observatory, sought to determine the elements which characterize the shape and the position of the apparent parabola followed by the meteoric current of the 10th of August. He then compared these astronomical elements with those obtained by calculating the orbits of the different comets. He was thus able to establish a very unexpected similarity between the orbit that he had just discovered for the swarm of shooting-stars of the 10th of August and that of the great comet seen in 1862.

Supposing that every one hundred and eight years these meteors have a frequency neither so sudden nor so short in duration as that of November meteors. but lasting twenty or thirty years, this period agrees with the duration of the revolution of the great comet of 1862, and may be, therefore, taken to represent that of the successive returns of the comet to its perihelion.

M. Schiaparelli then set to work to discover the elements of the orbit of the November swarm of shooting-stars. Observation in this instance supplied him with further data; the period of return for the great displays of November, indicated by Olbers in 1837, had just been confirmed in 1866, and might be fixed at thirty-three years and a fraction.*

* [Taking as data the observed directions, etc., of the November meteors, the researches of Professors Newton (New Haven, Conn.) and Adams have shown that their orbit must be an ellipse, the periodic time of which is about 33¼ years, agreeing exactly with observation. A small discrepancy has also been satisfactorily explained as the result of the attraction of the larger planets, especially Jupiter.—Ed.]

A swarm of shooting-stars, seen on the 10th of December, describes in space the same ellipse as the well-known Biela's comet, and the shooting-stars seen on the 20th of April move along the orbit of the first comet of 1861. Such researches have thrown a great light upon the question of shooting-stars. The comet which traces in space the same path as the swarm of meteors must be considered as an integral part of it. It is, in fact, merely a local concentration of the matter of the swarm —a concentration so intense that the mass of matter it forms is visible even at a great distance from the earth. According to this theory, shooting-stars are of the same nature as comets, consisting of small nebulous objects which move in space without being visible to us because of their smallness, and only becoming so when they penetrate into the atmosphere of the earth. Like comets, they seem to be gaseous.

A current of these meteors which encounters the orbit of the earth at a certain point, and the different parts of which take several years to pass this point of meeting, must be crossed by the earth each year at the same epoch. Hence the periodical showers of shooting-stars which are reproduced from year to year, with varying intensity, according to the greater or less concentration of the nebulous matter in the various parts of the current which the earth successively reaches.

Such are shooting-stars. Now we come to the *Bolides.* If shooting-stars are gaseous, there is an essential distinction between them and bolides, for the great majority of the latter are unquestionably solid. To give an idea of the meteoric phenomenon of the explosion of a bolide, I will cite, among the most recent falls, one that occurred by day and another that occurred at night, both in 1868.

This is the account of the fall of a bolide by day, which took place in the arrondissement of Casale, in Piedmont, on the 29th of February. It was half-past ten in the morning, but the sky was rather dark. Suddenly a loud detonation was heard, similar to the discharge of a heavy piece of artillery, or, perhaps, rather to the explosion of a mine. This was followed, at an interval of two seconds, by another report resulting from two distinct detonations, which succeeded each other so closely that the second seemed to be the continuation or the prolongation of the first. These detonations were heard as far off as Alexandrie, a distance of twenty miles. The sound had not yet died away when there became visible, at a considerable height above the ground, a mass irregular in shape and enveloped in smoke, thus resembling a small cloud. It left behind a long train of smoke; other spectators saw distinctly, and at a great height, not one but several spots like small clouds which disappeared nearly instantaneously. Some men at work in the fields saw several blocks fall through the air, and heard the noise which they made as they struck the ground. Every one whom it was possible to question on the subject was unanimous in affirming that there were a large number of these blocks, and that they must have occasioned a regular shower of aerolites of all sizes. Laborers at work felling trees in a wood three-quarters of a mile from Villeneuve, on the high-road from Casale to Vercelli, saw something like a hailstorm of grains of sand after these detonations, and a somewhat large fragment struck the hat that one of them was wearing. The aerolites found upon the ground consisted of: 1st, a piece weighing $4\frac{1}{4}$ lbs., which fell in a wheat-field 650 yards to the south-east of Villeneuve, and penetrated sixteen inches into the ground; 2d, a piece weighing $14\frac{3}{4}$ lbs., which fell in a newly-sown field to the north of Villeneuve, 7700 feet from the first, and entered the ground to a depth of $14\frac{1}{2}$ inches; 3d, the numerous fragments into which a third piece broke by falling upon the pavement in front of the inn of Molta dei Conti, at a distance of 10,335 feet from the first piece, and of 10,630 feet from the second.

The recital of the nocturnal fall will help to complete the comprehension of these singular occurrences. It took place in the arrondissement of Mauléon, in the Lower Pyrenees, on September 7, 1868, at half-past ten in the morning.

The sky was suddenly illuminated by a meteor, which looked like a burning ball with a long train of fire in its track. It emitted a bright light of a pale greenish hue, and lasted for six or ten seconds. Its disappearance was preceded by an explosion, and by the simultaneous projection of flaming fragments, while there remained for some time after a light and whitish cloud. This was followed by a continuous noise, like the distant rolling of thunder, then by three or four detonations of extreme violence, which were heard at points distant fifty miles from each other. Immediately after these detonations the inhabitants of Sanguis-Saint-Étienne heard a hissing noise like that made by red-hot iron when it is plunged into water, then a dull sound indicating the fall of a solid body to the ground. The mass had fallen at about thirty yards from the church of Sanguis, in the bed of a small stream, and was shattered into fragments, the largest of which was scarcely two inches long. The fall was witnessed by two men who were talking together, and who, terrified at the detonations and the hissing noise, had thrown themselves upon the ground just as the stone fell about twenty paces before them. The weight of the stone was estimated at from six to eight pounds.

These two instances, which I select from an immense number, give a sufficient idea of these downfalls from the sky, which were formerly looked upon as fabulous. It is only in the last half century that the facts have been credited and scientifically confirmed.

In contradistinction to the shooting-stars which become extinguished and lost in the upper regions, the bolides traverse all the atmospheric strata, and often reach the surface of the earth. This is the reason why the luminous phenomena that accompany them usually appear to us much more intense; because, in fact, the regions in which they occur are much nearer to us. But when seen from afar, as is the case with those whose directions prevent them from reaching the lower strata of the atmosphere, bolides present the same appearance to our eyes as shooting-stars. When they do reach the lower air, an explosion, simple or repeated, often takes place, followed in the majority of cases by a fall of fragments from the bolide that have become detached from the main mass by the effect of the explosion. Bolides, then, are solid bodies, like the fragments detached from them. The orbits described by these bolides, in their movement relative to the earth, have sometimes been found to be ellipses of such limited dimensions, that one would be led to suppose that the former were nothing but satellites of the earth, only visible during their passage through the atmosphere—a view adopted by Petit, of Toulouse. On the other hand, their orbits have sometimes been found to be hyperbolic arcs, nearly rectilinear, and traversed with great speed — a fact tending to show that bolides possessing such rapid movement must come from the stellar regions.

The *aerolites* are minerals that fall from the sky to the earth. They proceed from the explosion of a bolide.

Sometimes they plunge deeply into the soil upon which they fall. Thus the island of Lanaïä-Uawaï possesses an aerolite six or seven yards in diameter, which has remained imbedded in the ground in despite of all the efforts made to raise it to the surface. This aerolite fell at the beginning of the century. (Very recently, on the 9th of March, 1868, at 9·30 P.M., another bolide fell upon the same island.)

These stones, if touched immediately after their fall, seem to be burning hot; but they cool very rapidly—a fact indicating that their higher temperature was altogether superficial, and did not extend to the interior of their mass.

As to the shape of these aerolites, it is neither that of a ball, more or less round, nor that of an object with a rounded surface; they rather resemble polyhedra, with rough, irregular sides and ridges. The plane parts of their surface have often hollows analogous to those produced by the pressure of a round body upon a pasty substance. They are, moreover, enveloped in a black crust, generally of a dark hue, but sometimes lustrous, as if covered with very thin varnish.

Fig. 45.—Fall of a bolide in the day-time.

The light displayed in the movements of the bolides is due entirely to the heat produced by the compression of the air. Let us examine in what way the phenomena of explosion, and the falls of the aerolites which often succeed it, are produced.

The enormous compression of the air forced back by the bolide can not occur without this air reacting upon the anterior part of the surface of this body, and exercising a considerable pressure upon it. Attributing to the bolide a speed of four and a half miles per second—by no means an exaggerated estimate—M. Haidinger calculates the resisting pressure which the bolide meets with from the air at more than twenty-two atmospheres. Such a pressure evidently tends to crush the body which is exposed to it; and if this body, in its more or less irregular shape and constitution, offers portions of itself which are more opposed than the others to the action of this pressure, these portions may give way and become suddenly detached from the mass of the bolide.

Broken off and started in a direction contrary to that in which they were traveling a few moments before with the main mass of bolide, these fragments soon lose the speed with which they were endowed, and reach the terrestrial surface, still moving with very great velocity, but not with the rapidity of bodies falling to the earth from space.

We are inclined to look upon the bolides as being somewhat similar in origin and being to the planets which circulate in such great numbers around the sun, and as probably themselves forming part of our planetary system. Besides, the discovery recently made of a large number of planets of very small dimensions, induces us to believe that there exists a multitude of others still smaller which have escaped observation.

In consequence of the great difficulties that were encountered in attributing to the bolides a purely terrestrial origin, it was long ago suggested that they might be stones hurled to the earth *from the volcanoes of the moon.* This idea was taken up and developed, in 1795, by Olbers, and in the early part of the present century by Laplace, Lagrange, Poisson, and Biot; but serious objections of more than one kind soon appeared to render this theory untenable, and it was finally abandoned for that of Chladni, whose system consisted in regarding the bolides as bodies wandering freely in space, and penetrating every now and then the atmosphere of the earth.

Whatever may be the part played by the bolides in the universe, the possibility afforded us of examining the fragments which they leave in their passage is very useful in regard to the information which we are enabled to extract from them as to the constitution and nature of bodies foreign to the globe which we inhabit. Thus great pains have been taken of late years to collect from all quarters stones that have fallen from the sky after the explosion of bolides; and collections of this special kind of rock have been made, to which, in order to distinguish them from the terrestrial rocks, the special denomination of meteorites has been given. There are at various places beautiful and valuable collections of this kind; among others, that in the Museum of Natural

History in Paris, that in the British Museum, and that in the Mineralogical Museum at Vienna. The Paris collection, under the superintendence of M. Daubrée, contains at present specimens of 240 meteorites, while all the known falls do not exceed 255.

It is easy to understand that conflagrations may have been caused by the fall of aerolites, and that people may have been killed by them. Fourteen deaths have been ascertained to have taken place from this cause at various times.

The largest stones known to have fallen are as follows:

The aerolite that fell at Juvénas in the Ardèche, on June 15, 1821, weighed 212 lbs., exclusive of the fragments detached from it.

Fig. 46.—The Caille aerolite, weighing 12½ cwt.

The aerolite found in Chili, between Rio-Juncal and Padernal, in the Upper Cordilleras of Atacama, weighed 240 lbs., and was in the shape of a cone, measuring nineteen inches in length and eight inches in diameter. The miners who brought it home upon their mules had taken it for a block of silver. It was in the Paris Exhibition of 1867.

The meteoric stone of Murcia, which is in the Museum of Natural Sciences at Madrid, weighs 2¼ cwt.

The aerolite which fell in 1492, at Ensisheim, in the Upper Rhine, in the presence of Maximilian I., king of the Romans, weighs 2¾ cwt.: it is imbedded five feet in the ground, and was long venerated by the Church as a miraculous object.

The aerolite that fell on Christmas-day, 1869, at Mourzouk (latitude 26° N., longitude 12° E. of Paris), in the midst of a group of terrified Arabs, must weigh much more, for it is nearly a yard in diameter. It is to be taken to Constantinople, but will, unfortunately, have to be previously divided.

None of these, however, approach the Caille aerolite, in the Maritime Alps, which was used as a seat at a church porch, and which is now in the Paris Museum. It weighs 12¼ cwt. (see Fig. 46).

The aerolite that fell in 1810 at Santa-Rosa (New Granada) in the night of April 20, 21, weighs 14¾ cwt. When found, it was almost imbedded in the ground by the force of the fall.

Lastly, the most colossal of the known stones that have fallen from the sky is the aerolite brought back from the Mexico campaign, weighing more than 15¼ cwt. It had from time immemorial been lying at Charcas. Its shape is that of a truncated triangular pyramid, measuring a yard in height, and it is a fair specimen of the world that sent it to us. From several hundred analyses made by the most eminent chemists, it appears that the meteorites have added no single substance to the globe which it did not possess before. The elements up to this time discovered to be existent in them are twenty-two in number.

CHAPTER VIII.

THE ZODIACAL LIGHT.

To complete the panorama of the optical phenomena of the sky, we will now consider the nature of a nocturnal brightness which is seen in the heights of the atmosphere on certain clear nights. As in the case of shooting-stars and bolides, its origin is in the depth of space, and the explanation of it belongs to astronomy; but, as it reveals itself in our sky, it deserves notice in these pages.

After sunset in January, February, March, and April, and after sunrise in November, the celestial vault sometimes displays a band of light inclined toward the horizon and in the plane of the zodiac; that is, in the apparent path that, by its annual change of position, the sun seems to trace out in the sky. This light was not remarked till comparatively recently, and the discovery of it is due to Childrey, who speaks of it in his "Natural History of England," published about 1659. The earliest scientific researches with regard to this phenomenon were not, however, made until 1683; they are due to J. D. Cassini. When the zodiacal light first appears in the evening after sunset, it is interfered with near the horizon by the last traces of the twilight glimmer, and the union of these two lights presents the appearance of a cone. This oblique cone, at least in our climates, has its base upon the horizon and its summit at a certain height above.

Toward the equator this brightness rapidly loses its conical aspect as the last traces of twilight disappear, and when night has fully set in a band of light may be distinguished right round the sky, and making the zodiac luminous, so to speak; sometimes this band is visible uninterruptedly from sunset to sunrise. The parts nearest to the sun exceed in brilliancy the intensity of the Milky Way; the other parts are dim, and if they are visible at all in the intertropical zone, it is because of the great limpidity of the atmosphere in these regions.

The zodiacal light, when it is distinctly seen, as in the intertropical zone, is one of the most beautiful of the celestial phenomena. Its color is pure white. Certain observers in Europe have sometimes thought that they could discern a reddish tint in it. This tint has no real existence; for, if it had, it would be most distinctly discerned at the tropics, as the color would become more perceptible when the intensity of the light was increased. The last traces of twilight have been mistaken for it. In the tropics (in the months of January and February for the Tropic of Cancer) it rises perpendicularly to the horizon; then, when night has fully set in, there is seen rising in the west a beautiful white vertical column, the central axis of which equals and even exceeds in intensity the more brilliant parts of the Milky Way. Upon the edges of this column, the light gradually blends with the feeble glimmer of the sky. It differs in that respect from the Milky Way, the edges of which at certain points offer a noticeable contrast of light to the general darkness, as in the black hollow of the Southern Cross, called the *coal-sack*.

It is not visible in Europe during the summer. This is owing to its inclined position upon the southern horizon, which then grazes the part of the zodiac which is visible at night and during the twilights. In February its appearance is most complete. In warm countries, the shortness of twilights, and the elevation of the ecliptic, cause the phenomenon to be visible all the year round. There are, however, even in countries where this is the case, periodical maxima of beauty which depend upon the inclination of the plane of the zodiac to the horizon.

The observations of Cassini and of Mairan, who sometimes saw the zodiacal light at more than 100° from the sun, had long since indicated that this beautiful phenomenon extends beyond the terrestrial orbit. Humboldt and Brorsen had also remarked a luminous thread uniting the east and west.

Let us now consider what is the nature of this nebulosity which surrounds the sun. Several astronomers of the last century thought it was the atmosphere of that luminary, extending to an immense distance in the direction of its equator. From mathematical considerations, Laplace has shown that this hypothesis is inadmissible, and that the solar atmosphere can not extend beyond the limit at which the centrifugal force due to rotation would be in equilibrium with the attraction of the sun. It can easily be shown that at a distance from the sun equal to thirty-six times its semi-diameter, the centrifugal force developed by its rotation equals the weight of the atmospheric particles at that distance. It is mathematically impossible that the solar atmosphere can extend beyond this limit. It is not half the distance from Mercury to the sun, and but a sixth part of the distance at which the earth gravitates, for we

are situated at a distance of two hundred and fourteen times the semi-
diameter of this gigantic luminary from its centre.   Therefore the zodiacal light, which extends beyond the terrestrial orbit, is not an atmosphere of the sun.

Physicists have ascertained that all reflected lights acquire the properties peculiar to polarization, but that at the same time these properties may be lost in the event of the reflection arising, not from a gas or
a continuous surface, but from a series of distinct particles, as in the
clouds, which are composed of globules of water.   The zodiacal light
not being polarized, it results either that this light is not reflected, and
issues directly from matter luminous in itself, or, if it proceeds from the
sun, that it is caused by the reflection of the light of that luminary from
a multitude of corpuscles having no connection with each other, but obedient, like all matter, to the laws of universal gravitation.   These bodies
we must regard as circulating round the sun, and describing elliptical
orbits like the planets or the comets.   Now, if the zodiacal light proceeded from matter luminous in itself, this substance would still reflect
a certain quantity of the solar light, so that traces of polarization in the
zodiacal light would be perceived if it was not composed of distinct
corpuscles.   Therefore, in any case, we may consider as proved that it
is due to corpuscles with no connection between each other, and circulating in accordance to the laws of gravitation round the sun, from
which they receive their light.   Judging by the trifling intensity of
the light which they shed, it is improbable that they further possess a
proper light of their own.

It is possible that the aerolites, to the number of milliards upon milliards, distributed throughout the whole planetary system, and chiefly
in the general plane of movement—that is, in the plane of the ecliptic
— the bolides, the shooting-stars, corpuscles, solid, liquid, and gaseous,
form but one general kind of celestial fragmentary bodies, and that the
zone in which they chiefly gravitate is manifested to us by the reflection of the solar light, and constitutes the zodiacal light; and that, by
falling against the sun, these corpuscles cause the spots on its disk, and
help to keep up its immense heat.   If this whirlwind of corpuscles
does not circulate around the sun itself — a fact not proved — it circulates around the earth; and it is just possible that from afar it may
look like the ring of Saturn.

The appearance of the zodiacal light is somewhat rare in France; it
is scarcely ever seen distinctly more than once or twice a year, and then

in February. It was seen in Paris very clearly on the 20th of February, 1871, and lasted from 6·50 to 7·30. In the shape of a spindle, in which it is always seen, it measured 18° in width at its base, at the horizon, and, rising obliquely along the zodiac, terminated in a point before reaching the Pleiades. From the sun, which had set an hour and a half earlier, to the extremity of the spindle, it measured 86°; the part which was visible above the horizon measured 63°.

The determination of its intensity was all the more easy, as the atmosphere of Paris was scarcely lighted up at all, in consequence of there being no gas. Calm and motionless, this light was very different from the quivering gleam of the aurora borealis. This spindle was much more intense in the middle than at the edges, and at its base than at its apex. The tint, about half as brilliant again as the Milky Way, was rather more yellow. The smallest stars were visible through this veil; while in the case of the aurora borealis in October, 1870, the brilliancy of the stars in Ursa Major was eclipsed.

12

BOOK THIRD.

TEMPERATURE.

CHAPTER I.

HEAT: THE THERMOMETER—QUANTITY OF HEAT RECEIVED—TEMPER-
ATURE OF THE SUN—TEMPERATURE OF SPACE.

WE have, in the First Book, contemplated the earth as it is borne along in the midst of space by the force of universal gravitation, revolving in an orbit distant $91\frac{1}{2}$ millions of miles from the sun, which not only retains it, but also gives it beauty and life. From it we also derive heat, to the consideration of which we now proceed. Let us first see how heat, and its distribution over the surface of the globe, are to be estimated.

To measure the variations of temperature, the thermometer (θερμός, heat; μέτρον, measure) is used, just as the barometer was invented, as we have seen above, for ascertaining the variations in atmospheric pressure. Without discussing at greater length the employment of the thermometer, or the various forms of the instrument, than we did the above contrivance, it is, nevertheless, interesting to go back to its discovery, which also dates from the middle of the seventeenth century.

Our ancestors judged of temperature pretty much in the same way as we do in the present day, viz., by the principal effects resulting from it. Nowadays, science measures it more in detail and more uniformly by means of special instruments which permit of a comparison between the results obtained in different countries, or between those of one epoch and another. When the academicians of Florence established the fact that all bodies undergo a change in volume under the influence of heat, they laid the basis of thermometry. The instrument of which these savans made use consisted of a sphere soldered to a narrow tube, and containing colored alcohol. When this apparatus is transferred from one place to another warmer place, the liquid becomes dilated and the level rises, thus showing the augmentation of the temperature. This apparatus dates from 1660. In order that thermometers might be suitable for comparing with each other (that they might, that is to say, give the same indications under the same circumstances), the academicians of Florence had them all constructed, as nearly as was possible, upon one standard. A natural philosopher of Pavia, one Charles Renaldi, was

the first to suggest, about 1694, the means, still in use, for obtaining
thermometers suitable for making comparisons. The plan consists in
placing the instrument successively in two calorific positions, invariable
and easy of reproduction, viz., those corresponding to the melting of ice
and the boiling of water. Between these limits of temperature any
given body becomes dilated by the same fraction of its volume. As
a rule, 0 is marked at the point at which the liquid of the thermometer
stands in melting ice, and 100 at the point where it remains stationary
in the midst of boiling water. These two points being marked upon
the stem, the interval between them is divided into one hundred equal
parts. Newton, having conclusively demonstrated the fixity* of the
point at which water boils, the means adopted by Renaldi to render
thermometers capable of comparison was adopted by all physical philosophers. This is the Centigrade thermometer, the most convenient,
and the most in use.† Thirty years ago, Pouillet engaged in a series
of ingenious and patient experiments, with a view of determining the
quantity of heat transmitted to the earth by the sun, and the temperature of space—that is to say, the two constituent elements of the temperature of the globe.

The two contrivances made use of for the purpose were the *pyrheliometer* and the *actinometer*. The latter, being only used for researches
as to the temperature of the zenith, need not occupy our attention
here.

The pyrheliometer is in principle composed of a thin silvered box, A
(see Fig. 47), four or five inches in diameter, and holding, perhaps, three
or four ounces of water. Its surface, turned toward the sun, is blackened. A thermometer is introduced into the box and embedded in the
copper frame-work, B. The water in the box, at the same temperature
as the surrounding air, is exposed for five minutes to the sun. In order
to ascertain that the side of the box is quite perpendicular to the sun's
rays, care is taken to see that its shadow falls exactly upon the lower
disk, C, of the same diameter. By comparing its temperature with the
temperature of the air previous and subsequent to its exposure, the

[* That is to say, under the same atmospheric pressure. The boiling-point of water varies
every day with the height of the barometer; and, in fact, a method often used by explorers for
determining the height of the barometer (and therefore their own elevation above the sea-level)
is to find the temperature at which water boils at the place in question.—ED.]

[† The thermometer used in England is Fahrenheit's. The temperature of melting ice is
marked 32°, and that of boiling water (when the height of the barometer is 29·92 in.) 212°,
thus 180 graduations on the Fahrenheit scale correspond to 100 on the Centigrade.—ED.]

quantity of heat received from the sun in a minute by each square inch
of ground can be found and expressed in heat-units.*

Making allowance for the atmospheric strata traversed by the solar
rays, the experimentalist discovered that the pyrheliometer would be
raised 12°·1 Fahr. if the atmosphere were capable of transmitting in its
totality all the solar heat, without itself absorbing any, or if the apparatus could be placed at the limits of the atmosphere to receive at that
point, without any loss, the heat transmitted to us by the sun.

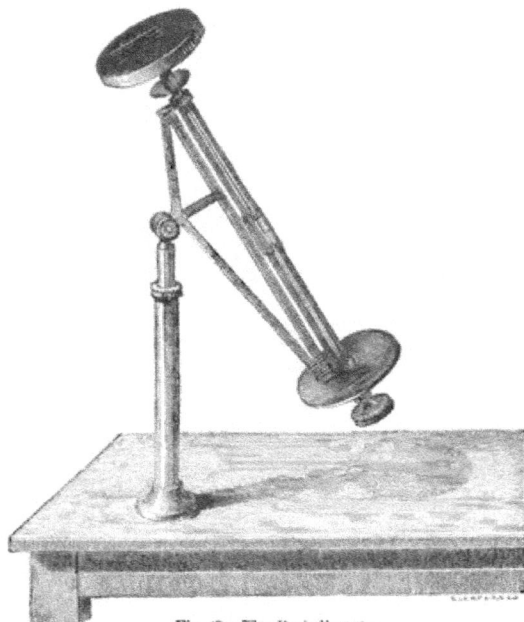

Fig. 47.—The Pyrheliometer.

We can thus tell the quantity of heat which the sun spreads in the
space of a minute over a square inch at the limit of the atmosphere, and
which would also be received at the surface of the ground, were it not
that the air of the atmosphere absorbed some of the rays as they passed
through it.

From these data and the law in accordance with which transmitted

[* The heat-unit generally adopted in English works is the quantity of heat necessary to
raise one pound of ice-cold water one degree Centigrade, viz., to raise one pound of water from
0° to 1° C.—ED.]

heat diminishes in proportion as the obliquity increases, it is easy to calculate the proportion of incident heat which arrives each instant upon the lighted hemisphere of the globe, and the proportion absorbed in the corresponding half of the atmosphere. The calculation shows that when the atmosphere is, to all appearance, quite still, it is absorbing nearly one-half of the total quantity of heat which the sun emits to us, and that it is only the other half of this heat which reaches the ground. Since the sun, as has been calculated, transmits every minute to each square yard of the ground that it shines perpendicularly upon a degree of heat equal to about 35,200 heat-units, it is easy to conclude therefrom the total quantity of heat which the terrestrial globe and its atmosphere together receive in a year. The result is more than 2,660,000,000,000,000,000,000 heat-units! This heat would raise, if such were possible, by 2315 degrees, a body of water three feet three inches deep, and enveloping the whole. By transforming this quantity of heat into a quantity of melted ice, the following result is arrived at: If the total quantity of heat which the earth receives from the sun in the course of the year was uniformly distributed over all parts of the globe, without any loss in melting ice, it would be sufficient to melt a coat of ice enveloping the whole globe to a depth of about one hundred feet. Such is the simplest way of expressing the total quantity of heat which the earth receives each year from the sun.

It is this gigantic quantity of heat which sets in motion the mechanism of terrestrial action, which lets loose the tempests over the ocean, and, in a word, sustains the vast aërial life of this planet. The same fundamental data permit of our ascertaining the total amount of heat which is emitted from the sun in a given time.

Let us consider this luminary as the centre of a vast sphere, the radius of which is equal to the mean distance of the earth from it; then it is evident that over the surface of this sphere each square yard receives every minute from the sun precisely as much heat as the square yard of the earth—that is to say, 35,200. Consequently, the total quantity of heat which it receives is equal to its entire surface, expressed in yards and multiplied by 35,200.

The same thing may be expressed by stating that the terrestrial globe, with its 8000 miles of diameter, only intercepts, in this sphere of $91\frac{1}{2}$ million miles of radius, $\frac{1}{2300000000}$ of the total heat that leaves that luminary, and that the heat emitted by the sun is 2,300,000,000 greater than that received by the earth.

Transforming into the quantity of melting ice, we obtain the following result:

If the total amount of heat emitted by the sun were exclusively employed in melting a coat of ice placed right around the body of the sun, it would be capable of melting in a minute a thickness of nearly $39\frac{3}{4}$ feet—that is, a thickness of more than $10\frac{1}{2}$ miles in the twenty-four hours!

One part of this immense source of "energy" is employed in heating the terrestrial rind to a certain depth; but as the soil and the atmosphere radiate into space, and as the terrestrial globe does not seem to lose or gain in reference to the mean temperature, at least during long periods of years, all this part of the sun's radiation may be considered as maintaining the equilibrium of the temperature of our planet. Another part is transformed into molecular movements, in chemical action and reactions, which are the source whence the life of animals and vegetables derives unceasingly the wherewithal of their perpetuation and sustenance. Heat, which thus seems necessary for these beings, is but an emanation from our luminary. As Tyndall remarks, "It is thus we are, not merely in the poetical sense, but practically, children of the sun."

The American engineer, Ericsson, the inventor of the solar steam-engine, has calculated that the mechanical effect of the solar heat which falls upon the roofs at Philadelphia would keep in motion more than 5000 steam-engines, each of 20-horse power.

The work done in raising the temperature of a pound of water by one degree Fahrenheit is exactly as great as that required to raise a weight of one pound to the height of 772 feet.

Solar heat is the source of the only natural works that man has yet been able to divert to his profit, and among them we must include water-courses and the winds.

Moreover, the combustible matter of manufacture is derived from the same luminary; as wood, it is carbon absorbed by the vegetables breathing in the air under the influence of the sun; as coal, it is still carbon that has been in earlier ages fixed by the same influence in the large antediluvian forests.

The sun's rays, after having traversed either the air, a pane of glass, or any transparent body, lose the faculty of retreating through the same transparent body to return toward celestial space. It is by a procedure founded upon this physical law that gardeners accelerate in spring the vegetation of delicate plants by covering them with a glass bell, admitting the solar rays, which have great difficulty in effecting their egress. If the gardener places two or three of these bell-glasses one upon another, he invariably burns up the plant underneath them, and even in the mild weather of March or April he is often obliged to raise one of the edges of the glass to prevent the plant from being injured by the sun at noon. By means of an apparatus composed of a box blackened inside, and of several pieces of glass laid one upon the other, Saussure was enabled to raise water to boiling-point; and Sir John Herschel, during his stay at the Cape of Good Hope, in the burning heat of the last days of December, was enabled to cook a piece of " bœuf à la mode " of very fair dimensions, by means of two blackened boxes placed one inside the other, and each provided with one single glass, with no other source of heat than the solar rays which were ingulfed without possibility of escape in this kind of trap. "There was," says M. Babinet, "sufficient to regale the whole of his family and their guests at this meal, prepared with a stove of such a novel kind."

Herschel's box, closed only by two panes of glass, reached successively 80, 100, and 120 degrees of heat.

Although this oven appears so novel, it may almost be said to be taken from the Greeks. We find, indeed, that a century before the Christian era, Hero of Alexandria described in his "Pneumatics" a large number of ingenious contrivances devised by the ancients, and, no doubt, by the learned hierophants of Egypt. One of these, which seems to have been constructed by Hero, draws water from a reservoir by the sole effect of the dilatation and condensation of air under the influence of the sun alternately shining on and concealed from the apparatus.

At the close of the sixteenth century, the Neapolitan savant, J. B. Porta, set forth in his "Natural Magic" the mechanical applications of solar heat. If, he says, a hollow copper globe is placed upon the summit of a tower, and if from it there descends a pipe into a reservoir of water, by heating the globe above, either by means of fire or the sun's rays, the rarefied air escapes. Soon after, when the sun declines, the copper globe cools, the air becomes condensed, and the water rises up the pipe.

The concentration of solar heat in a glass-covered inclosure is an experiment so easy that the observation of it must have followed very closely upon the invention of glass. Nevertheless, despite the different proofs of this fact and the applications of its principle to which I have alluded, there is no complete scientific study of the phenomenon earlier than that of Saussure. Subsequent to his work and that of Herschel, the subject had been considered in various lights by different philosophers. This curious problem is just now in perhaps its most interesting phase, viz., that which gives on the one hand serious results, and on the other allows the imagination to guess at others in the future still more important.

It is a natural question to ask, What is the temperature of the sun? To this we can give no satisfactory answer. Two estimates have been made by Secchi and Zöllner, which, however, differ enormously, the former giving about 19,000,000° Fahr., while the latter only amounts to about 49,000° Fahr.

To determine the temperature of the sun, an apparatus has been used which exposes the thermometer to its rays in an inclosed place, the temperature of which is previously ascertained. Reading the indication given by the mercurial column, the number is multiplied by the ratio of the surface of the celestial sphere to the apparent surface of the sun. As the solar disk has a mean diameter of 31′ 3″·6, the ratio of the whole celestial sphere to this is 183,960. The apparatus in question is as follows: Two concentric cylinders soldered together form a kind of double caldron, the annular interval of which may be filled with water or oil at a given temperature. A thermometer passes by means of a small tube through the annular space and penetrates to the axis of the cylinder, where it receives the solar rays, which are introduced by means of a diaphragm, the orifice of which is scarcely larger than the ball of the thermometer. The interior cylinder and its thermometer are covered with lamp-black; a second thermometer gives the temperature of the annular space and, consequently, that of the inclosure. The whole apparatus is mounted upon a stand having a parallactic movement, corresponding to the diurnal motion of the sun.

The apparatus being exposed to its rays, the two thermometers are noticed, the difference of their temperature gradually increases, and at the end of a certain time becomes constant. The two temperatures are then marked and the difference calculated.

One word must be added as to the interior heat of the earth. Mairan, Buffon, and Bailly estimated, so far as France is concerned, the heat which escapes from the interior of the earth at twenty-nine times as much in summer, and four hundred times as much in winter, as that which reaches us from the sun. Thus, the heat of the luminary which

gives us light would, if this were true, form but a small fraction of that
of the globe. This idea was developed with great eloquence in the
"Époques de la Nature," but the ingenious romance to which it forms
a basis is dispelled like a phantom before the stern evidence of mathematical calculations. Fourier having discovered that the excess of the
temperature of the terrestrial surface over that which results from the
mere action of the solar rays has a necessary relation to the increase of
the temperatures at different depths, succeeded in deducing from the
amount of this increase, as found by experiment, a numerical determination for the excess in question—that is, for the thermometrical effect
which the central heat produces upon the surface. And, instead of the
high figures given by Mairan, Bailly, and Buffon, he obtained as his result only the thirtieth part of a degree!

The surface of the globe, which, at the beginning of the world, was
probably incandescent, has cooled down, in the lapse of ages, so much
as to retain scarcely a trace of its primitive temperature. Nevertheless,
we know that the temperature increases as we descend into the interior
of the earth at the rate of 1° to about 112 feet, on the average, and that
the heat must be very great underneath volcanoes. Upon the surface
(and the phenomena of the surface can alone alter or compromise the
existence of human beings) all changes are limited to about the thirtieth part of a degree. The terrible congelation of the globe, which
Buffon fixed for the epoch when the interior heat should be entirely
dissipated, is therefore a mere dream.*

---

* [M. Flammarion concludes this chapter with a discussion of the temperature of space, and
he states that the mechanical theory of heat shows that there is an *absolute zero* of temperature
at —459° Fahr. (—273° Cent.), so that no body can be colder than this; it being, in fact, the
temperature of a body totally devoid of heat, and therefore the temperature of space. I should
merely have contented myself with the omission of this portion of the chapter without remark,
only it appears to me that the error reproduced by M. Flammarion is sufficiently wide-spread to
make it worth while to call attention to the matter. In point of fact, we have no evidence for
asserting that the temperature of space is —459°. We know that gases, at ordinary temperature, expand equally by heat, so that if a thermometer were made in which the fluid was air
kept at a constant pressure, its reading would be the same as if any other gas were used, the
pressure being the same. Consider, therefore, a thermometer composed of air contained in a
long straight tube, so arranged that the pressure of the gas is kept constant whatever its volume may be, and suppose the freezing and boiling points determined as usual, and the intervening space divided into 180 equal parts, as in Fahrenheit's scale, then it follows, assuming
Boyle and Mariotte's law, that if the graduations were continued right down to the end of the
tube, the last division would be marked very nearly —459°, so that it is clear that no temperature, however low, can correspond to —459° of the thermometer (*i. e.*, the air thermometer can
never read so low as —459°), as in that case the air would have been compressed into nothing;

but as it is clearly convenient to start from the end of the tube, this point can very well be taken as our zero, merely to measure from. There are other reasons, of a more strictly scientific character, derived from thermo-dynamical considerations, that also lead to approximately the same point as the absolute zero of temperature; but they do not, in the very slightest degree, imply that this is the temperature of space; in fact, such an assertion would be unintelligible, even if true, without much explanation. The lowest artificial temperature observed is —220° Fahr. (—140° Cent.), obtained by Natterer, by exposing to evaporation a mixture of nitrous oxide and carbonic disulphide.—Ed.]

CHAPTER II.

HEAT IN THE ATMOSPHERE.

It now becomes necessary to ascertain what part of the immense calorific radiation which is incessantly emanating from the sun is at work
in the atmosphere.

Meteorology is nothing but a great physical problem. We have to
determine what are the laws which regulate the manner in which heat,
barometrical pressure, vapor of water, and electricity, are distributed in
our atmosphere, in relation to the movements which the solar heat en-
genders in the solid, liquid, and gaseous superficial stratum of our
globe. This problem, vast as it is, says Father Secchi, is in reality but
an application of the best known laws of physics; the difficulties of
solving it are owing rather to the large number of disturbing causes,
and to the incalculable reactions of effects upon causes, than to any real
deficiency in the general theory. Hence the necessity of numerous experimental data in order to arrive at a complete solution.

The atmosphere is in reality an immense machine, to the action of
which is subordinated every thing upon our planet that has life.
Though there are neither fly-wheels nor pistons in this machine, it none
the less does the work of millions of horses—a work the aim and effect
of which is the sustenance of life.

All the movements of the atmosphere are the consequence of the
property which gases possess of being expanded by heat. The variations of volume, and, consequently, of density, are, at each instant, disturbing the equilibrium which would be tending to establish itself in
the atmosphere. The air, heated in the equatorial zones, rises into the
upper regions to fall again near the poles; there it becomes cool, returns to the equator, and recommences its circulating movement. The
work thus performed in the atmosphere is enormous. To this property
of gas must be added another, not less important—that of dissolving*

[* The air and the vapor of water form, as it were, different atmospheres: that is to say,
that the vapor atmosphere could still remain if all the air were removed. Water, placed under the air-pump, evaporates till the space under the receiver is filled with aqueous vapor to an
extent dependent on the temperature.—Ed.]

the vapor of water which, as it rises in prodigious quantities about the equator, is thence distributed all over the earth in the shape of rain. Thus is effected another and scarcely less potent work—the distribution of rain over the surface of the globe. The running waters which set our machines in motion were originally raised into the air by this mighty agency; from thence they pour down on the mountains in the shape of rain, run into the rivers, and so make their way again into the ocean from whence they started.

The sun is the power that regulates all the movements of the planetary system; not only their motions in their orbits, but also the physical or physiological phenomena which take place upon their surface. On the earth, in particular, the atmospheric movements and those of the waters, the development of vegetation, the production of the force which results from combustion and the nutrition of animals—all these phenomena are due to the influence of the sun's heat rays.

What may seem to us still more perfectly organized is the way in which this calorific power is, so to speak, stored up in the vegetables; not only in those which, still alive, serve for our use and nourishment, but also in those which, buried for ages in the bowels of the globe, at length emerge therefrom to warm us and supply our machines with the required motive power. Each plant is a veritable machine, in which are elaborated the extremely combustible substances which serve to furnish us, in the absence of the sun, with heat and light, or to produce, in providing us with nutriment, the force and vital warmth which we stand in need of. It is, therefore, on the sun, as Father Secchi again remarks, that depend entirely all the phenomena of nature, and our existence itself.

In the solar radiation what is at first so striking is the light which gives us day, and the heat which warms us; but, besides these two orders of phenomena, there is a third of equal importance, viz., the chemical actions which accompany the two others. Thus three classes of action must be distinguished in the solar work—the *luminous* rays, the *calorific* rays, and the *chemical* rays. It is well known that, to analyze a sunbeam, it is passed through a triangular prism of glass, on emerging from which the ray is decomposed into a colored ribbon, as we have already seen in our study of the rainbow. But the visible spectrum is not the only component part of a sunbeam. The many-hued ribbon is continued at each end by an invisible ribbon. The waves—the length of which is included between ·0000167 and ·0000266 of an inch—are capable of causing our optical nerve to vibrate, and thus producing the sensation of *light*, the diversity of colors being dependent only upon the length of the waves, the longest of which belong to the red rays, and which gradually diminish toward the violet. To the left of the red extremity of the spectrum, there are long and slow waves of heat. To the right of the violet end, there are short and rapid waves of chemical action. The eye sees neither the first nor the second of these, but they may be recognized by the use of suitable apparatus. In reality, however, there exists in nature but one single series of waves, the lengths of which continually decrease from the extremity of the obscure calorific spectrum to the extremity of the invisible chemical spectrum. Between these two extremes there is but a very limited part which has the power of giving sensation to the optical nerve.

Fig. 48 shows the relative extent and intensity of these different actions, separated from each other as they are made manifest to us by the

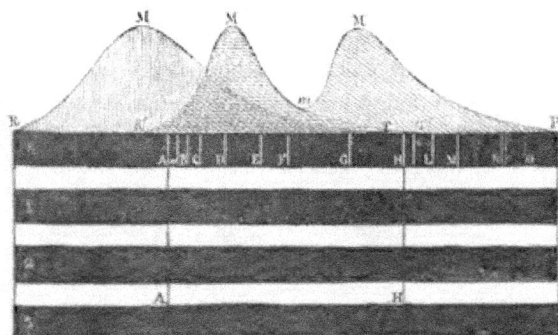

Fig. 48.—Relative Intensity of the calorific, luminous, and chemical rays of the sun.

dispersive action of a prism. The band which forms the basis of this figure indicates the length of the solar spectrum. From A to H is the *luminous* part; to the right, from H to P, is the *invisible* chemical part; to the left, from A to S, is the *calorific* part, also *invisible*. The curves traced above show the relative intensities at each point of the spectrum. The intensity of the light is represented by the curve R' M' T', that of chemical action by m M″ P, that of calorific radiations by R M T. It has been attempted to represent the three respective intensities by the three bands, 1 (light), 2 (heat), and 3 (chemical action).

Thus we do not see all that goes on in nature. The luminous rays are the only ones which we can behold; but the calorific and chemical rays take effect without being visible to us.

The illuminating power of the different rays consists in their greater or less capacity for giving an impulse to the optical nerve. It is probable that the faculty of perceiving luminous phenomena has not the same scale for every individual, and that it is much more extended in the case of certain animals than with man, both at the red and violet ends of the spectrum. Pure water possesses a very considerable absorbing power for thermal rays. The moisture contained in the eye differs very little from pure water, and it is this fact which very likely renders the organ of sight insensible to calorific rays. The extent of the luminous waves which are sensible to the eye ordinarily corresponds to what is called in acoustics an octave, so that man is only placed in relation with nature by a very small part of the solar rays.

Gases possess the faculty of absorbing heat rays, and consequently our atmosphere absorbs a very considerable portion of the rays which are transmitted to us. The longest waves are those which are most easily absorbed; thus a large number of the less refrangible rays which fall upon our atmosphere are stopped, and do not reach us at all.

The absorption produced by the simple gases, oxygen and nitrogen, is extremely small; but this does not hold good of the compound gases existent in our atmosphere, such as carbonic acid, vapor of water, ammonia, etc. Professor P. M. Garibaldi,[*] of Genoa, has proved by conclusive experiments that, for a pressure of 29·92 inches, these gases have absorption-powers represented by the following figures:

Atmospheric air	1
Carbonic acid	92
Ammonia	516
Vapor of water	7937

A quantity of vapor of water capable of producing a pressure of 0·4 inch exercises an absorption a hundred times greater than that of atmospheric air. Thus, a considerable portion of the dark heat rays proceeding from the sun are intercepted by the vapor of water contained in the air, and are unable to reach the surface of the earth.

The luminous rays have been separated from the heat rays by Professor Tyndall. To effect this, a pencil of solar rays was made to pass

[* Professor Tyndall has made an elaborate and careful series of experiments on the absorptive power of different gases. He concludes that on an average day the water present in the air absorbs about sixty times as much heat as the air itself.—ED.]

through a solution of iodine in carbonic disulphide. The rays become
invisible without losing their calorific power, and if the vessel containing the solution is made in the shape of a convex lens,* the temperature
in the focus of the lens is increased to such an extent that combustible
bodies may take fire there. Professor Tyndall on one occasion placed
his eye at the focus, and the retina received no luminous influence.
The calorific rays were, however, so powerful that a sheet of metal was
immediately made red-hot at the very spot where the eye had received
no impression. The ratio of luminous radiations to obscure radiations
is equal to $\frac{13}{720}$ for incandescent platinum. For sunlight, the heat which
accompanies the luminous portion of the spectrum is only $\frac{1}{9}$ of that
which is found in the obscure part.

The action of the atmosphere is to raise the temperature of the earth,
for it allows the calorific rays to reach the earth, and then prevents them
from making their way back into space. The greater portion of heat
rays sent out from the earth are no longer able to traverse the atmosphere, while only a few of the sun's rays, which are of high temperature, are stopped.

Further, the nocturnal radiation is considerably diminished by the
presence of the atmosphere, and in this way the cooling of the globe
and of the plants which it nourishes becomes modified and diminished.
The vapor of water acts very efficiently, and a moist stratum, only a
few yards thick, arrests the nocturnal process of cooling as completely
as the whole atmosphere.

But the most striking fact in connection with this matter is the great
absorption of heat which accompanies the transformation of water (or
any other liquid) into vapor. The heat so absorbed is termed latent
heat, from its not being spent in raising the temperature of the vapor.
Water evaporates in large quantities, especially in the equatorial regions,
and thus absorbs a large quantity of heat which remains latent. As
much heat is necessary to vaporize one pound of water (at the temperature of the boiling-point) as to increase by $1°$ (Centigrade) the heat of
537 pounds of water. The vapor of water absorbs this enormous proportion of heat, which it, however, restores in its entirety when it returns to the liquid state as rain. This heat is destined to be transported to the most distant latitudes, and to establish in the atmospheric en-

[* The solution must be inclosed in a lens of rock-salt, a substance which allows heat to
pass through it without absorbing scarcely any; it is therefore termed *diathermanous*, an ad-
jective having the same reference to heat that *transparent* has to light.—ED.]

velop which surrounds the globe an equality of temperature which would not otherwise be produced. The quantity of heat which thus passes from the equator to the poles is beyond conception.

Thus, for instance, numerous and rather exact observations have taught us that in the equatorial regions evaporation each year removes a body of water at least sixteen feet deep. Let us suppose that in the same regions there is an annual rain-fall of rather over six feet, there still remains a quantity of water represented by a depth of nearly ten feet, and which must pass, in the form of vapor, into the countries nearer to the poles. The surface over which the evaporation takes place may be estimated at seventy million geographical miles; and, starting from this datum, it will be seen that the depth of ten feet represents a volume of water equal to twenty-five thousand billions of cubic feet (25×10^{15}). This enormous mass of heat passes *incognito*, so to speak, from the equator to the poles, transported by the action of the vapor, and this latter, as it becomes transformed into water and ice, sets free all the heat which it had absorbed, thus contributing to make milder the climate of these desolate regions. In this way the heat is distributed in the atmosphere, and thus are created clouds and rain, which will be explained below.

The thickness of the strata of air traversed by the solar rays has a notable influence upon heat and light. The rays do not fall upon the earth perpendicularly, but obliquely, and the loss is greater the more they are inclined to the vertical.

This diminution has been submitted to different calculations: the two formulæ which seem to be most in harmony are those of Bouguer and Laplace. Making use of them, the following results are arrived at, as to the thickness of the strata of air for the different heights of the sun:

Height above the Horizon.	Zenith's Distance.	Thickness of the Strata of Air.
90 deg.	0 deg.	1·00
70 "	20 "	1·06
50 "	40 "	1·30
30 "	60 "	1·99
20 "	70 "	2·90
15 "	75 "	3·80
10 "	80 "	5·51
5 "	85 "	10·21
4 "	86 "	12·15
3 "	87 "	14·87
2 "	88 "	18·88
1 "	89 "	25·13
0 "	90 "	35·50

Thus, if the thickness of the atmosphere traversed by a ray of the sun at the zenith be represented by 1, the thickness traversed by the sun's rays at the horizon is more than thirty-five times greater. This difference is much larger than can be indicated in the annexed illustration (Fig. 49). The first result of this inequality is, that the sunlight becomes feebler in proportion as the sun sinks toward the horizon. At the zenith and in the higher regions of the sky the sun is dazzling,

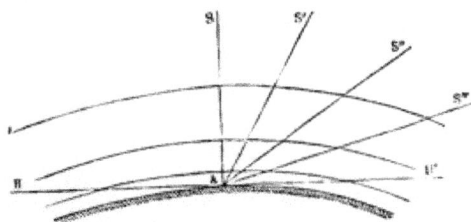

Fig 49.—Inequality of the thickness of air traversed by the sun, according to its position above the horizon.

and no human eye can withstand its blaze. At sunrise and at sunset we are able to fix our eyes upon its reddened disk without inconvenience. The smaller stars do not become visible till they reach a certain height, and we can only witness the rising and setting of those of the first magnitude. According to the researches of Bouguer, if 10,000 be taken to represent the luminous intensity of the sun as it would be seen from a point external to the atmosphere, its intensity at the different altitudes above the horizon may be thus stated:

At 50 degrees		8123
" 30 "		7624
" 20 "		6613
" 10 "		5474
" 5 "		3149
" 4 "		1201
" 3 "		802
" 2 "		454
" 1 "		192
" 0 "		467

That is to say, that, at sunrise and sunset, this luminary has only $\frac{1}{1134}$ of its apparent brilliance when at the zenith, and $\frac{1}{1300}$ of its brilliance when at its midday elevation over our horizon during the summer solstice. These comparisons are made on the supposition that the sky is clear, and consequently vary with the more or less misty state of the atmosphere. Heat varies, like light, with its angle of incidence. The most accurate observations prove that the atmosphere absorbs, of vertical rays, ·28 of the heat which falls upon its surface, and the total absorption in the illuminated hemisphere is about equal to

three-fifths of the incident heat; the transmitted part at different heights
being represented as follows:

Height.	Amount transmitted.
At the zenith...........................	0·72
" 70 degrees.........................	0·70
" 50 "	0·64
" 30 "	0·51
" 10 "	0·16
" 0 "	0·00

As was remarked above, it is not the air itself—that is to say, the
mixture formed of oxygen and nitrogen—which absorbs the most heat,
but the vapor of water, which always exists in the air, but in very varying proportions.

The luminous rays pass almost in their entirety, and reach the ground;
the heat rays are, on the contrary, absorbed to a large extent. If, therefore, the atmosphere prevents a great part of the solar heat from reaching the surface of our globe, it makes up for it by retaining for us the
part that we do receive. Without the atmosphere and the vapor of
water contained in it, since the radiation of the soil goes on almost without obstacle toward the interplanetary space, the loss would be enormous, as indeed is the case in the higher regions. No sooner has the
sun set than a rapid coldness succeeds the intense heat of the sun's direct rays; in a word, there is an enormous difference between the maxima and minima of temperature, either daily or monthly. This occurs upon the lofty plateaux of Thibet, and explains the severity of
the winters, and the decline of the isothermal lines in these regions.
Tyndall says very truly: "The suppression, for a single summer's night,
of the vapor of water contained in the atmosphere over England (and
the proposition holds true for all the countries in similar latitudes) would
be accompanied by the destruction of all the plants which are killed by
frost. In the desert of Sahara, where *the ground is fire and the wind a
flame*, the cold at night is often very difficult to support. In this hot
country ice is seen to form in the course of the night."

Moisture is not distributed in equal proportions at all elevations of
the atmosphere. We shall see, further on, that it decreases in amount
beyond a certain height. Heat traverses air the more easily, the less
moisture it contains. After the lower regions of the atmosphere have
been passed, and (say) an altitude of 6000 feet attained, it is impossible
to avoid noticing the very considerable increase in the heat of the sun

relatively to the temperature of the surrounding air. This fact never struck me so strongly as in an aeronautical ascent on June 10, 1867, on which occasion I noted, at 7 A.M., at a height of 10,000 feet above the ground, that there was, for half an hour, a difference of 27° Fahr. between the temperatures of my feet and head ; or, to speak more accurately, between the temperature of the interior of the car (shade), and that of the exterior (sun). The thermometer marked 46° Fahr. in the shade, and 73° Fahr. in the sun. While our feet felt the effects of this relative cold, a hot sun scorched our necks and faces, and those parts of the body directly exposed to the solar radiation. The effect of this heat is, of course, further augmented by the absence of the slightest current of air.*

In a subsequent ascent, I experienced at the same time the remarkable difference of 36° Fahr. between the temperature in the shade and that in the sun, at an altitude of 13,500 feet.

The influence of altitude upon the intensity of the sun's calorific influence at points nearly vertically above one another has recently been studied very carefully by M. Desains and a colleague at the Schweitzerhoff, Lucerne, and at the Righi-Culm Hotel, about 1500 yards above the lake. These experiments demonstrated that at the same hour, and under equal conditions, the solar radiation was more intense upon the summit of the Righi than at Lucerne, but that it was less capable of being transmitted through water. It was found that the solar rays in their passage—at an angle of about seventy degrees with the vertical—through the stratum of air comprised between the level of the Righi and that of Lucerne, underwent a loss of about seventeen per cent.

This shows that the terrestrial temperatures depend not only on the quantity of heat received from the sun, but also and especially upon the *absorbing power* of the air in regard to the rays of heat. Such also, no doubt, is the case in the other planets ; and the influence of the atmospheres is such that, in spite of its close proximity to the sun, it is possible that Mercury may possess a much lower temperature than that of the earth, and the surface of Jupiter may present a climate far warmer than ours, in despite of its greater distance from the sun.

Recent spectroscopic experiments of M. Janssen render the existence of planetary atmospheres generally similar to our own probable. As-

---

[* My experiences at high elevations were quite opposed to those stated in the text. At great heights I observed no difference in reading between a thermometer with the sun shining full on its bulb, and another in which the bulb was carefully shaded.—ED.]

tronomical observations also long since pointed to the same conclusion with regard to some of the planets.

After having appreciated the action of solar heat throughout the atmosphere and upon the surface of the globe, to which, in fact, almost every movement that takes place there can be traced, we must now complete the account by noticing that the amount of this heat diminishes as we ascend higher into the atmosphere. We have seen that the pressure of the air diminishes in proportion as we rise higher into it. The temperature is subject to an analogous decrease, which may be estimated, though not nearly so accurately as in the case of the diminution of the atmospheric pressure. Corresponding to the indications of the barometer, the following are those given by the thermometer:

When an ascent in a balloon is made with the sky cloudy, the temperature generally declines until the clouds have been reached; once above them, a rise of several degrees always takes place, but the temperature soon begins to fall again. With a clear sky, the initial temperature is, *cæteris paribus*, higher than in the preceding case by a quantity about equal to the rise observed after emerging from the clouds. The diminution of heat is never absolutely regular, as strata of hot air are nearly always encountered in the atmosphere, sometimes five or six in succession, at very great elevations. These alternations, and this variability of the sky, do not prevent the manifestation of one general fact—that of the decrease in the temperature with an increase of elevation.

The following is the result of a series of observations upon this point which I have made in the course of my various ascents :

The decrease in the temperature of the air, which plays so important a part in the formation of the clouds and in the elements of meteorology, is far from following a regular and fixed law. It varies according to the hours, the seasons, the state of the sky, the direction of the winds, the condition of the vapor of water, etc., etc. It is only after a great number of observations that it is possible to deduce any fixed rule, several secondary causes which must be first ascertained and eliminated being always at work.

From several observations, taken in very dissimilar conditions (which are, however, less unfavorable than those under which observations are taken on mountain sides), it follows that the decrease in the temperature of the air differs, in the first instance, according to whether the sky is clear or cloudy, being more rapid in the first case than in the second.

With a clear sky, the mean fall in the temperature has been found to be 7° Fahr. for the first 1600 feet from the surface of the ground; 13° at 3280 feet; 19° at 4900 feet; 23° at 6560 feet; 27° at 8200 feet; 31° at 9840 feet; 34° at 12,500 feet—an average of 1° Fahr. per 340 feet.

With the sky cloudy, the fall in the temperature is 5½° Fahr. for the first 1500 feet; 11° at 3000 feet; 16° at 4900 feet; 19° at 6560 feet: 29° at 9840 feet; 32° at 12,500 feet—of average of 1° Fahr. per 350 feet.

The temperature of the clouds is higher than that of the air immediately above or beneath them. The decrease is more rapid near the surface of the ground, and more gradual at greater elevations. It is also more rapid in the evening than in the morning, and also in warmer than in colder weather. Regions hotter or colder than the mean temperature for their altitude are sometimes met with in the atmosphere, crossing it like aërial rivers. Notwithstanding these variations, the general law enunciated above is the expression of the true state of things. The difference between the indications of the thermometer in the shade and in the sun augments with elevation. Thus, the general result of these aërial ascents tends to show that the temperature decreases about 1° Fahr. for an elevation of 345 feet. The result of the well-known and numerous aërostatical observations taken by Glaisher differs but little from the above. The ascents of mountains have furnished a certain number of important data, among which may be quoted the following:

Humboldt found that the decrease, in a southern atmosphere, was 1° Fahr. to 344 feet in the mountains, and to 440 feet upon the table-lands. A series of places in Southern India gave 320 feet; in the north of Hindoostan, on the other hand, the decrease was 1° in 410, an amount approaching to that noted by Humboldt upon the table-lands of America. Everywhere analogous differences of level are remarked; in Western Siberia, 1° in 450 feet is the result arrived at; and this number is converted into 440, if the comparison includes the elevated regions of Northern India. In the United States the decrease is 1° to 400 feet. The configuration of the country seems to be the most important element in the calculation. If there is a gentle rise in the ground, or if the country is made up of successive gradients, the decrease in the temperature is much more gradual than upon the sides of steep mountains. In the first case, 1° may be taken to represent a difference in level of 420 feet; in the second, of 350 only.

Schouw remarked in Italy, upon the southern slopes of the Alps, a decrease of 1° to 300 feet; less on Mount Ventoux, a steep and isolated mountain in Provence (lat. 44° 10′ N., long. 2° 56′, height 6270 feet above the level of the Mediterranean). Martins found, after nineteen observations, taken under dissimilar conditions, a decrease of 1° to 340 feet in winter, and 230 feet in summer, or an average of 260 feet. The observations of Ramond, made between 43° and 44° of latitude, give an average of 1° to 265 feet.

.

CHAPTER III.

THE TEMPERATURE OF THE AIR: ITS MEAN CONDITION — DAILY AND
MONTHLY VARIATIONS OF THE TEMPERATURE — TEMPERATURE OF
EACH SUMMER, WINTER, AND YEAR AT PARIS AND AT GREENWICH
SINCE THE LAST CENTURY — DAILY AND MONTHLY VARIATIONS OF
THE BAROMETER.

WE have seen that the earth, by its annual revolution round the
sun, and by its daily rotation upon its axis, produces a variation in the
obliquity of the solar rays which find their way to it. By its annual
revolution, they become more and more vertical during six months of
the year—from December 21 to June 21—and less and less so during the
other six months. By its rotation the horizon each morning is brought
into the presence of the sun, causing the heat-giving luminary to reign
in the heights of the heavens during the day, and in appearance to sink
again to the horizon on each evening. Thus, it is evident that, by these
two movements of the earth, there are two general principles in regard
to solar heat upon our planet; the one annual, the other diurnal.

Let us consider the latter first. To determine it exactly, the thermometer must be consulted hourly, night and day, for several years together, in order to distinguish and eliminate the effects due to the rotation of the earth from those due to the numerous other causes which influence change of temperature. Few meteorologists have been willing
to undertake so arduous a task. Ciminello of Padua made such observations for nearly sixteen consecutive months. I say *very nearly*, because the observations at midnight, and at the hours of one, two, and
three in the morning, were replaced by two, taken during the same interval at different hours. He was the first to make hourly series of
thermometrical observations. Since that time others have been made
by Gatterer, a contemporary of his; by the artillery officers at Leith;
by Neuber at Apenrade, in Denmark; by Lohrmann at Dresden; by
Koller at Kremsmunster; by Kaemtz at Halle; and at the Observatories of Milan, St. Petersburg, Munich, and Greenwich. Such observations are now continuously recorded at the Roman Observatory, and
some others by means of a self-registering apparatus.

The result of these observations, and of many others which have been made every two or every three hours, shows that the hottest moment of the day is two in the afternoon, and the coldest about half an hour before sunrise. These two limits vary but little from one month to another. The difference of temperature between the hottest moment and the average coldest period of the twenty-four hours is about 14° at Paris. This amount, however, varies with the time of the year.

The average yearly maximum temperature at the Paris Observatory is 58° at 2 P.M.; the average minimum is 44°·8 at four in the morning; and the average mean temperature of the year, as taken at 8·20 A.M. and 8·20 P.M., is 51°·3.

The interval of time, between the minimum in the morning and the maximum in the afternoon, is only ten hours; and the interval is fourteen hours, viz., from 2 P.M. to 4 A.M. between the time of maximum and the next minimum. The minimum of the diurnal variation, as a rule, takes place just before sunrise; in the early part of the year it is just before 6 A.M., and occurs earlier as the days lengthen. After the month of February it occurs at about 5 A.M., then at 4 A.M., afterward oscillating between three and four in the morning during the longest days. In the beginning of August the minimum is again at 4 A.M., returning to about 6 A.M., when the days are at their shortest. It is even somewhat later than this for a short period, but soon afterward resumes the annual progress given above.

The *mean* temperature of a day, in the mathematical acceptation of the term, represents the average of the temperatures corresponding to every instant of the day. If the duration of these instants be a minute, it would be necessary to divide the sum of the 1440 thermometrical observations taken between two consecutive midnights by 1440 (the number of minutes in twenty-four hours), and the quotient would give the required mean temperature. Again, by dividing by 365, the sum of the 365 mean temperatures of every day in the year, we should obtain the mean annual temperature.

It would seem, from the preceding definition, that to obtain the mean temperatures accurately, observations at short intervals would be indispensable; but the change of temperature under ordinary conditions is, fortunately, of such a nature that the half sum of the maximum and minimum temperatures (at 2 P.M. and sunrise) is found to differ but little from the mean of observation taken at every hour. So early as 1818, Arago pointed out that the average temperature at 8·20 A.M. was nearly the same as the average temperature of the year. Numerous thermometrical observations taken under his direction were based upon the fact of the mean temperature of the day, occurring twice in the course of the day. But it has since been found that this method is defective; for from 8 to 9 A.M., and also from 8 to 9 P.M., the temperature often varies very rapidly. The averages were afterward formed by taking the temperatures at 4 A.M. and 10 A.M., and again at 4 P.M. and 10 P.M., adding and dividing by four. The arithmetical mean of the observations taken at 6 A.M., 2 P.M., and 10 P.M., also gives about the same value, the difference being about two-tenths of a degree. Since meteorology has been more methodically followed, still greater accuracy has been acquired, and it has been found that the twenty-four hourly observations may be replaced by eight tri-hourly observations, taken at 1 A.M., 4 A.M., 7 A.M., and 10 A.M.; and at 1 P.M., 4 P.M., 7 P.M., and 10 P.M.

Let us now consider the *annual* movement of the temperature.

The various causes which influence the action of the sun's heat vary but little throughout the year in the regions near the equator, whether situated in the northern or the southern hemisphere, the tropical regions, as they are called, and which form the torrid zone. The day has about the same length all the year round; the meridian height of the sun undergoes but little variation there; and the four seasons differ very little, in regard to temperature, the one from the other. For an entirely opposite cause, the seasons are very dissimilar both to the north and to the south of the equator in the regions where the length of the day varies very much in course of the year, or, to express the same thing in other words, where the meridian height of the sun at one solstice is very different from that of the other.

We have considered the general condition of the seasons in our latitudes. Let us now consult the figures themselves. The table appended gives the mean temperature at the Paris Observatory.

It shows that, whether the average maximum or the average minimum of each month be taken into account, or, indeed, if we merely take the mean temperatures alone, the heat follows an ascending scale from January to July, and a descending scale from July to December. The hottest month is that of July, which follows upon the summer solstice, and the coldest that of January, which comes after the winter solstice. The average of the minima is only once (in January) below 32°; the coldest months are December, January, and February, constituting

the real climatological winter; spring is made up of the months of
March, April, and May; summer of the three hottest months, June,
July, and August; and the other three months, September, October,
and November, form the true autumn.

TABLE OF THE MEAN TEMPERATURES AT PARIS (ARAGO, 1806–1851).

Months.	Maxima.	Minima.	Means.
	Degrees.	Degrees.	Degrees.
January	41·0	30·5	35·8
February	45·1	33·3	39·2
March	50·0	37·7	43·9
April	55·6	43·7	49·6
May	65·1	51·3	58·1
June	70·0	56·5	63·1
July	72·9	59·7	66·2
August	72·3	58·3	65·3
September	65·9	53·8	59·9
October	58·3	45·1	51·8
November	49·5	39·0	44·2
December	44·3	32·5	38·5
Annual Temperatures	57·5	45·1	51·3

The above averages are those which Arago arrived at after forty-six
years of observations (1806–1851). Since then, further observations
have given a result still more in conformity with the secular mean temperature of Paris, representing, as it does, a longer series of years.

The heat received from the sun by the earth varies inversely as the
square of its distance from the sun; and as the earth does not move in
a circular orbit, there is, in addition to the monthly variation caused by
the inclination of the solar rays, a variation due to the distance of the
sun. In fact, we are farther from the sun during the summer than we
are in the winter; and the difference is considerable. The following
are the deviations, taking as unit the mean solar distance, and regarding the heat as reciprocal to the square of the distance:

	Distance.	Solar Heat.
Mean distance	1,000,000	1·0000
In Perihelion (least distance)	983,208	1·0345
In Aphelion (greatest distance)	1,016,792	0·9673

Thus, before even reaching our atmosphere, the solar heat rays are
subject to a variation of nearly one-fifteenth; that is to say, that the
solar heat during winter is, in respect to our globe, about one-fifteenth
greater than during summer.

This difference is sufficiently great to be taken into account.

The diurnal and monthly variations of temperature increase as the distance from the equator increases. From the equator to 10° north latitude, the mean temperatures of the various months scarcely differ more than 4° to 6°. At 20° north latitude they vary from 10° to 12°. At 30° distance the regular mean monthly variation is found to reach 22°. In Italy the regular curve at Palermo, in Sicily, extends from 51° to 74°; and this range is moreover diminished by the contiguity of the sea. At Paris the mean curve varies from 35½° (January) to 66° (July), and the changes become much greater between the frosts of winter and the heat of summer. At Moscow the mean monthly range extends from 12° (January) to 75° (July); showing a difference of 63° of mean temperature. Lastly, we may add to this scale of variations that of Boothia Felix, a northern country of America, situated beyond 72° of north latitude. There the range varies from −40° (72° below the freezing-point of water) in February to 41° in July; exhibiting a difference of 81° between the mean monthly temperatures of the year.

The diurnal variation also gives rise to remarkable curves in its successive temperatures. The range of thermometrical oscillation is greater in warm climates and inland countries than it is in colder lands and in the neighborhood of the sea. Apart from the equalizing influence of the sea, which remains about the same all the year round, the distance from the equator acts in an opposite way upon the annual and the diurnal oscillations of the thermometer. While the first increases on account of the length of the nights in winter and of the days in summer, the second decreases because in the southern countries the heat of the sun's rays is greater and the sky clearer during the night. Thus, for instance, at Padua the diurnal variation in July is about 16°; at Paris it is on an average about 13°; at Leith it is about 9°.

These are the mean values. But if the changes of temperature in a given district be constantly recorded, it will be found that, apart from these regular mean variations caused by the sun, there are other variations of a much larger amount which exercise the greatest influence upon the public health; these are the diurnal variations that occur in the space of twenty-four hours. These differences are very interesting, especially if we notice the reading of a thermometer with its bulb placed in the full rays of the sun by day, and of another with its bulb exposed fully to the clear sky at night. There are also often very great differences between the maximum and the minimum temperatures of the air of the same day, especially in the months of May and June—differences which reach, in Paris, to as much as 45° to 55°.

The following are some of the maxima observed at Montsouris, between 1 and 4 P.M., with a thermometer with green bulb, exposed to the sun at a height of about four inches above grass, as also some of the minima taken from the same thermometer between one and four the following morning. I select those that exhibit the greatest differences:

	Maximum.	Minimum.	Difference.
	Degrees.	Degrees.	Degrees.
May 11, 1870	87·5	39·4	47·9
" 16, "	86·4	42·8	43·6
" 17, "	90·9	44·4	46·5
" 18, "	102·9	53·8	49·1
" 19, "	106·7	57·9	48·8
" 20, "	107·5	55·2	52·3
" 21, "	111·2	60·8	50·4
" 25, "	86·0	41·0	45·0
" 27, "	87·4	43·0	44·4
" 30, "	94·6	50·4	44·2
June 8, "	86·9	42·8	44·1
" 12, "	89·6	46·4	43·2
" 13, "	92·5	47·3	45·2
" 14, "	107·4	53·6	53·8
" 16, "	106·5	61·0	45·5
" 23, "	105·4	53·1	52·3
" 29, "	95·2	48·2	47·0
" 30, "	95·0	44·8	50·2
July 2, "	86·0	42·8	43·2

This shows how great at times are the diurnal variations in these latitudes. The mean temperature of a place is that found by adding up the annual mean temperatures and dividing their sum by the number of years during which the observations have been taken. This mode of operation is only applicable to a limited number of stations. It was necessary, therefore, to seek a method of obtaining, by means of experiments which could be readily made, approximate mean temperatures, with a fair approach to accuracy. We know that the surface of the soil undergoes daily variations of temperature, that lower down there is a stratum which experiences only small annual variations, and that at a greater depth still, at about seventy to eighty feet, there is a stratum with constant temperature which is found to be very nearly the same as the average of a long series of the daily temperatures of the atmosphere made at the same place. By finding the temperature of this stratum at a sufficient depth, or, which comes to the same thing, by ascertaining the constant temperature of springs or wells in a certain district, or even of tunnels, we may thus succeed in obtaining for the temperature of each place a number differing but slightly from that which would be found by taking a long series of annual temperatures at that place. In the equinoctial regions, a thermometer simply sunk in the earth to the depth of thirteen inches in sheltered spots will continue to mark the same degree of temperature with a difference of $0°\ 2'$ or $0°\ 4'$ of a degree at most. For this purpose a hole is dug under the tents of the Indians or inside a shed, in a place where the ground is protected from the heat caused by the direct absorption of the solar rays, from nocturnal radiation, and from the infiltration of rain.

By taking the temperature of springs as that of the highest stratum of constant temperature, there will be found a great similarity in respect to the zone comprised between 30° and 55° latitude, provided that the places are not more than 3000 feet above the level of the sea.

In respect to latitudes above 55°, the difference between the temperature of the air and of springs increases to a marked extent.

Toward the peak of the Swiss Alps, above an elevation of from 4600 to 4900 feet, as in the high latitudes, the springs are nearly 6° Fahr. warmer than the air.

In Southern countries the temperatures of springs and of the ground are less than the mean temperatures of the air, as may be gathered from the accounts of Humboldt and Leopold von Buch.

In our latitudes this temperature is equal to that of the soil near the surface, and is a little higher than the average of the particular place.

It is worth our while to complete this general study of the meteorology of our climate by enumerating *the mean temperatures at Paris and at Greenwich since the beginning of the present century.* They are furnished from the archives of the Observatories at Paris and at Greenwich.

MEAN TEMPERATURES AS DETERMINED AT THE PARIS OBSERVATORY AND AT THE ROYAL OBSERVATORY, GREENWICH.

Years.	Winter. (Dec., Jan., Feb.)		Summer. (June, July, August.)			
	Paris.	London.	Paris.	London.	Paris.	London.
	Degrees.	Degrees.	Degrees.	Degrees.	Degrees.	Degrees.
1800	—	34·6	—	60·7	50·4	48·3
1801	—	38·7	—	60·5	51·3	49·0
1802	—	36·0	—	59·6	50·0	48·0
1803	—	35·8	—	60·5	51·1	48·2
1804	41·0	41·1	65·5	60·5	52·0	49·5
1805	36·0	36·3	65·1	58·4	49·5	47·7
1806	40·8	40·5	65·5	60·8	52·4	50·5
1807	42·8	41·2	67·3	61·6	51·4	48·3
1808	35·8	36·6	66·6	62·1	50·7	48·1
1809	40·8	38·5	62·4	58·7	51·1	48·0
1810	35·6	38·0	63·5	60·0	51·1	48·7

Years.	Winter. (Dec., Jan., Feb.)		Summer. (June, July, August.)			
	Paris.	London.	Paris.	London.	Paris.	London.
	Degrees.	Degrees.	Degrees.	Degrees.	Degrees.	Degrees.
1811	39·2	37·2	64·6	59·0	53·6	49·6
1812	39·4	38·6	63·0	56·1	48·2	46·5
1813	35·1	37·0	61·7	57·5	50·4	47·2
1814	33·8	32·5	63·3	57·7	50·2	45·8
1815	39·7	38·1	62·8	59·4	50·9	49·0
1816	36·0	36·8	59·5	55·2	48·9	46·4
1817	41·4	39·9	62·8	57·4	50·9	47·7
1818	38·3	37·4	66·6	61·2	52·3	50·8
1819	39·6	39·6	64·8	60·6	52·0	49·3
1820	35·4	35·2	63·5	58·0	50·6	47·4
1821	36·5	37·8	63·0	57·8	52·0	49·3
1822	42·8	42·5	67·5	62·1	53·8	51·0
1823	34·7	35·4	62·8	58·0	50·7	47·3
1824	39·9	37·8	64·0	59·2	52·2	48·3
1825	41·0	39·4	66·0	62·0	53·1	49·6
1826	38·7	38·3	68·4	63·9	52·5	49·9
1827	34·7	35·6	64·8	60·0	51·4	48·5
1828	42·8	41·4	64·6	60·3	52·7	50·1
1829	35·2	38·2	63·7	58·9	48·4	46·6
1830	28·8	31·9	63·1	58·8	50·2	47·8
1831	38·7	36·8	64·6	62·3	53·1	50·4
1832	38·3	38·7	66·6	60·5	51·4	49·1
1833	38·7	39·8	63·9	59·5	51·6	49·0
1834	43·3	43·1	66·6	62·5	54·1	51·0
1835	39·9	40·1	66·8	62·6	51·3	49·2
1836	35·4	36·3	65·5	60·3	51·3	48·1
1837	39·0	39·0	66·0	59·8	50·0	47·3
1838	33·3	34·3	63·3	59·1	48·6	46·4
1839	37·8	38·3	65·1	59·3	51·6	47·7
1840	39·6	38·9	65·1	59·9	50·5	47·8
1841	33·3	34·1	62·1	58·3	52·2	48·7
1842	37·2	38·1	60·4	62·8	51·8	49·6
1843	39·4	40·3	64·2	59·8	52·3	49·4
1844	37·8	39·4	62·2	59·9	50·4	48·7
1845	32·7	34·7	62·6	59·3	49·5	47·6
1846	42·4	43·4	69·1	64·3	53·1	51·3
1847	35·1	34·5	65·1	61·8	51·4	49·6
1848	37·9	40·3	64·8	59·5	52·5	50·2
1849	42·8	42·4	64·8	61·0	52·3	49·9
1850	38·8	39·2	65·3	61·1	51·1	49·3
1851	39·6	41·2	64·9	60·4	50·9	49·2
1852	39·0	41·1	66·7	61·6	53·1	50·6
1853	41·4	41·1	63·9	59·5	50·2	47·7
1854	37·5	37·5	63·0	58·9	51·6	49·0
1855	35·8	35·2	64·0	60·4	49·1	47·8
1856	39·4	39·0	66·0	61·1	51·4	49·0
1857	37·8	38·7	66·7	64·0	52·3	51·1
1858	36·3	39·1	66·6	62·5	50·7	49·2
1859	40·1	41·5	67·1	64·3	52·5	50·8
1860	36·5	37·4	61·2	56·7	48·6	47·0
1861	36·0	37·4	65·5	61·1	51·3	49·4
1862	39·0	40·4	62·4	58·3	51·3	49·6
1863	41·2	42·5	65·5	60·3	52·5	50·3
1864	37·6	38·6	63·1	59·6	49·8	48·5
1865	36·1	37·8	65·1	61·3	52·5	50·3
1866	40·5	41·9	64·2	60·4	52·0	49·8
1867	41·2	40·6	63·7	59·8	50·9	48·6
1868	36·9	39·2	67·3	64·4	53·2	51·6
1869	43·7	44·1	63·3	60·2	51·3	49·5
1870	36·5	37·5	66·0	62·5	50·4	48·7

The preceding table shows that the coldest winter of the present century in Paris was that of 1830; the mildest, that of 1869; the coldest summer, that of 1816; the hottest, that of 1842; the coldest year was 1829; and the hottest, 1834.*

This list gives the mean annual temperature of winter and summer, as ascertained at the Paris Observatory. We shall see farther on that there have been more severe frosts and greater heat in France than those given, but they have been observed at different places.

I have already stated that, taking the mean temperatures of each day of the year at Paris, it would be seen that there is an increase in heat from the first week in January to the middle of July, with a continuous decrease from the latter date until the close of the year. The general phenomenon, however, exhibits certain discontinuities which can not be treated so simply.

It is true that, generally speaking, it is the movement of the earth which gives rise to the grand phases of the temperature, and which produces in our climates, for instance, a minimum in January and a maximum in July. But the curve which unites these two extreme points is not regular. There are unmistakable departures from continuity which seem subject to periodical returns.

In its more general aspect, the question may be put in the following manner:

What is, for a given locality, the mean departure which the temperature of each day in the year exhibits in relation to the supposed regular march of these temperatures between the annual extremes?

[* M. Flammarion has given this table for Paris only; I have added the corresponding values for Greenwich, as taken from my paper in the Philosophical Transactions for the year 1848, supplemented by subsequent results. I may remark that I have altered some values in M. Flammarion's table as seemed to be necessary by comparison with the tables in the "Annuaire" for 1872.

This table shows that the coldest and warmest winters, the coldest summer, and the coldest year, were the same at Paris and at Greenwich, and that the warmest summer and the warmest year at Greenwich was 1868.

It also shows that the most severe winter of all was at Paris, and that the winter temperature of Paris is frequently lower than at Greenwich, although generally it is higher.

The mean temperature of the winter at Paris from all the years is 38°·4, while that at Greenwich is 37°·1.

The mean temperature of the summer months has in every case been warmer at Paris than at Greenwich.

The mean of summer at Paris is 64°·7; at Greenwich, 60°·4.

The mean temperature of every year is higher at Paris than at Greenwich: the mean at Paris is 51°·3; that at Greenwich, 48°·9.—En.]

Is this departure about the same for each year, or for a small group of years—or does it, on the contrary, vary from one year to another, or from one group of years to another. so as to present a certain periodical recurrence?

The questions, which are secondary to the first general question, are very numerous, inasmuch as the quantities of light which enter into the atmosphere, the electric state of the air, and its so-called ozonometrical properties, its hygrometrical condition, as also the variations in the atmospheric pressure, the displacement of the air, or the winds and tempests—in a word, all the atmospheric phenomena are intimately bound up with the distribution of heat over the surface of the globe.

Lastly, a very natural and important addition consists in the influence of these thermometrical perturbations upon the health of men, animals, and of plants.

Four epochs in the year are remarkable for a fall in the temperature, and atmospheric perturbations caused thereby, viz., about the 12th of February, the 12th of May, the 12th of August, and the 12th of November.

The periodical cold of the month of May is a popular tradition ; horticulturists term St. Mamert, St. Pancras, and St. Servais, whose anniversaries are on the 11th, 12th, and 13th of May, the three *ice-saints.*

In February there are the same indications, but they are even more marked. The fall after the 7th of February is very sudden, and continues to the 12th, which gives but a single minimum even in the middle of the ice-saints of February. As February with us represents Northern climates, every thing will be extreme, the rise as well as the fall ; in August, on the other hand, which gives us an idea of the tropical climates, the changes are less sudden, and the slight movement corresponding to that of the 10th to the 14th in May, or, in another form, of the August ice-saints, continues until the 16th.

In November, as in August, the decline of the temperature is seen to be struggling against influences which tend to an abnormal return of heat ; the points of inflection correspond precisely to those of the three other months, and one of the last of them produces, on the 14th, the Martinmas summer.

The careful examination of a large number of years shows that, at London and Berlin, as at Paris, there is a certain agreement between the four days of the same date, as exhibited in their mean temperatures. M. Ch. Sainte-Claire Deville ascertained that these curious periods are to be found in the most ancient of known meteorological documents; for instance, in the observations of the pupils of Galileo, and of the Academy of Cimento. These observations extend over fifteen years (1655-1670). The minimum of the ice-saints is found to occur on the 12th, with a remarkable regularity.

It is certain for the last two centuries, in this part of Europe, that the periodic anomalies of the temperature, some of which were proverbial among our ancestors, have manifested themselves in the same manner stated above.

Certain astronomers, Erman and Petit among the number, have attributed these frigorific phenomena to masses of asteroids which, in their orbit, sometimes come between the sun and the earth.

The action of the sun produces, therefore, in the temperature of the air, variations according to the hours of the day and the month of the year. This same solar action produces a diurnal and a monthly variation in the readings of the barometer, which, perhaps, had better be considered here, as it is a consequence of the temperature.

The atmospheric pressure increases and decreases twice each day with regularity, in a manner dependent on the sun's position. The reading of the barometer, which shows the weight of the atmosphere, gradually increases from 4 to 10 A.M. This atmospheric tide is not due, like that of the sea, to the attraction of the moon and the sun, since it takes place every day at the same hour, and does not follow the course of the moon. It is due to the expansion produced by the solar heat, and to the increase in the vapor of water, also produced by this same heat.

This barometrical variation is not great, for it never attains so much as one-tenth of an inch. It was about the year 1722 that the *diurnal variations* of the barometer were first ascertained by a Dutchman, whose name has not reached us. Since that epoch, several observers have endeavored to determine their amounts and their periods for different parts of the earth. Humboldt proved, by a long series of observations, that, at the equator, the maximum of elevation corresponds with 9 A.M.; after that hour, the barometer reading decreases until 4 or 3·30 P.M., when it attains its minimum. It afterward increases again until 11 P.M., when it reaches a second maximum, and, lastly, decreases again until 4 A.M. Thus there are each day two minima, at 4 A.M. and 4 P.M.; and two maxima, at 9 A.M. and 11 P.M. The movements are so regular that a simple glance at the barometer suffices to ascertain the hour, especially during the day, without any probability of being more than a quar-

ter of an hour in error. They are so permanent that neither tempest,
nor storm, nor rain, nor earthquake af-
fects it; they main-
tain themselves as
steady in the warm
regions of the coast
of the New World
as upon table-lands
more than 13,000
feet high, where the
mean temperature
falls to 44½°. The
amount of the os-
cillations diminish-
es as the latitude in-
creases, in the same
manner as the mean
temperature of a
place is, in general,
higher the nearer it
is to the equator.

At the Antilles, it
is found that there
is a distinctly-mark-
ed maximum for the
diurnal oscillation
along the northern
coast of America,
which is situated op-
posite to the sea of
the Antilles. The
stations upon this
coast-line give, on
an average, an am-
plitude of 0·11 inch-
es; whereas at all
the other stations
the amount is small-

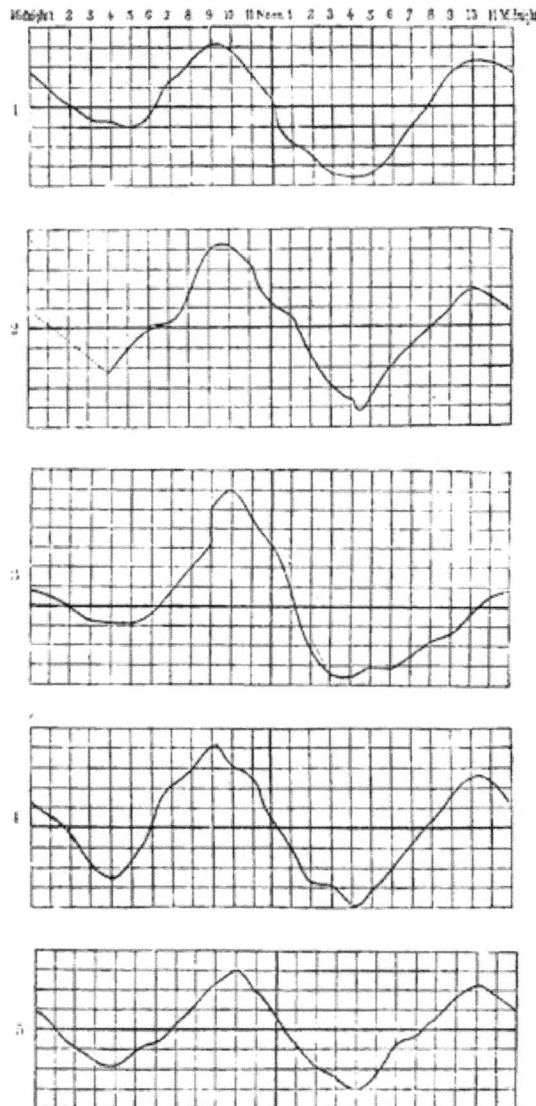

Fig. 56.—Regular diurnal oscillation of the Barometer: 1. Ascension Isl-
and; 2. Port d'Espagne; 3. Acapulco; 4. Cumana; 5. Basse-Terre.

er, whether they are situated to the north or south of the littoral region in question.

The northern coasts of Venezuela and New Grenada are exactly those which the thermal equator follows, rising in this district to the twelfth degree of north latitude, whence it descends again toward the equator, on both sides of the continent. The place of the maxima oscillations of the barometer is, therefore, the same as that of the maxima temperatures, and the two phenomena follow a similar march in the intertropical American zone. This is, moreover, quite in accord with the causes which influence the distribution of temperature over the different hours of the day.

Various observations have made it evident that the amplitude of the total oscillation diminishes with increased altitude. It may be stated as a general rule that this amplitude is dependent on the mean temperature of the place, and that it decreases with it not only according to the vertical co-ordinate of the altitude, but according to the two co-ordinates of latitude and longitude.

The diurnal oscillation of the barometer varies with the latitude as follows:

Places.	Latitude.	Mean Height.	Diurnal Oscillation.
	Degrees.	Inches.	Inches.
Lima	12·3 S.	29·202	0·107
Caraccas	10·31 N.	26·848	0·085
Payta	5·6 S.	29·841	0·082
Santa-Fé de Bogota	4·36 N.	29·918	0·080
Bagné	4·28	25·934	0·076
Calcutta	22·35	29·877	0·072
Cumana	10·28	29·770	0·070
Rio de Janeiro	22·54 S.	30·117	0·067
Mexico	19·26 N.	22·958	0·063
Cairo	30·2	29·816	0·061
Rome	41·54	29·971	0·040
Bâle	47·34	29·087	0·033
Brussels	50·50	29·806	0·032
Paris	48·50	29·757	0·028
Frankfort	50·8	29·626	0·022
Dresden	51·7	29·310	0·019
Berlin	52·33	29·869	0·013
Cracow	50·4	29·228	0·012
Edinburgh	55·55	29·406	0·008
Königsberg	54·42	29·956	0·007
St. Petersburg	59·56	29·895	0,005

The last column of this table shows that at 60° of latitude the diurnal barometrical oscillation is very small.

In our climates these hourly variations are so masked by accidental variations that to discover and measure them was a work requiring the greatest sagacity and precision. It is only by the average of many years' observations taken with care and at suitable hours that the hourly periods can be arrived at.

The following table gives the diurnal and monthly atmospheric variation due to the expansion of air by solar heat, as found from observations at the Paris Observatory:

Month.	Mean Heights of the Barometer reduced to the Temperature of 0.			
	At 9 A.M.	At Noon.	At 3 P.M.	At 9 P.M.
	Inches.	Inches.	Inches.	Inches.
January	29·813	29·810	29·786	29·799
February	29·798	29·782	29·768	29·782
March	29·773	29·764	29·740	29·761
April	29·705	29·690	29·678	29·694
May	29·738	29·727	29·707	29·725
June	29·787	29·777	29·758	29·773
July	29·786	29·773	29·761	29·777
August	29·780	29·766	29·749	29·768
September	29·773	29·761	29·741	29·761
October	29·756	29·745	29·725	29·745
November	29·738	29·727	29·711	29·729
December	29·816	29·796	29·795	29·813
Means of the Year	29·772	29·759	29·743	29·760

This table shows that the morning maximum attains on an average 29·772 inches, and the afternoon minimum 29·743 inches: the difference is 0·029 inch. It, moreover, shows that there is not only a diurnal variation of the barometer, but also a monthly variation.

The atmospheric pressure decreases gradually from January to April, increases a little up to July, decreases a little until November, and then increases in December and January. This movement of the atmospheric pressure is almost the exact opposite of that of the temperature; it is much more marked in the tropical regions, as may be seen by consulting the curves which M. Deville traced in the Antilles. The amplitude of the monthly oscillation is there on an average (29·81 − 29·69 =) 0·12 inch, between January and April, according to observations taken at noon. The nearer one approaches the tropics, the greater it is: correspondents at the Calcutta Institute inform me that 0·7 inch represent the amplitude between January and July, and at Benares 0·6 inch.

The observations at Brussels show that the diurnal and monthly variations in our climates are distinct. By comparing them it is seen that the diurnal maxima of temperature are pretty constant during the year, occurring about 10 A.M. and 10 P.M. As to the minima, the interval between them is greater in summer than in winter; the two quantities also exhibit a greater deviation in the summer months. During the

Jan. Feb. Mar. Apr. May. June. July. Aug. Sept. Oct. Nov. Dec. Jan.

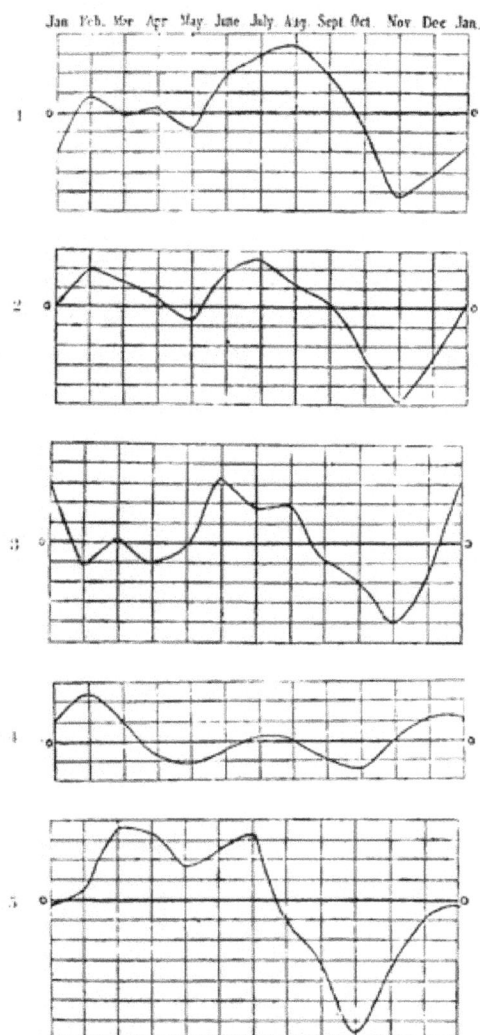

shortest days (November, December, January), there are only eight hours between the minima, which occur at 6 A.M. and 2 P.M., whereas during the other months the interval between them is longer.

The time at which the first minimum takes place varies more than two hours, being at 8·30 A.M. in June, and 6·22 A.M. in December.

There is an equally great change in the time of the first maximum. The extreme limits take place at 10·50 A.M. in February, and at 8·40 A. M. in June. Local causes exercise a certain influence upon the epochs of these extreme limits.

The epoch of the second minimum varies still more, as it occurs at 2·15 P.M. in January, and at 5·30 P.M. in June, showing a difference of time of three and a quarter hours. The limits within which the barometrical epoch varies are, in the case of the first max-

Fig. 54.—Regular monthly oscillation of the Barometer: 1. Cayenne; 2. Guiana; 3. Trinidad; 4. Santa-Fé de Bogota; 5. Guadaloupe.

imum and the first minimum, about two hours. The interval of time which elapses between the first maximum and the second minimum deserves especial attention, there being a separation of four hours only in January, which increases to eight in June, the latter being the double of the former. The results show that the total diurnal variation is made up by the combination of two waves: the one scarcely perceptible, which, in the space of twenty-four hours, has a maximum and a minimum of 0·009 inch only; the other much greater, with two maxima and two minima of 0·01 inch.

Such are the regular variations of the barometer, due to diurnal and annual action of solar heat. These are the least important variations. The atmosphere is unceasingly in movement by influences which acquire a greater intensity, although they have the same origin. The irregular variations are larger, and increase from the Equator to the Poles. While the extreme differences of the barometer do not exceed upon an average two-tenths of an inch in the equinoctial regions (exclusive of the cyclones, which will be alluded to hereafter), they reach to two and three inches in our latitudes.

The greatest barometrical variations occur in winter, the smallest in summer.

At all times of the year the barometer reading is higher during the minima of temperature than during the maxima.

It is especially during autumn and winter nights that the differences of temperature have the greatest effect on the reading of the barometer. In spring this influence is much less, and is to a great extent disguised by other causes.

CHAPTER IV.

REMARKABLE SUMMERS—THE HIGHEST KNOWN TEMPERATURES.

THE first summer of the present century, or, to speak more exactly, according to chronology, the summer of the last year of the eighteenth century, was remarkable for its high temperature, and I might commence the series with it but for the fact that Europe experienced an exceptional degree of heat at a date which will remain famous, that of 1793.

The summer of this year was memorable for the intense and unexampled heat, which occurred in July and August. According to Cassini IV., then director of the Observatory, the results for Paris were:

Great heat (77° to 88°)	36 days.
Very great heat (90° to 95°)	9 "
Extraordinary heat (95° or higher)	6 "

The highest temperatures occurred as follows:

Valence, July 11	104·0	degrees.
Paris, July 8	101·1	"
Paris, August 16	99·1	"
Chartres, August 8	100·4	"
Chartres, August 16	100·6	"
Verona, July and August	96·1	"
London, July 16	89·1	"

The great heat began in Paris about the 1st of July, and increased very rapidly. The sky remained continually clear and cloudless; the wind was in the north and generally very gentle, and the barometer remained very high. The hottest days were the 8th and 16th of July. On the 9th, a fearful thunder-storm raged at Senlis and the immediate neighborhood. Hailstones as large as eggs destroyed the crops; a tremendous wind blew down more than 120 houses. This tempest was followed by very heavy rain, and the water, collecting in the fields, swept off cattle, furniture, women, and children. At Bougueval, in the Oise, a woman, after rescuing her nine children, was swept off by the flood. On the 10th of July, to complete the destruction, there came a second hailstorm.

The extreme heat of the month of July continued through part of

August. On the 7th of this latter month it was very great, very general, and most oppressive. The sky was still clear, and the wind, from the north-east, was so scorching that it seemed as if emitted from the mouth of a furnace. It came by whiffs, and was as severe in the shade as in the sun. This was experienced not only throughout Paris, but in the country districts as well. The suffocating heat paralyzed the breathing, and was felt far more severely on that day, with the thermometer at 86°·6 Fahr., than on the 8th of July, when it was 101°·2.

The dryness of the ground became very great. The level of the waters of the Seine fell to the lowest water-mark of 1719, at the end of August and in the middle of September. There fell in Paris but 10·9 inches of rain in the year. In the country, trees and shrubs were generally burned up; and fruits, including apples among them, showed signs of having been burned. There was a great scarcity of vegetables. The land, dried up, hardened and cracked: it was impenetrable both to the plow and the spade. Workmen engaged in sinking a well in a place exposed to the sun, found the soil dried up to a depth of more than five feet. By the 1st of September the trees had lost nearly all their leaves; the dryness and the heat had caused the bark to crack and the branches to look dead. Very many of the trees did, in fact, die.

In Burgundy the vintage began on the 23d of September; the wine was abundant, but of inferior quality, as the vines had been affected by a cold rain which fell in that district. The summer was dry and hot in the neighborhood of Toulouse, and the maize crop was a complete failure. It will be remembered that 1793 was a year of great scarcity in France.

1800.—The summer was marked by extreme heat, which extended over part of Europe. From the 6th of July to the 21st of August the temperature decreased but five times below 74°·2; there were, according to Bouvard, of

Great heat	25 days.
Very great heat	5 "
Extraordinary heat	2 "

The highest temperatures were as follows:

Bordeaux, August 6	101·8 degrees.
Nantes, August 18	101·8 "
Montmorency, August 18	100·2 "
Limoges	99·5 "
Paris, August 18	95·9 "
London, August 2	88·0 "

Conflagrations were very numerous in the early part of April. A whole village in the department of the Eure, the forest of Haguenau, and part of the Black Forest, became a prey to the flames. Myriads of grasshoppers alighted in the neighborhood of Strasbourg. In the night of July 20th the ancient monastery of the Augustins, in Paris, was struck by lightning. In the south there occurred numerous cases of sudden madness.

1811.—The summer of 1811 was in many respects one of the most memorable known in Northern Europe.

The following is the table of the maxima temperatures:

Augsburg, July 30	99·5 degrees.
Vienna, July 6	96·3 "
Avignon, July 27	95·0 "
Hamburg, July 19	94·6 "
Naples, July 20	94·3 "
Copenhagen, July	92·8 "
Liége	92·7 "
Strasbourg	91·4 "
St. Petersburg, June 27	88·0 "
Paris, July 19	87·8 "

In Burgundy the vintage began on the 14th of September. A hail-storm that occurred on the 11th of April spoiled two-thirds of the crop: but the summer was very favorable for vines, and the small crop yielded wine of an excellent quality, which was long famous as the *comet wine*.

1822.—The summer of 1822 was remarkable throughout France for high mean temperature.

At Paris there were of

Great heat	55 days.
Very great heat	3 "

The maxima of temperature were as follows:

Malines, in July	101·8 degrees.
Joyeuse, June 25	99·1 "
Alais, June 14 and 23	97·7 "
Liége	95·0 "
Maëstricht, June 11	93·2 "
Paris, June 10	92·8 "

The drought was very great in France during the warm season: from the 21st of August to the 26th of September the Seine was nearly continuously below the mark of zero at the Pont de la Tournelle. As early as the month of March there was a scarcity of water: for cattle in the south of France water had to be brought from great distances upon the backs of mules, and the spring temperature in that country was as high as that generally experienced in August. The harvest was finished in Languedoc by the 23d of June, and though there was very little straw the ears were well filled. In Burgundy the sky was unusually clear. The vintage began on the 2d of September, but, according to the vine-dressers, it might have been begun on the 15th of August, and in the neighborhood of Vésoul the grapes were gathered on the 19th of August. There was an average quantity of wine, and the quality was exceptionally good; the grain crops were, as a rule, less abundant than usual.

1826.—A very hot and dry summer: thirty-six days of great heat in Paris, seven of very great, and two of extraordinary heat. The mean of the summer was very high, $69\frac{1}{2}°$ Fahr. Crops were destroyed, and forests burned, in Sweden and Denmark.

The highest temperatures observed were—

Maëstricht, August 2	101·8 degrees.
Épinal, July 1	97·7 "
Paris, August 1	97·2 "
Metz, August 3	97·0 "
Strasbourg	93·6 "

1834.—This year, though not remarkable for any very great heat, is noticeable for the very high mean temperature of the spring and summer throughout France. Vegetation was very forward, and there fell, in many places, rain distributed in such a manner as to be most favorable to the crops. In Paris there were forty-three days of great, and three days of very great heat.

The mean average of the summer, 69°, is the highest of the century, next to 1826, 1842, and 1846. The drought was very great in August. The maxima temperatures of 1834 are thus distributed:

Avignon, July 14	95·0 degrees.
Geneva, July 18	93·9 "
Liège	92·3 "
Metz, July 12	91·4 "
Strasbourg	91·0 "
Paris, July 12 and 18	90·7 "

In the south, the temperature was lowered by plentiful rains, and was very mild. Burgundy was this year celebrated for the superior quality of its wine, though the quantity ran very short. Such was also the case in the Bordeaux district. The harvest was almost universally good in France.

1836.—The summer of this year was memorable for the stormy nature of the month of June and the early part of July, and for the number of fatal accidents caused by the heat in the south of France. In Denmark, Russia, and Spain, the temperature also produced some remarkable effects.

The drought in the month of August was intense; the Seine fell about ten inches below the low-water mark of 1719. There was an average crop of wine in the south of France, the quality being fairly good. The vintage did not begin in Burgundy till the 6th of October. The corn harvest was bad.

1842.—The summer of this year was the hottest during the first half of this century, especially in Paris and the north.

In Paris there were of

Great heat	51 days.
Very great	11 "
Extraordinary	1 "

The mean temperature of the season in Paris was 69°·4.

The following is a list of the highest temperatures recorded:

Paris, August 18	99·0 degrees.
Agen, July 4	98·6 "
Bordeaux, July 16	94·6 "
Toulouse, July 17	93·9 "

Many accidents caused by the heat were registered; wheels of several mail-carts took fire. At Badajoz, in Spain, three laborers died of the heat on the 28th of June, and a lady expired from its effects in a diligence. At Cordova several reapers fell down asphyxiated, and frequent cases of madness were attributed to the same cause.

In Burgundy the vintage began on the 21st of September, the crop being abundant and the quality good. The grain crop was below the average.

1846.—The temperature this summer was very remarkable, and there were periods of intense heat in France, Belgium, and England. In Paris there were of

Great heat	18 days.
Very great	9 "
Extraordinary	2 "

The mean summer temperature was 69°·2.

The maxima of this year are as follows:

Toulouse, July 7	104·0 degrees.
Quimper, June 19	100·4 "
Rouen, July 5	98·2 "
Paris, July 5	97·7 "
Orange, July 13	97·7 "
Angers, July 29	95·0 "
Metz, August 1	94·6 "

Accidents occurred in Brittany. At the Pont-de-Croix fair several persons had fits, occasioned by the heat; at Benzee, a little girl, left in the sun, was found dead a few minutes afterward. The temperature during June was also very high at Toulouse, Toulon, and Bordeaux. In the Landes, farmers obtained a second crop of rye. Near Niort, early in July, three laborers died while at work.

The vintage in Burgundy began on the 14th of September, the quality being exceptionally good, though there was only a half crop. The corn harvest, too, was much below the average.

1849.—The heat was very great in the south of France, and the maximum at Orange is the highest temperature in the shade yet recorded in France.

The table gives the following figures:

Orange, July 9	106·5 degrees.
Toulouse, June 23	99·7 "
Bordeaux, July 7	94·3 "
Gand	93·9 "
Metz, July 8	92·5 "

1852.—This was a remarkable summer in Russia, England, Holland, Belgium, and France. There were in Paris of

Great heat	30 days.
Very great	6 "
Extraordinary	1 "

The summer mean in Paris was 67°. The mean of July was 72¼°. There was an unusual continuance of great heat: July 9th, 88°·0; the 10th, 92°·3; the 11th, 87°·8; the 12th, 90°·5; the 13th, 92°·8; the 14th, 93°·6; the 15th, 93°·6; the 16th, 95°·2.

The highest temperatures throughout Europe were—

Constantinople, July 27	101·3 degrees.
Rouen, July 5	97·0 "
Versailles, July 16	96·3 "

Orange, August 25	95·5 degrees.
Dunkerque, July 7	96·3 "
Paris, July 16	95·2 "
Verviers, July 18	95·2 "
London, July 12	95·0 "

At Amsterdam, a thermometer rose, on July the 12th, to 102°·2. At Alphen, near Leyden, two peasants, asphyxiated by the heat, were found dead in a field; at Alkenaer an engine-driver became insane, after congestion of the brain produced by sun-stroke. In the centre of France the thermometer stood for more than 10 days at over 86°. Many domestic animals perished from the heat. At Thourotte, in Belgium, there fell a disastrous hailstorm on the 11th of August: many of the hailstones weighed from two to three ounces, and were from two to three inches in diameter.

In France the harvest was mostly over by the middle of July, and was an average. On the other hand, the vintage did not begin till the early part of October, and the wine crop was small in most vineyards, and of inferior quality.

1857.—This summer was hotter than usual in France, and the months of July and August were nearly everywhere distinguished for extreme heat. The highest temperatures observed were—

Montpellier, July 29	101·5 degrees.
Orange, July 18	100·9 "
Les Mesneux, August 1	98·6 "
Toulouse, July 27	98·2 "
Clermont, July 14 and 15, and August 3	98·2 "
Blois, in August	97·7 "
Paris, August 4	97·2 "
Metz	96·1 "

There were three distinct streams of summer heat. The first, on the 27th of June, passed over the highest and the most southerly stations in France, and reached, on the 28th, the northern frontier; the second extended over the north-west, from the 14th to the 16th of July; the 3d, and the most intense, moved slowly and in the same direction, traveled from south to north in the interval between July the 27th and August the 4th. The drought this summer was very great throughout nearly the whole of France; fortunately, in the middle of August, some rain fell.

In Burgundy the vintage began on the 16th of September, and the crop was passable as to quantity and good as to quality. The corn crops were, generally speaking, up to the average.

1858.—This summer was remarkable for great drought, and prolonged rather than intense heat, in England, Belgium, the centre and a part of the south of France, and Algeria. In the north the heat was less than in 1857, but greater in the south. The maxima temperatures were—

Montpellier, June 20	100·9	degrees.
Orange, July 19	100·9	"
Vendôme, June 15	97·0	"
Tours, June	96·8	"
Clermont	96·4	"
Lille, June 15	95·9	"
London, June 16	94·8	"
Paris, June 3	89·6	"

The drought was very great throughout nearly all France in the spring and part of the summer, and was very inimical to the rearing of stock; during June the sky was remarkably clear, but in July and August some rain fell, at least in the north, so that the meadows that had been scorched up, owing to a want of moisture dating from the year before, partially recovered themselves. The harvest, which terminated on the 1st of July in the south, and the 1st of August in the north, was an average crop in respect to quantity, and a rather more than average one in respect to quality. The vintage, begun in Burgundy on the 18th of September, yielded a remarkable crop, both in respect to quantity and quality.

During recent years I must mention 1865 and 1868 as having been marked by a long series of hot days. The former, as is well known, was very favorable for the vintage.

1865.—The mean monthly temperatures at the Paris Observatory were—

January	38·5	degrees.	July	67·8	degrees.
February	36·1	"	August	63·9	"
March	36·0	"	September	66·6	"
April	60·4	"	October	54·0	"
May	61·3	"	November	46·4	"
June	64·4	"	December	36·1	"

The extreme heat in Paris was 91°·9 on the 6th of July. The average of the three summer months was 65°·3. Adding to them September, the average of the four months was 65°·5; an average that rarely

15

lasts so long. The mean of the year was 52°·5, being 1°·2 above the average. The month of January was relatively warm. In April, after the 4th, the weather was exceptionally fine, and the thermometer readings were very high: from the 8th the temperature was that of June. In May and June the temperatures were above their normal points. July and August were cold. In September the temperature was higher than in August. October and November were warm. The highest temperatures were—

Nimes, July 5	100·2 degrees.
Nice, July 10	95·5 "
Perpignan, July 4	95·4 "
Aix, August 28	94·5 "
Montpellier, July 26	93·2 "

1868.—The mean monthly temperatures at the Paris Observatory were—

January	32·0 degrees.	July	70·2 degrees.
February	41·7 "	August	65·7 "
March	44·6 "	September	63·7 "
April	50·9 "	October	50·9 "
May	64·2 "	November	40·8 "
June	64·4 "	December	47·5 "

The maximum temperature in Paris was 93°·2 on the 22d of July. The average of the three summer months was 66°·9. This summer is notable in the annals of meteorology for its thermometrical elevation, and its combination of circumstances favorable to the crops, both as to their quantity and quality. The averages of the temperatures of May, June, and July were very high in the south. Thus at Tours the average of May was 65°·1; that of June, 67°·6; that of July, 71°·2. The highest temperatures observed in France are appended :

Nimes, July 20	106·5 degrees.
Perpignan, July 25	99·0 "
Draguignan, July 24	98·4 "
Montauban, July 20	98·1 "
Toulouse, July 19	95·0 "
Montpellier, July 20	94·3 "
Aix, July 20	93·2 "

The temperature rose higher in 1859, without giving so high an average. This latter was due less to the height of the diurnal maxima than to that of the nocturnal minima. In fact, notwithstanding the almost uninterrupted serenity of the nights, the cold caused by nocturnal radiation was at no time very remarkable. Nearly every morning before

sunrise a slight fog, indicating a somewhat elevated hygrometrical condition, covered the soil, moistening the plants, and modifying the effects
of the great heat during the day. The vapor of water prevents the
radiation of the obscure heat; the air which was lying over our part
of the country, and the somewhat elevated hygrometrical condition of
which increased the transparency for the stellar light, nullified the effects of nocturnal radiation, which is so potent even in the tropical regions when it has only to traverse an air devoid of moisture.

This remarkable summer affected the temperature of the soil to
the depth of more than a yard. During the summers of 1864, 1865,
1866, and 1867, the heat at the depth of 39 inches was 57°·7, 58°·5,
57°·2, and 57°·6. In 1868 it was 60°·6, nearly 61°·0.

Such are *the memorable summers* of the present century. The following are the highest temperatures of the air (*in the shade*, and to the
north) observed in France since they have been truthfully ascertained
by the thermometer. I have recorded all those which have reached or
exceeded 37°, and those only, except in the case of Paris, where there
are several readings. The towns are given in the order of latitude from
N. to S.

Places.	Latitude.	Longi-tude.	Elevation above the Sea.	Dates.		Extreme Maxima.
	Deg.	Deg.	Feet.			Deg.
Saint-Omer............	50·45	0·05	75	August	10, 1777	99·5
Cambrai...............	50·11	0·54	177	"	4, 1783	99·5
Rouen.................	49·26	1·15	127	"	18, 1800	100·4
Les Mesneux..........	49·13	1·37	279	"	4, 1857	99·5
Metz.................	49·07	3·50	597	"	4, 1781	100·6
Montmorency..........	49·00	0·02	469	"	18, 1800	98·6
				"	26, 1765	104·0
				"	14, 1773	102·9
				"	19, 1763	
				"	5, 6, 1705	102·2
				July	16, 1782	101·7
				"	8, 1793	101·1
Paris.................	48·50	0·00	213	"	10, 1766	101·0
				August	18, 1842	99·0
				July	31, 1803	98·1
				"	5, 1846	97·7
				"	19, 1825	97·3
				August	4, 1857	97·2
Haguenau..............	48·48	5·25	439	July	16, 1782	102·9
Nancy................	48·42	3·51	656	"	26, 1782	99·7
Chartres..............	48·27	0·51	518	"	16, 1793	100·6
Quimper..............	48·00	6·26	19	June	19, 1846	100·4
Montargis.............	48·00	0·23	380	1777	and 1778	99·5
Angers................	47·28	2·54	154	July	17, 1784	100·4
Tours.................	47·24	1·39	180	August	1840	100·4
Nantes	47·13	3·53	144	"	18, 1800	101·8
Chinon................	47·10	2·06	269	July	21, 1783	100·6
Seurre (Côte-d'Or) ...	47·01	2·48	492	"	6, 1783	102·2

Places.	Latitude.	Longitude.	Elevation above the Sea.	Dates.		Extreme Maxima.
	Deg.	Deg.	Feet.			Deg.
Nozeroy.....................	46·47	3·42	492	July,	1787	99·5
Luçon.....................	46·27	3·30	265	"	21, 1777	101·8
La Rochelle............	46·09	3·30	82	"	4, 5, 1836	102·2
Saint-Jean d'Angély.	45·57	2·52	78	"	1787	99·5
Limoges	45·50	1·05	941	"	23-25, 1800	99·5
Valence..................	44·56	2·33	419	"	11, 1793	104·0
Bordeaux..............	44·50	2·55	59	August	6, 1800	101·8
Joyeuse (Ardèche)....	44·32	2·00	482	June	23, 1822	99·1
Agen.....................	44·12	1·43	141	July	4, 1842	98·6
Orange	44·08	2·28	150	"	9, 1849	106·5
Avignon	43·57 N.	2·28	118	{August	14, 1802 16, 1803	} 100·6
Nimes....................	43·51	2·01	374	July	20, 1868	106·5
Manosque.............	43·49	3·35	1312	"	18, 1782	101·8
Arles.....................	43·41	2·18	55	August	20, 1806	99·5
Toulouse...............	43·37	0·54	649	{July 30, 31, 1753	7, 1846	} 99·9 104·0
Montpellier............	43·37	1·32	98	"	29, 1857	101·5
Béziers.................	43·21	0·52	252	"	1847	98·6
Sorèze...................	43·19	0·43	1640	"	12, 1824	99·5
Pau......................	43·18	2·43	672	August	4, 1838	101·8
Perpignan.............	41·42	0·34	137	July	29, 1857	101·5

The greatest heat in the shade and in the north is 106½° for France (Orange July 9th, 1849, and Nîmes July 20th, 1868); 96° for Great Britain; 102° for Holland and Belgium; 99½° for Denmark, Sweden, and Norway; 102° for Russia; 103° for Germany; 105° for Greece; 104° for Italy; 102° for Spain and Portugal. In non-European countries, the highest temperatures, as given by Arago, are as follows:

Tunis... 112·5 degrees.
Manilla.. 113·5   "
Nubia.. 115·2   "
Am-Dize (Egypt)... 116·1   "
Esné (Africa)... 117·3   "
Bagdad (Asia).. 120·0   "
Near Suez (French Expedition to Egypt)............... 126·5   "
Near Port Macquarie (Archipelago)..................... 129·0   "
Near Syène (Africa)...................................... 129·2   "
Murzouk (Africa)... 133·4   "

These are the maxima of the temperatures of *the air* in the shade.

In presence of such elevations of temperature, it may be asked to what point human organism can support it without incurring the danger of sudden death. The mean temperature of the human body is about 96° (it is easily ascertained by placing the bulb of a thermometer under the tongue). That of birds is higher, and reaches 111° with certain kinds. That of fish is lower, and about 37°.

CHAPTER V.

AUTUMN—WINTER: WINTER LANDSCAPES—COLD—SNOW—ICE—HOAR-FROST, RIME, ETC.—REMARKABLE WINTERS—THE LOWEST KNOWN TEMPERATURES.

Take, in the first place, this winter landscape which is represented in the preceding page. It is the same as that which we saw, full of color and movement, on a fine summer's day. It is now transformed beneath the gray and sombre sky of winter. The green foliage has disappeared from the trees, the meadow is covered with a pall of snow, the rivulet is frozen over, and the laborer's cottage seems as lifeless as Nature herself. With the progressive decline of temperature the thermometer has fallen to 32°, a remarkable point, at which water ceases to preserve its liquid condition and becomes solid. It then may assume various forms, becoming either massive in the shape of ice, light in the shape of hoar-frost, or falling slowly as snow-flakes. It is, as a rule, in this latter form that winter begins to manifest itself, for snow is produced as soon as the temperature is at or about 32°. If this temperature extends from the clouds to the surface of the earth, the water reaches the ground as snow. If snow in falling has only a thin stratum of air above 32° to traverse, and if it be abundant, it still reaches the ground and preserves its consistency. This occurs sometimes in summer.

Snow, in covering the earth as a carpet, forms at once a covering and a screen; a screen, because, possessing but little conducting power, it obstructs the passage of heat from the earth, and thus prevents the earth from becoming as cold as the air. Snow also adds its influence in favor of the fertilizing of the soil. Like rain and mists, it contains a considerable proportion of ammonia, which exists in a volatile state in the atmosphere, and which it conveys to the soil, afterward preventing it from becoming volatile again, as is the case after rain, especially after warm rain.

In the origin—that is to say, in the frozen clouds high up in the atmosphere—the snow appears to be formed of very slender fibres of ice. When the small drops of water, which form mists and ordinary clouds, become congealed, it is probable that these drops do not preserve their spheroidal shape, but that they fall an instant and take the shape of a filament which freezes concurrently with its physical transformation. By virtue of the laws of crystallization, these small filaments of ice become cohesive at angles of 60°, and form the figures which, though so numerous, still appertain to the same geometrical order.

Glaisher, in his ascent of June 26, 1863, encountered at 13,000 feet an immense cloud of snow, extending to a thickness of nearly one mile. It was a truly wonderful sight. This snow was composed entirely of small and perfectly-formed crystals, of an extreme delicacy. The points were visible, separate from each other, following two systems of crystallization; for the angular intervals were some at 60°, and others at 90°.

The construction of snow-flakes has long attracted the attention of observers. Kepler speaks of their structure with admiration, and other natural philosophers have endeavored to determine their cause; but it is only since the laws of crystallization in general have been ascertained that it has been possible to throw any light upon this subject.

In a circle, of all the polygons which can be inscribed, there is but one whose sides are equal to its radius; that is, the regular hexagon, or figure with six sides. This hexagonal figure is traced upon the flowers of the field, and we meet with it also in the crystallization of ice and snow, in the analysis of all the forms presented to our notice. The tendency of ice to take a crystalline shape is made evident by the fern-like leaves noticeable on window-panes during winter, when water becomes congealed upon them.

The examination of the figures of snow leads to impressions not less marked as to the existence of geometry, Number and Beauty, in the works of nature. It is not merely a few ice-flowers, such as the above, which have been remarked and designed in the slender snow-flakes, but there are many hundred different kinds, all constructed upon the same fundamental angle of 60°.

The snow sometimes falls in such compact flakes that behind the first planes it forms a white, cloudy veil, which hides the landscape. These heavy falls of snow are mostly met with upon the lofty table-lands of Asia or the Andes, where the caravans have often to encounter them. The routes soon become concealed beneath the pall that covers them; it becomes difficult to find one's way; and just as, in the rarer falls of snow in our countries, travelers wander over St. Bernard, and even

Fig. 52.—Snow Crystals.

crowd, are in no danger when standing upon ice eleven or twelve inches thick.

In very severe Russian winters the ice in the rivers is more than one
yard thick; but in France it has never exceeded more than about two
feet. Its power of resistance is so great that, in 1740, a large palace of
ice was constructed at St. Petersburg, 55½ feet long, 17 feet wide, and
21 feet high, the weight of the top and of the higher parts of the edifice
being readily supported by the foundations. In front of the building
were placed six guns in ice, with their carriages made of the same material. They were made to fire ball; and each piece pierced, at a distance of sixty yards, a plank two inches in thickness. The guns were
not more than four inches thick; they were loaded with a quarter of a
pound of powder, and not one of them burst. The Neva supplied the
materials for this singular edifice.

I have said that water when congealed increases in volume. One
consequence and one proof of this expansion is the bursting of the vessels containing it—a fact which occurs all the more readily when the
process of freezing is rapid and the vessel narrow in the neck.

I will complete this chapter by a notice of some of *the hardest winters
upon record*—considering those as hard winters in which the cold has
been of sufficient length and severity to freeze certain sections of large
rivers, such as the Seine, the Saône, and the Rhine—to congeal wine,
to destroy the tissues of certain trees, and to be followed by very grave
consequences for both the vegetable as for the animal world.

The following among the remarkable winters are the severest during
the last hundred years. Let me, in the first place, mention that the
hardest winters of past centuries were those of 1544, 1608, and 1709, in
which latter year the thermometer at the Paris Observatory fell as low
as −9°·6 Fahr. The winter of 1776 next comes as an exceptionally
cold one. The Tiber, the Rhine, the Seine, and even the Rhône, rapid
as it is, were nearly entirely frozen over.

After 1776, we come to the winter of 1788–1789, precursor of the
Revolution. This was one of the severest and longest winters that have
ever prevailed in Europe. In Paris the cold commenced on the 25th
of November, and lasted, with the exception of Christmas-day, when it
did not freeze, for fifty consecutive days. The thaw began on the 13th
of January, and the snow was found to be twenty-six inches deep. In
the great canal at Versailles, in the ponds and in several streams, the
ice was two feet thick. The water also froze in several very deep wells,

and wine became congealed in cellars. The Seine began to freeze as
early as November 26th (1788), and for several days its course was impeded, the breaking up of the ice not taking place until the 20th of
January. The lowest temperature observed at Paris was —7°·2 Fahr.,
on the 31st of December. The frost was equally severe in other parts of
France and throughout Europe. The Rhône was quite frozen over at
Lyons, the Garonne at Toulouse; and at Marseilles the sides of the docks

Fig. 53.—Winter.—The Seine full of floating ice.

were covered with ice. Upon the shores of the Atlantic the sea was
frozen to a distance of several leagues. The ice upon the Rhine was so
thick that loaded wagons were able to cross it. The Elbe was covered
with ice, and also bore up heavy carts. The harbor at Ostend was
frozen so hard that people could cross it on horseback; the sea was
congealed to a distance of four leagues from the exterior fortifications,
and no vessel could approach the harbor. The Thames was frozen as
low as Gravesend, and during the Christmas holidays and the early

part of January the stream in the neighborhood of London was covered with shops.

The following are the lowest temperatures that were noted in different places:

Bâle (Suisse), December 18	−35·5 degrees.
Bremen (Germany), December 16	−32·1 "
St. Alban's, December 31	−28·8 "
Warsaw (Poland), December 18	−26·5 "
Dresden (Germany), December 17	−25·8 "
Eosberg (Norway), December 29	−24·3 "
St. Petersburg, December 12	−23·1 "
Berlin (Prussia), December 28	−19·8 "
Strasbourg, December 31	−15·3 "
Tours, December 31	−13·0 "
Lons-le-Saulnier, December 31	−11·2 "
Troyes, December 31	−10·8 "
Orléans, December 31	− 8·5 "
Lyons, December 31	− 7·4 "
Rouen, December 30	− 7·2 "
Paris, December 31	− 7·2 "
Grenoble, December 31	− 6·2 "
Angoulême, December 31	− 1·7 "
Marseilles, December 31	+ 1·4 "

The cold of this winter was very fatal to men and animals, and injurious to vegetables. In the Toulouse district the bread was nearly everywhere frozen, and it was impossible to cut it until it had been laid before the fire. Several travelers perished in the snow; at Lemberg, in Galicia, thirty-seven persons were found dead in three days toward the end of December. The birds that belong to the extreme north were seen in several parts of France. Fish were killed in nearly all the ponds by the great depth to which the ice penetrated.

1794-95.—This was a remarkably long and severe winter throughout Europe. In Paris there were forty-two consecutive days' frost; and January 25th (1795) was the coldest day ever known, the thermometer falling to −10°·3, or 42°·3 below the freezing-point of water. In London the minimum temperature, 8°·1, occurred upon the same day; and at midnight, on the banks of the Rhône, near Geneva, it was 6°·8. The Maine, the Scheldt, the Rhine, and the Seine were so frozen over, that carriages and army corps crossed them in several places. The Thames was frozen over in the beginning of January, near Whitehall, in spite of the height of the tide. Pichegru, then in the north of Holland, sent detachments of cavalry and infantry about the 20th of Jan-

uary, *with orders to the former* to cross the Texel, and to capture the
enemy's vessels caught at anchor by the frost. The French horsemen
crossed the plains of ice at full gallop, approached the vessels, called on
them to surrender, captured them without a struggle, and took the crews
prisoners.

1798–99.—This was a very cold winter all over Europe. In Paris
there were thirty-two consecutive days' frost, and the Seine was completely frozen from the 29th of December to the 19th of January, from
the Pont de la Tournelle to beyond the Pont Royal, but not sufficiently so to admit of its being crossed on foot. The lowest temperature
remarked was $+0°\cdot3$, or $31°\cdot7$ below the freezing-point of water in
Fahrenheit's scale, on December 10th, 1798. An Alpine eagle was
shot at Chaillot. The Meuse, the Elbe, and the Rhine were frozen
more completely than the Seine. Carriages crossed the Meuse; at the
Hague and at Rotterdam fairs were held upon the stream. A regiment
of dragoons, starting from Mayence, crossed the Rhine upon the ice instead of by the bridge at Cassel, which it had been found necessary to
raise.

1812–13.—This winter will ever be remembered for the terrible disasters which attended the retreat of the French army through Russia,
after the capture and conflagration of Moscow. The frost set in early
all over Europe. The retreat of the army began on the 18th of November; Napoleon left the capital of the Muscovite Empire on the
19th, and the evacuation of the city was complete on the 23d. The
army marched toward Smolensk, the snow falling without intermission. The cold became very intense after the 7th of November, and
on the 9th the thermometer marked $5°\cdot0$ (Fahr.). On the 17th the temperature fell to $-15°\cdot2$ (Fahr.) according to Larry, who had a thermometer suspended from his button-hole. The army corps commanded by
Ney escaped from the Russian troops, by whom it was surrounded, according to Arago, by crossing the Dnieper, which was frozen over, on
the night of the 18th–19th of November. The day before some Russian
troops, with their artillery, had crossed the Dwina upon the ice. The
cold diminished, and a thaw began on the 24th, but did not last; so that
from the 26th to the 29th, during the fatal passage of the Berezina,
the water contained numerous blocks of ice without offering a passage
at any part to the troops. The cold soon set in again with fresh intensity; the thermometer fell again to $-13°\cdot0$ (Fahr.) on the 30th of
November; to $-22°$ (Fahr.) on December 3d; and to $-35°$ on the 6th

at Molodeczno, the day after Napoleon left Smorgoni, and published the bulletin (No. 29) which informed France of a part of the disasters incurred during this terrible campaign.

1819–20.—This was also a very severe winter throughout Europe, although the extreme cold did not last so long. In Paris there were forty-seven days' frost, nineteen of which were consecutive, from the 30th of December, 1818, to January 17th, 1819. The minimum temperature occurred on the 11th of January, viz., −14°·3. The Seine was entirely frozen over from the 12th to the 19th of January. The Saône, the Rhône, the Rhine, the Danube, the Garonne, the Thames, the Lagoons of Venice, and the Sound, were so far frozen that it was possible to walk upon the ice. The lowest temperatures observed in different towns are as follows:

St. Petersburg, January 18	−25·6 degrees.
Berlin, January 10	−11·9 "
Maestricht, January 10	− 2·7 "
Strasbourg, January 15	− 1·8 "
Commercy (Meuse), January 12	− 1·8 "
Marseilles, January 12	+ 0·5 "
Metz, January 10	+ 2·7 "
Mons, January 11–15	+ 3·9 "
Paris, January 11	+ 6·3 "

In France the intensity of the cold was heralded by the passage along the coast of the Pas de Calais of a great number of birds coming from the farthest regions of the north, by wild swans and ducks of variegated plumage. Several travelers perished of cold; among others a farmer near Arras, a gamekeeper near Nogent (Haute Marne), a man and a woman in the Côte d'Or, two travelers at Breuil, on the Meuse, a woman and a child on the road from Etain to Verdun, six persons near Château Salins (Meurthe), and two little Savoyards on the road from Clermont to Châlons-sur-Saône. In the experiments made at the Metz School of Artillery on the 10th of January, to ascertain how iron resisted low temperatures, several soldiers had their hands or their ears frozen.

1829–30.—This was the earliest and longest winter of the first part of the nineteenth century; its duration was especially injurious to agriculture in southern countries. The cold, without being extremely rigorous, extended all over Europe; a great number of rivers were congealed, and the thaw was accompanied by disastrous inundations; many men and animals perished, and field labor was for a long time interrupted. The following are the principal temperatures observed:

St. Petersburg, December 19............................	−26·5 degrees.
Mulhouse, February 3....................................	−18·6 "
Bâle, February 3...	−16·6 "
Nancy, February 3.......................................	−15·3 "
Épinal, February 3.......................................	−14·1 "
Aurillac, December 27...................................	−10·5 "
Strasbourg, February 3	−10·1 "
Berlin, December 23.....................................	− 5·8 "
Metz, January 31...	− 4·9 "
Pau, December 27..	+ 0·5 "
Paris, January 17...	+ 1·0 "

In Switzerland the winter was severe in the great altitudes. At Freiburg there were one hundred and eighteen days' frost, sixty-nine of which were consecutive, and the minimum was −1°·3 Fahr. In the plains, at Yverdun, among other places, the effects of radiation were felt very intensely: the thermometer fell in a few hours from +14° to −4°. The snow termed *polar* snow, the crystallization of which is very close, and which is peculiar to very low temperatures, also fell there.

The length of time during which the Seine was frozen and its subsequent thaw, excited public curiosity to the highest degree. The river remained frozen from December 28th to the 26th of January; that is, for twenty-nine days, on the first occasion. It was frozen over afterward from the 5th to the 10th of February, making in all thirty-four days, or as long as was the case in 1763. It was frozen over at Havre from the 27th of December, and a fair was established upon the ice at Rouen on the 18th of January. On the 25th of January, after six days' thaw, the ice from Corbeil and Melun blocked up the bridge at Choisy, forming a wall sixteen and a half feet high.

1840–1841.—During this winter there were fifty-nine days' frost, twenty-seven of them consecutive in Paris. The cold began on the 5th of December, and lasted, with an intermission from the 1st to the 3d of January, until the 10th of that latter month. There was another frost from the 30th of January to the 10th of February. On the 3d of February the thermometer still marked 16°·2 Fahr. From the 16th of December the Seine was full of blocks of ice, and one of the arches of the Pont Royal was obstructed. Upon the evening of the same day the current was stopped at the Pont d'Austerlitz, and was frozen from Pont Marie to Charenton. The next day it was frozen at the bridge of Notre-Dame, and on the 18th people crossed from Bercy to the railway station. In several places the blocks of ice forced together were as much as seven to eight feet thick. On the 15th of December the ashes of Napoleon, brought back from St. Helena, entered Paris by the Arc de Triomphe. The thermometer, in places exposed to nocturnal radiation, had that day marked +6°·8 Fahr. An immense crowd, the National Guard of Paris and its suburbs, and numerous regiments, lined the Champs Élysées from the early morning until two in the afternoon. Every one suffered severely from the cold. Soldiers and workmen, hoping to obtain warmth by drinking brandy, were seized by the cold, and dropped down dead of congestion. Several persons perished, victims of their curiosity: having climbed up into the trees to see the procession, their extremities, benumbed by the cold, failed to support them, and they were killed by the fall. I append some of the temperatures noted during this winter.

Mount St. Bernard, January 22	−9·9 degrees.
Geneva, January 10	+0·0 "
Metz, December 17	+4·5 "
Paris, December 17	+8·2 "
Paris, January 8	+8·4 "

1853–54.—This was a severe winter in the temperate regions of Europe. It lasted from November to March, and caused several rivers to be frozen over. The cold was intense in many places, yet it proved rather beneficial to agriculture than otherwise.

The principal temperatures were as follows:

Clermont, December 26	− 4·0 degrees.
Châlons sur Marne, December 26	− 4·0 "
Lille, December 26	− 0·4 "
Kehl, December 26	+ 0·3 "
Metz, December 27	+ 0·5 "
Brussels, December 26	+ 3·0 "
Lyons, December 30	+ 5·7 "
Paris, December 30	+ 6·8 "
Bordeaux, December 30	+14·0 "

The next winter was also severe, especially in Southern Russia, Denmark, England, and France, and was of unusual length. The frosts commenced as early as October in the east of France, and lasted until the 28th of April. The Loire was blocked with ice on the 17th of January, and its course was arrested the next day. The Seine, though full of blocks of ice on the 19th of January, was not frozen over. The Rhône was impeded on the 20th, and the Saône on the same day. The

Rhine was completely frozen over at Manheim on the 24th, and people crossed it on foot. The appended table gives the lowest temperatures:

Vendôme, January 20	− 0·4 degrees.
Clermont, January 21	+ 1·4 "
Brussels, February 2	+ 1·9 "
Turin, January 24	+ 2·3 "
Metz, January 29	+ 3·2 "
Strasbourg, January 29	+ 3·2 "
Montpellier, January 21	+ 3·2 "
Lille, February 2	+ 7·2 "
Paris, January 21	+11·7 "
Toulouse, January 20	+12·7 "

The winter of 1857–1858 was the type of the average severity of a winter in the temperate zone. The Seine contained blocks of ice on the 5th of January, and the small arm of the stream by the *Cité* was covered with ice on the 6th. The Loire, the Cher, the Nièvre, the Rhône, the Saône, and the Dordogne were stopped in several places. The Danube and the Russian ports in the Black Sea were frozen in January. The lowest temperatures were:

Le Puy, January 25	+ 6·1 degrees.
Clermont, January 7	+ 6·8 "
Bourg, January 29	+ 9·5 "
Vendôme, January 6	+12·2 "
Lille, January 7	−14·0 "
Paris, January 7	−15·8 "

The winter of 1864–1865 was more severe. The Seine was frozen over at Paris, and people crossed it by the Pont des Arts. The extreme temperatures were:

Haparanda, February 7	−28·1 degrees.
St. Petersburg, February 9	−19·8 "
Riga, February 4	−14·4 "
Berne, February 14	+ 5·0 "
Dunkirk, February 15	+10·4 "
Strasbourg, February 11	+12·2 "

Lastly, the winter of 1870–1871 will also be classed among severe winters, because of the extreme cold in December and January (notwithstanding the mild weather of February), and also because of the fatal influence which the cold exercised upon the public health at the close of the war with Germany. The great equatorial current, which generally extends to Norway, stopped this year at Spain and Portugal, the prevailing wind being from the north. On the 5th of December there was a temperature of 21° Fahr.; and on the 8th, at Montpellier, the thermometer stood at 17°·6 Fahr. A second period of cold set in on the 22d of December, lasting until the 5th of January. In Paris the Seine was blocked with ice, and seemed likely to become frozen over. On the 24th there were 21°·6 of frost; and at Montpellier, on the 31st, 28°·8. It is well known that many of the outposts around Paris, and several of the wounded who had been lying for fifteen hours upon the field, were found frozen to death. From the 9th to the 15th of January a third period of cold set in, the thermometer marking +17°·6 Fahr. at Paris, and +8°·6 Fahr. at Montpellier. The most curious fact was that the cold was greater in the south than in the north of France. At Brussels the minima were +11°·1 in December, and +8°·2 Fahr. in January. There were forty days' frost at Montpellier, forty-two in Paris, and forty-seven at Brussels, during these two months. Finally, the winter average (December, January, and February) is 35°·2 in Paris, whereas the general average is 37°·9. In the north of Europe this was also a very hard winter, though the cold set in at a different time from what it did in France. There were 40° of frost at Copenhagen on the 12th of February, or the temperature was −7°·6 Fahr. By the documents which M. Renou has furnished me with for France, I discover a minimum of −9°·4 Fahr. at Périgueux, and of −13° Fahr. at Moulins! I find by the documents supplied me by Mr. Glaisher, that he also considers the winter of 1870–1871 as appertaining to the class of winters memorable for their severity.

For the Seine to freeze in Paris there must be a temperature about +16° Fahr., lasting several days. We have seen above how this is brought about. Since the beginning of the century it has been entirely frozen over eleven times: January, 1803; December, 1812; January, 1820, 1821, 1823, 1829, 1830. and 1838; in December, 1840; in January, 1854; and in January, 1865.

M. Renou has noticed that the severest winters seem to recur about every forty years: 1709, 1749 (less severe), 1789, 1830, 1870.

The following are the lowest temperatures observed in France since they have been carefully noted by the thermometer. They are inscribed, like the previous list of the highest temperatures, in geographical order from north to south. I have taken all those that have reached 20° of frost, and only those, except in the case of Paris, where there are several means of comparison.

Places.	Latitude.	Longi-tude.	Alti-tude.	Date.	Minimum.
	Deg.	Deg.	Feet.		Deg.
Douai....................	50·22	0·44	78	January 28, 1776	− 5·1
Arras....................	50·17	0·26	219	December 30, 1788	−10·1
Amiens...................	49·53	0·02	118	February 27, 1776	− 4·5
Saint-Quentin.............	49·50	0·57	341	January 28, 1776	− 5·1
Vervins..................	49·55	1·34	574	December 31, 1788	− 7·4
Montdidier	49·39	0·14	324	January 29, 1776	− 8·5
Rouen...................	49·26	1·15	121	December 30, 1788	− 7·2
Clermont (Oise)............	49·23	0·05	282	" 26, 1853	− 4·0
Les Mesneux...............	49·13	1·37	278	January 19, 1855	− 4·4
Metz....................	49·07	3·50	597	" 31, 1830	− 4·9
Montmorency..............	49·00	0·02	690	" 1795	− 4·0
Châlons-sur-Marne..........	48·57	2·01	269	December, 1788 / " 26, 1853	− 5·1 / − 4·0
Goersdorff................	48·57	5·26	747	" 27, 1853	− 7·2
Paris....................	48·50	0·00	213	January 25, 1795 / " 13, 1709 / December 31, 1788 / February 6, 1665 / January 22, 1716 / " 29, 1776 / December 30, 1783 / January 20, 1838 / " 17, 1830	−10·3 / − 9·6 / − 7·2 / − 6·2 / − 3·5 / − 2·4 / − 2·2 / + 1·0
Haguenau	48·48	5·25	213	December, 1788	− 6·7
L'Aigle..................	48·43	2·00	446	" 30, 1788	− 7·2
Nancy	48·42	3·51	656	February 1, 1776 / " 3, 1830	− 8·7 / −15·3
Strasbourg................	48·35 N.	5·25	472	December 31, 1788 / February 3, 1830	−15·3 / −10·1
Étampes..................	48·26	0·10	416	December 31, 1788	− 7·4
Mayenne.................	48·18	2·57	334	" 1788	− 4·0
Troyes...................	48·18	1·45	360	" 31, 1788	− 9·4
Saint-Dié................	48·17	4·37	1125	" 31, 1788	−14·8
Épinal...................	48·10	4·07	1118	February 3, 1830	−14·1
Colmar..................	48·05	5·01	639	December 19, 1788	−14·1
Neufbrissac..............	48·00	5·00	649	" 18, 1788	−22·4
Orleans..................	47·54	0·26	403	" 31, 1788	− 8·5
Mulhouse.................	47·49	5·00	751	January, 1784 / February 3, 1830	− 8·3 / −18·6
Beaugency................	47·46	0·46	328	December 31, 1788	− 8·5
Tours...................	47·24	1·39	180	" 31, 1788	−13·0
Dijon...................	47·19	2·42	807	February 1, 1776	− 4·0
Chinon..................	47·10	2·06	268	December, 1788	−10·8
Bourges.................	47·05	0·04	511	January, 1789	− 9·4
Pontarlier...............	46·54	4·01	2749	December 31, 1788 / " 14, 1846	−10·8 / −24·3
Lons-le-Saulnier...........	46·40	3·13	846	" 31, 1788 / January 16, 1838	−11·2 / −12·1
Poitiers..................	46·35	1·60	387	December, 1788	− 4·0
Moulins	46·34	1·00	744	December 31, 1788 / " 22, 1870	− 8·7 / −13·0
Roanne..................	46·02	1·44	938	" 31, 1788	− 5·1
Limoges.................	45·50	1·05	941	" 1788	−10·7
Lyons...................	45·46	2·29	967	" 31, 1788 / January 16, 1838	− 7·4 / − 4·0
Grande-Chartreuse..........	45·48	3·23	6660	December 30, 1788	−15·3
Grenoble.................	45·11	3·24	698	February, 1776	− 6·9
Périgueux................	45·11	1·36	321	December, 1870	− 9·4
Aurillac.................	44·56	0·06	2040	December 27, 1829	−10·5

The greatest cold yet experienced has been −24° in France; −5° in

England; —12° in Holland and Belgium; —67° in Denmark, Sweden, and Norway; —46° in Russia; —32° in Germany; zero in Italy; —10° in Spain and Portugal. As to other countries, not European, more observations must be taken before one can speak with any degree of certainty upon the point. It is, nevertheless, certain that at Fort Reliance. in British North America, there have been —70° of cold, and at Semipalatinsk —76°. Quicksilver freezes at —39°. There are inhabited points of the globe where it remains congealed for several months of the year—on Melville Island, for instance. Captain Parry, moreover, asserts that a person sufficiently wrapped up may safely expose himself to the open air in —50°, or 82° below freezing-point of water—that is, if there is no wind. In this latter event the skin is rapidly affected. Frozen mercury looks like lead, but it is not so hard, is more fragile, and less coherent. If touched it burns like hot iron. Small statuettes can be made with it which melt when the temperature is higher than —39°.

Such are the greatest frosts that have been experienced. If they are compared with the extremes of heat noticed in the previous chapter (165° upon the surface of the soil of Africa), it will be seen that the extremes of temperature upon the globe may attain a scale of nearly 240 degrees.

CHAPTER VI.

CLIMATE: DISTRIBUTION OF TEMPERATURE OVER THE GLOBE — ISO-
THERMAL LINES — THE EQUATOR — THE TROPICS — THE TEMPERATE
REGIONS — THE POLES — THE CLIMATE OF FRANCE.

IF two lines parallel to the equator be traced upon the globe, at the distance of 23° 28′ in each hemisphere, they will mark two circles between which the sun is seen to pass across the zenith at certain epochs of the year; these are the tropics. That of the northern hemisphere is known as the Tropic of Cancer, because, during the summer solstice, the sun passes at its zenith and is in the zodiacal sign of Cancer. That of the southern hemisphere is known as the Tropic of Capricorn, because the sun passes at its zenith during the winter solstice in the zodiacal sign of Capricorn. The zone included between these two circles is the hottest part of the earth, inasmuch as it comprises the places over which the sun rises to its greatest altitudes; it is termed the torrid or intertropical zone.

If two other circles, distant 23° 28′ from the pole, or at 66° 32′ from the equator, be drawn upon this same terrestrial globe, they will mark the points below which the sun may remain for several days together, and above which it remains at its least altitudes; these are the *polar* circles. During one half of the year the sun rises spirally above them to the height of 23° 28′, and during the other half descends below them to the same amount. Between these two zones is the *temperate zone*, in respect to which the sun rises and sets each day, without ever reaching so high as the zenith, attaining an increasing elevation, and giving a greater length of day, so far as our hemisphere is concerned, from the solstice of December to the solstice of June, corresponding with which there is an inverse rate of progress in the other hemisphere.

The two glacial zones form 82 thousandths of the surface of the earth; the two temperate zones represent 520 thousandths; and, finally, the torrid zone, composed of the two regions comprised between the tropics and the equator, is 398 thousandths of the whole surface of our planet.

The length of the longest and the shortest days, in the different latitudes of our hemisphere, from the equator to the polar circles, gives the following scale:

Latitudes.	Names of Places.	Longest Day.	Shortest Day.
Deg.		hrs. min.	hrs. min.
0	(Quito)....................................	12 0	12 0
5	(Bogota)	12 17	11 43
10	(Gondar, Madras)........................	12 35	11 25
15	(St. Louis)...............................	12 53	11 7
20	(Mexico, Bombay)	13 13	10 47
25	(Canton).................................	13 34	10 26
30	(Cairo)...................................	13 56	10 4
35	(Algiers).................................	14 22	9 38
40	(Madrid, Naples)	14 51	9 9
45	(Bordeaux, Turin)........................	15 26	8 34
50	(Dieppe, Frankfort).......................	16 9	7 51
55	(Edinburgh, Copenhagen).................	17 7	6 53
60	(St. Petersburg, Christiania).............	18 30	5 30
65	(Archangel)...............................	21 9	2 51
66·32	(Polar Circle)............................	24 0	0 0

It is of course the same in the southern hemisphere. Beyond the polar circles the length of the day varies from 0 to 24 hours in that part of the year during which the sun rises or sets. The number of days during which the sun is constantly above or constantly below the horizon in different latitudes, from 66° 32' to 90°, is given in the following table, the phenomena being just the reverse for the two glacial zones:

Latitudes.	The Sun does not set in the Northern Hemisphere nor rise in the Southern during	The Sun does not rise in the Southern Hemisphere nor set in the Northern during
Deg.	Days.	Days.
66·32	1	1
70	65	60
75	103	97
80	134	127
85	161	153
90	186	179

In this theory of the climates we suppose the sun to be reduced to a point at its centre; we have, moreover, neglected to take into account the phenomena of starlight produced by the refraction of light. As the diameter of the sun is about 32', the latitude at which it would disappear altogether must be placed at 16' farther back. The refraction, too, raising it by 33' at the horizon, the absolute polar circles must be also placed that distance farther back. Lastly, night is not complete until the sun has descended to about 18° below the horizon; and this circumstance must also be taken into account, whence it results that near the poles day is hardly ever extinct, and night, in the absolute sense of the term, almost unknown.

The seasons are exactly opposite in the two hemispheres, as we have already stated; they are indeed neither more nor less than the intervals of time which the earth takes to traverse the four parts of its orbit comprised between the equinoxes and the solstices. In consequence of the eccentricity of the earth's orbit, and by virtue of the *law of superficies*, the lengths of the seasons differ; they are represented by the following figures, which show that the sun is, in the course of each year, about eight days longer in the northern hemisphere than in the southern hemisphere:

	Days.	Hrs.	Min.
Autumn (September 22 to December 21)	89	18	35
Winter (December 21 to March 21).................	89	0	2
Sojourn of the sun in the southern hemisphere...	178	18	37
Spring (March 21 to June 21).......................	92	20	59
Summer (June 21 to September 22)................	93	14	13
Sojourn of the sun in the northern hemisphere...	186	11	12

The sun being, in fact, the source of heat for the surface of the earth, it follows that the hottest countries are those over which it remains the longest, and upon which it darts its rays the most vertically, that is, the regions situated along the equator, and upon each side of it, as far as the tropics. Thus, these warm regions are known by the generic appellation of "the torrid zones." In proportion as one recedes from the equator and approaches the poles, it is seen that the sun attains a lesser elevation, and that for six months the nights are longer than the days; these are the temperate regions, where the seasons lend a far greater variety to the products of nature, but where the mean of the annual temperature gradually diminishes according to the diminution in the apparent height of the sun at noon. Lastly, when we pass beyond 66° of latitude, the glacial polar region is reached over which the sun, even on the finest days, scarcely rises sufficiently high to melt the eternal ice subsisting in these regions.

It is superfluous for me to mention that the south pole is cold like the north pole, notwithstanding the idea attaching to this direction. Some few poets still talk of traveling

"Du pôle *brûlant* jusqu'au pôle *nord;*"

but such metaphors are no longer admissible. The equator is to the south of our position, and the winds blowing from there toward us are hot. The equator is to the north of the other hemisphere, and the winds which reach it from the equator are also hot, though they come from the north. In respect to meteorological direction, as in regard to the seasons, the inhabitants of Australia, the Cape of Good Hope, Cape Horn, Buenos Ayres, and Santiago feel and speak just contrary to what we do.

Latitude—that is, the angle at which the solar rays reach the surface of the ground—being the great cause of the succession of climates from the equator to the poles, the diminution would be progressive and regular if the earth was a globe, perfectly regular in shape, instead of being divided into earth and water, and broken by mountains, table-lands, and valleys. The quantity of heat, estimated at 1000, for instance, at the equator, would follow a constantly descending scale, marking 923 at each of the tropics, 720 at the latitude of Paris, and 500 at the polar circle. But the earth is not a smooth and undisturbed sphere, and revolutions more or less harmonious are constantly occurring.

We shall see in this work that the atmosphere is in a perpetual state of circulation, and that there are general winds which periodically traverse different countries of the globe. These regular currents modify the normal distribution of climates. Thus the trade-winds, which establish a double current between the equator and the poles, temper at once the cold of the high latitudes over which they pass and the heat of the tropical regions, heating the former and cooling the latter.

A second cause is added to this by which the temperature along the same circles of latitude is varied. The globe is divided into oceans and continents. Water has a greater capacity for heat than land, whence it results that the sea is cooler than the land in summer, and warmer in winter. The winds which blow from the sea prevent the coast-line from being as cold as the country farther inland. As the south-west wind is that which blows oftenest, the western coasts of Spain, France, Scotland, and Norway are warmer than the inland country in the same latitudes. The great marine current known as the Gulf Stream adds still further to this modifying cause.

Water becomes less readily heated upon the surface than earthy matter, because the latter has a specific heat much below that of water; so that the quantity of solar heat required to raise its temperature by 10°, for instance, is much less than that which would raise the temperature of a liquid surface the same number of degrees.

It must, furthermore, be mentioned that the solar rays which become absorbed in a very thin layer of earth penetrate, at least in part, to a very considerable depth in the water; that at sea, especially, they do not become totally extinguished until they have reached a depth of more than three hundred yards; so that the heat arising from absorption, instead of being concentrated upon the surface, is spread over a great mass of water, and must be the less in proportion as this mass is larger.

Evaporation, which is, as we have seen, a very great cause of cold, is the greater according as this phenomenon occurs upon a larger scale. And, where the liquid mass continually furnishes the means of evaporation, there exists a cause of cold which does not exist at all, or in a much less degree, upon dry land. From these three causes (specific heat, diathermacy, and evaporation) it follows that water, and the atmosphere which is in contact with it, must be less heated than continental regions situated in the same latitude. In winter, on the other hand, it is warmer, and that for a reason which it is easy to comprehend.

I have already stated that the superficial moleculæ, rendered cold by their radiation toward the cold regions of space, are precipitated toward the bottom by reason of the excess of their specific weight; consequently the surface of the sea must preserve a higher temperature than that of the surface of continents, since in this case the superficial moleculæ that have become cold do not plunge into the ground.

These consequences, deduced from a minute examination of the action of the solar rays upon a liquid and a continental surface, are confirmed by observation.

Thus, at Bordeaux, the mean winter temperature is 42°·8: whereas, in the same latitude, the temperature of the Atlantic never falls below 51°·3.

In latitude 50° the ocean has never been known to be less than 48°·2.

The mass of observations collected show that, in the northern hemisphere and in the temperate zone, the mean temperature of an island situated in the midst of the Atlantic would be higher than the mean temperature of a spot similarly situated upon the main-land—that the winter would be warmer and the summer cooler. This has been especially remarked in the Island of Madeira.

The sea serves to equalize the temperatures. Hence there is an important difference between the climate of islands or coast-lands peculiar to all continents that abound in gulfs and peninsulas, and the climate of the interior of a great and compact mass of dry land. In the interior of Asia, at Tobolsk, Barnaul-upon-Obi, and Irkoutsk, the summer is the same as at Berlin, Münster, and Cherbourg; but these summers are followed by winters when the temperature is as low as $-0°\cdot4$ or $-4°$. During the summer months the thermometer will remain for weeks together at $86°$ or $88°$. These *continental climates* have been very appropriately termed *excessive* by Buffon, and the inhabitants of countries in which they prevail seem to be condemned, like the spirits alluded to by Dante,

"A sofferir tormenti e caldi e gieli."

The climate of Ireland, of Jersey and Guernsey, of the peninsula of Brittany, of the coasts of Normandy, and the south of England, countries in which the winters are mild and the summers cool, contrasts very strikingly with the *continental* climate of the interior of Eastern Europe. In the north-east of Ireland ($54°\cdot56$), in the same latitude as Königsberg, the myrtle grows in the open ground just as it does in Portugal. The temperature of the month of August in Hungary is $69°\cdot8$; in Dublin (upon the same isothermal line of $49°$) it is 61 degrees at most. The mean temperature of winter descends to $36°\cdot3$ at Buda. In Dublin, where the annual temperature is only $49°$, that of the winter is, nevertheless, $7°\cdot7$ above the freezing-point, or nearly four degrees higher than at Milan, Pavia, Padua, and all Lombardy, where the mean heat of the year reaches $55°$. In the Orkney Islands, at Stromness, a little to the south of Stockholm (there is not one degree difference in latitude), the mean winter temperature is $7°$, or higher than that of London or Paris. Stranger still, the inland waters of the Färoe Islands never freeze, situated in $62°$ of north latitude, beneath the mild influences of the west wind and the sea. Upon the coast of Devonshire, one part of which has been termed the Montpellier of the North, because of the mildness of its climate, the Agave Mexicana has been known to flower when planted in the open air, and orange-trees trained upon a wall to bear fruit, though only scantily protected by a thin matting. There, as at Penzance, Gosport, Cherbourg, and the coast of Normandy, the mean temperature of winter is $42°$, being but $18°\cdot5$ below that of Montpellier and Florence.

The mean annual temperature of London, as deduced by Glaisher from one hundred years' observations (1771–1870), is $48°\cdot5$. The mean summer temperature is $60°\cdot2$, and that of winter $38°$. The winter, therefore, is warmer at London than at Paris, and the summer and the year cooler. Although Cherbourg is one degree of latitude north of

Paris, its mean temperature is, notwithstanding, higher, being 52°·3, while that of Paris is only 51°·3. The difference between the winter climates of the two towns is much greater, since the winter mean is 43°·7 at Cherbourg, and 37°·8 at Paris. Thus fig·trees, laurels, and myrtles, which would perish in the neighborhood of Paris, are found to flourish in the former place. The enormous fig-trees which grow at Roscoff, in Brittany, are almost equal to those of Smyrna.

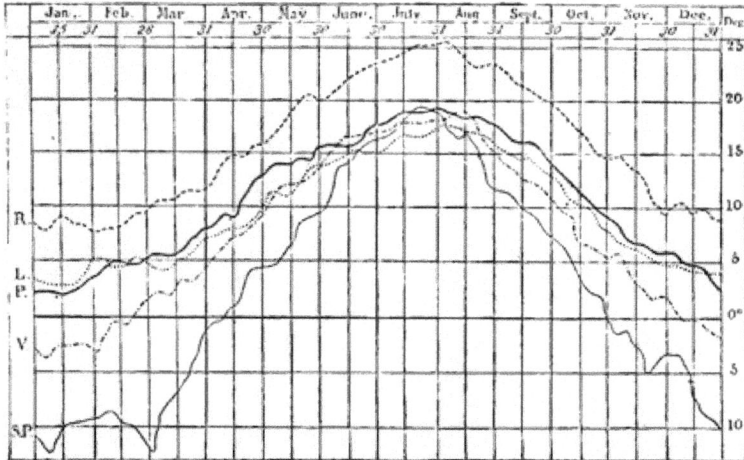

Fig. 64.—Comparative temperatures of the European capitals of Rome, London, Paris, Vienna, St. Petersburg.

These comparisons are sufficient evidence as to how the same mean annual temperature may be distributed in many different proportions over the various seasons, and how great an influence these diverse modes of distribution of heat may exercise in the course of the year upon vegetation, agriculture, the ripening of fruit, and the comfort of man.

The same relations of climate, which are remarked as existing between the peninsula of Brittany and the rest of France, the mass of which is more compact, and where the summers are hotter, and the winters colder, are reproduced to a certain extent as between Europe and the continent of Asia. Europe owes the mildness of its climate to its abundantly indented configuration, to the ocean which washes its western shores, to the sea which separates it from the polar regions, and, above all, to the existence and to the geographical situation of the African continent, the intertropical regions of which radiate excessively and cause the ascent of an immense current of hot air, whereas the regions situated to the south of Asia are, for the most part, oceanic.

Europe would become colder if Africa were submerged—if the fabled Atlantides, emerging from the ocean, were to unite Europe to America —if the warm waters of the Gulf Stream did not flow into the northern seas, or if a new land, upheaved by volcanic agency, were to become inserted between the Scandinavian Peninsula and Spitzbergen. In proportion as we advance from west to east, along the same latitude, in France, Germany, Poland, Russia, and as far as the Ural Mountains, we find that the mean annual temperatures follow a uniformly descending scale. But as we penetrate inland, the form of the continent becomes more and more compact, its breadth increases, the influence of the sea diminishes, and that of the west wind becomes less perceptible. Therein lies the chief cause of the progressive decline in the temperature.

The mean temperature of the equator is 81°·5. Owing to the causes which I have specified, and to the absence of vegetation, that of inland Africa is 86 degrees with the thermometer placed in the shade and protected from hot winds; but there are points at which the action of the burning breeze, and the absence of clouds, combine in producing an intolerable degree of heat. Thus, in the interior of Abyssinia, and in the neighborhood of the Red Sea, it is by no means rare to meet with a summer temperature of 118 to 122 degrees in the shade. That of the soil is higher still. In the afternoon the valleys of Abyssinia are regular furnaces; M. d'Abbadie having observed the temperature of the soil at 160 degrees nearly, while Colonels Ferret and Galinier met with a temperature of 167 degrees. The air is stagnant in the midst of all this heat, and there is no refreshing breeze. The air in the depths of these ravines is often mephitic, and to repose therein after or before the rainy season is fatal. It is necessary at that period to travel by night, as plains have to be crossed which afford no place of shelter.

"Sometimes in crossing these deserts," says M. d'Abbadie, "one is assailed by the *karif*, a sort of aërial hurricane, a phantom of burning dust which appears upon the horizon, and seems to grow in size as it approaches. The wind which wafts it blows like a hurricane; men and animals are obliged to turn their backs, and are enveloped in a dry and black cloud, which covers them as with a hideous mantle. Fortunately this storm of fire lasts only a few minutes, and after it has passed the intense heat which is peculiar to these regions is felt as a relief.

"At other times one is overtaken by the simoom (the *poison*), a wind of flame which begins to blow without any premonitory sign. The camel is then seen to lay his head upon the ground, seeking coolness from it, though it is itself like a furnace. The hardiest of the natives are struck down in despair. The prostration is so great, that I was myself unable to lift a small thermometer placed within my reach, in order to ascertain, at all events, the temperature of this remarkable wind. Its duration was five minutes: it causes death when it continues for a quarter of an hour.

"If one happens to meet with a small stream in these regions, it soon disappears, absorbed by the sand. These miniature oases, composed of a few trees and some grass, are very rare.

"These same valleys are the theatre of a very extraordinary phenomenon; a sudden irruption of water, which at certain periods of the year causes inundations, to which those occurring in European countries are trifling. And, strange to relate, they take place during the summer.

"One may be traveling with a full sense of security, when a native, hearing a strange and distant noise, commences to shout at the top of his voice, 'The torrent!' and climbs as fast as possible to the nearest elevated point. In a few seconds the hollow of the valley is hidden by a deep body of water, which carries with it trees, rocks, and even wild animals. These torrents, formed in an instant, disappear in the course of the same day, and leave no trace of their passage, save débris and muddy deposits.

"How is this strange phenomenon to be explained? The barrenness of the mountains accounts for these sudden down-pours. From the hollow of the ravine in which the traveler is journeying, he is unable to see the narrow clouds which suddenly dissolve into water, with an abundance unknown, save in tropical regions. There is very little soil, and still fewer roots of trees, to absorb this sudden rain, which consequently runs off at once, leaping from rock to rock, as down a roof, flowing from each minor valley into the principal ravine, and there forming a stream which, though short-lived, is of mighty dimensions."

M. d'Abbadie further relates how he was just too late on one occasion to witness, in all its grandeur, one of these sudden inundations. He found a native, who was regarding the wet ground with the air of one who had been stunned. "Peace be with you," said M. d'Abbadie; "what news? Where are your arms? Surely you can not be without your lance and shield?" "Peace be with you," replied the African,

"the torrent has carried off my lance, my shield, my camel, and all my
fortune, my wife and my children."

It will thus be seen that various causes influence the climates of different countries of the globe; and it would involve great errors were
we to take into account the distance from the equator only in calculating the decrease of the temperature toward the poles. We have seen
that the average temperature of the equator is 81°·5; the mean temperature of Paris is 51°·3; that of regions within the polar circle about 5°.

To establish a correct table of the distribution of temperature over
the surface of the earth, Humboldt marked upon a map all the points
at which reliable thermometrical observations had been taken, noted
the degrees recorded, and then traced lines passing respectively through
all the places where the mean temperature was the same. These he
termed isothermal lines (from ἴσος, equal, and θερμός, heat). During
the fifty years since, observations have been multiplied and the maps
made more perfect.

We see in diagrams of isothermal lines, or lines of equal temperature
running along the western shores of Europe, that the line of 50°, for instance, touches the fortieth degree of latitude south-west of New York,
and reaches as far as 55° near England; so that Dublin and London
have nearly the same mean temperature as New York, although they
are situated much farther north; the same temperature then falls again
toward the south, passing to Vienna, Astrakhan, and Pekin, and descending even below the fortieth parallel of latitude. The greatest heat
line, called the thermic equator, is nearly entirely to the north of the
equator, and its temperature varies, according to situation, from 81° to
86°. Within the polar regions the mean temperature of different places
decreases to as much as 1°, which has, as yet, scarcely been traced, in
consequence of the difficulty of traveling in these inhospitable regions.

Humboldt has pointed out that, notwithstanding these great differences, the mean temperature decreases almost uniformly at the rate of
nearly a degree of the thermometer to each degree of latitude. But as,
on the other hand, the heat diminishes by 1° for an increase of height
of about three hundred feet, it follows that an elevation of about one
hundred yards produces the same effect upon the temperature of the
year as an approach of 1° of latitude toward the north. Thus, the mean
annual temperature of the monastery of Mount St. Bernard, situated at
a height of 8173 feet, in latitude 45° 50', is the same as that of low
ground in 75° 50' latitude. By studying the distribution of heat over

the surface of the globe, and by tracing a system of isothermal lines, Humboldt demonstrated the causes which raise the temperature of a particular spot, and those which lower it. The augmenting causes are as follows:

The proximity of the ocean on the west in the temperate zone. The configuration peculiar to continents which are cut up into numerous peninsulas. The Mediterranean, and the gulfs penetrating far inland. The direction, that is to say, the position of a country in respect to a sea free from ice, which extends beyond the polar circle, or in regard to a continent of considerable extent, situated upon the same meridian, at the equator, or at least in the interior of the tropical zone. The south-westerly direction of the prevailing winds in the case of the western fringe of a continent situated in the temperate zone, the chains of mountains acting as a rampart and a protection against the winds which blow from colder countries. The scarcity of pieces of water, the surface of which is covered with ice during the spring, and up to the beginning of summer. The absence of forests on a dry and sandy soil, the constant serenity of the sky during the summer months, and, lastly, the near neighborhood of a maritime current, whose waters are warmer than those of the surrounding ocean.

The decreasing causes are: the height above the level of the sea of a region which does not possess extensive table-land. The distance of the sea to the west and the south in our hemisphere. The compact shape of a continent, upon the coasts of which there are no bays; a great extent of land toward the pole, and toward the regions of eternal frost, except in the case of there being between the land and this region a sea that is free of ice during the winter; a geographical position such that the tropical regions in the same longitude are covered by the sea: in other words, the absence of any tropical land upon the meridian of the country whose climate is being studied; a chain of mountains which by its shape or direction prevents the access of warm winds, or indeed the presence of isolated peaks, because in both these cases currents of cold air make their way down the slopes; forests of great extent, for these prevent the solar rays from acting upon the soil; the leaves cause the evaporation of large quantities of water, by reason of their organic activity, and increase the superficies liable to be rendered cold by radiation. The forests act, therefore, in three ways: by their shade, by their evaporation, and by their radiation. The numerous pieces of water which, in the north, are regular receptacles of ice up to the middle of summer. A cloudy sky in summer, because it intercepts a portion of the sun's rays; a very clear sky in winter, because it facilitates the radiation of the heat.

To the general conditions of climates must be added the influence which local circumstances may have upon the state of the temperature. It is far more difficult than is generally supposed to ascertain exactly the temperature of a given spot upon the surface of the globe, and especially of an inhabited spot; for ten thermometers, identically the same, and carefully compared, will not mark the same point at the same moment in ten different streets of the same town. The principal remark to be made in reference to this is, that in consequence of the radiation of dwelling-houses, and on account of the obstacles which an agglomeration of buildings puts in the way of free circulation of air, the temperature of large towns is always less marked and higher than that of the country around them. Howard showed that the mean temperature of London exceeds by 2° that of the surrounding district.* The thermometers of the Paris Observatory are never so high as those in the heart of the city, but are higher than those placed in the open air in the field adjoining. Every one has noticed that it is cooler in summer and warmer in winter in the narrow streets of old Paris than upon the modern squares and boulevards. There is frequently a difference of several degrees. Even in the open country, at the same altitude and in the same frontage, the temperature differs according to the distance from woods. These latter act upon the temperature of the air, which is lower in than it is outside them. The mean maxima outside of woods are higher than inside. The mean temperature of summer is also higher in the former case than in the latter. These facts are clearly shown, according to MM. Becquerel, by the results of more than fourteen thousand observations made during the last few years.

The hours of maxima and minima are not the same inside the trees (even when they stand alone) as in the open air. They vary according to the kind and the diameter of the tree. The variations of temperature among the leaves are about the same as those in the surrounding air; in the young branches they occur more slowly, and so on to the trunk, where they are very gradual. I am excluding from the question the

[* In a paper published in the Philosophical Transactions, Part II., for 1850, I proved that those parts of London situated near the river Thames are somewhat warmer upon the whole year than the country, but that those parts of London which are situated at some distance from the river do not enjoy higher temperatures than those due to their latitudes.—ED.]

special heat of the trees which results from the various reactions which take place in their tissues, and that which they derive from liquids absorbed by the roots, because it is very slight as compared to that caused by solar or nocturnal radiation, as is proved by the maxima and minima of temperature which correspond with the maxima and minima of the air, though occurring at different hours of the day. This special heat of trees plays an important part in winter, by preventing a decline in temperature which would be fatal to them. In a tree twenty to twenty-four inches in diameter, the maximum temperature occurs in summer about 10 or 11 P.M.; in winter, toward 6 P.M.; whereas in the air it is at 2 or 3 P.M., according to the season. From this difference between the hours of maxima, it results, as experience has proved, that the temperature of the air may be lowered by some cause, such as the passage of a cloud, a change in the direction of the wind, etc., and yet rise in the interior of trees, because of the heat acquired by the outer surface, which is transmitted slowly to the inner portion of the tree, owing to its non-conductibility. The abundance of forests and moisture tend to lower the temperature, while clearing away timber and causing dryness of atmosphere produce a contrary effect; the difference in some cases for the mean temperature of the year being as much as four degrees nearly.

The numerous observations taken by MM. Becquerel in the Loiret have been particularized by them as follows:

1st, in summer, the mean temperatures of the air outside of woods are higher than they are inside.

2d, in winter, the reverse is the case.

3d, the difference between the mean annual temperature of the air at several miles from woodland and that inside a wood is about 3°.

The mean temperatures of the air in summer being about 2½° higher outside than they are inside a wood, and the reverse being the case in winter, it follows that the woodland climate is not so extreme as that of the open plain; it partakes, therefore, of the nature of a warm climate in respect to temperature. Local conditions modify more or less the general type of climates. The greatest local action is always exercised by unevenness of soil. The mountain chains divide the surface of the earth into large basins, into deep, hollow, or circular valleys. These valleys, often shut in, as between ramparts, individualize local climates (in Greece, for instance, and in part of Asia Minor), and place them in special conditions in reference to heat, moisture, the transparency of the air, and the frequency of winds and storms. After having studied the general condition of climates, and before coming to the poles in the course of this short geographical review, it is interesting to endeavor to form a correct idea of the extreme differences of temperature throughout the world.

In no place of the globe, and in no season, has the thermometer at an elevation of two or three yards above the soil, and sheltered, reached 135°.

In the open sea the temperature of the air has never exceeded 86°.

The most extreme degree of cold ever recorded upon a thermometer suspended in the air is 72° below zero.

The extreme difference in the temperatures of the atmospheric air is, therefore, 207°.

Comparing together the most extreme temperatures recorded, Arago constructed the remarkable table appended. The places are given according to their decrease in latitude.

Places.	Latitude.	Longitude.	Highest Temperature observed.	Lowest Temperature observed.	Difference.
	Deg.	Deg.	Deg.	Deg.	Deg.
Melville Island.............	74·47 N.	113·8	+ 60·1	−54·9	115·0
Port Félix...................	70·0	94·13	+ 70·0	−59·4	129·4
Nijnei-Kolymsk	68·32	158·34	+ 72·5	−65·0	137·5
Reikiavik	64·8	24·16	+ 68·9	− 4·0	92·9
Drontheim	63·26	8·3	+ 83·7	−10·7	74·4
Jakoutsk...................	62·2	127·23	+ 86·0	−72·4	158·4
Abo........................	60·27	19·57	+ 95·0	−32·8	127·8
St. Petersburg.............	59·56	27·58	+ 90·0	−37·8	127·8
Upsal.....................	59·52	15·18	+ 86·0	−25·1	111·1
Stockholm.................	59·20	15·43	+ 99·5	−28·7	128·2
Nijnei-Taguilsk	57·56	57·48	+ 95·0	−60·7	155·7
Kasan	55·48	46·47	+ 96·8	−40·0	136·8
Moscow....................	55·45	35·14	+ 94·1	−46·7	140·8
Hamburg	53·33	7·38	+ 95·0	−22·0	117·0
Berlin.....................	52·31	11·3	+ 102·7	−19·8	122·5
London....................	51·31	2·28	+ 95·0	+ 5·0	100·0
Dresden...................	51·4	11·24	+ 101·8	−25·8	127·6
Brussels	50·51	2·1	+ 95·0	− 6·0	101·0
Liége	50·39	3·11	+ 99·5	−11·9	111·4
Lille......................	50·39	0·4	+ 96·1	− 0·4	96·5
Dieppe....................	49·49	1·12	+ 92·3	− 3·6	95·9
Rouen	49·26	10·15	+ 100·4	− 7·2	107·6
Metz	49·7	3·50	+ 100·6	− 6·3	106·9
Paris.....................	48·50	0·0	+ 104·0	−10·3	114·3
Strasbourg................	48·35	5·2	+ 96·6	−15·3	111·9
Munich (1765 feet)	48·8	9·14	+ 95·0	−19·8	114·8
Bâle......................	47·33	5·15	+ 93·2	−35·5	128·7
Buda......................	47·29	16·43	+ 96·3	− 8·5	105·3
Tours.....................	47·24	1·39	+ 100·4	−13·0	113·4
Dijon.....................	47·19	2·42	+ 96·1	− 4·0	100·1
Quebec	46·49	73·36	+ 99·5	−40·0	139·5
Lausanne (1732 feet).....	46·31	4·18	+ 95·0	− 4·0	99·0
Geneva	46·12	3·49	+ 97·2	−13·5	110·7

Places.	Latitude.	Longitude.	Highest Temperature observed.	Lowest Temperature observed.	Difference.
	Deg.	Deg.	Deg.	Deg.	Deg.
St. Bernard (8172 feet)...	45·50	4·45	+ 67·4	−22·4	89·8
Gr.-Chartr. (6660 feet)...	45·18	3·23	+ 81·5	−15·3	96·8
Grenoble......................	45·11	3·24	+ 95·0	− 6·9	101·9
Turin...........................	45·4	5·21	+ 99·7	− 0·0	99·7
Le Puy (2493 feet)	45·0	1·33	+ 93·6	− 3·8	97·4
Orange	44·8	2·28	+ 106·5	− 0·4	106·9
Toulouse......................	43·37	0·54	+ 104·0	+ 4·3	99·7
Montpellier	43·37	1·32	+ 101·5	− 0·4	101·9
Marseilles....................	43·18	3·2	+ 98·4	+ 0·5	100·7
Perpignan	42·42	0·34	+ 101·5	+15·1	86·4
Rome...........................	41·54	10·7	+ 100·4	+19·6	80·8
Naples.........................	40·51	11·55	+ 104·0	+23·0	81·0
Pekin..........................	39·54	114·9	+ 109·6	+ 3·9	105·7
Lisbon.........................	38·42	11·29	+ 101·8	+24·7	77·1
Palermo.......................	38·7	11·1	+ 103·5	+32·0	71·5
Algiers........................	36·5	0·44	+ 99·5	+27·5	72·0
Havana........................	23·9	84·43	+ 90·1	+45·1	35·0
Vera-Cruz....................	19·12	98·29	+ 96·1	+60·8	35·3
Curaçao	12·6	71·16	+ 92·0	+75·0	17·0
Pulo-Penang Island.......	5·25	97·59	+ 90·0	+75·9	14·1
Quito (9540 feet)..........	0·14 S.	81·5	+ 71·6	+42·8	28·8
St. Louis de Maranam,....	2·31	46·36	+ 91·9	+75·0	16·9
Isle of Bourbon.............	20·52	53·10	+ 99·5	+60·8	38·7

Generally speaking, the differences between the highest and the lowest temperatures are less the farther one travels from the pole toward the equator.

Let us now deal with the limits of climates, the extremity of the world, the icy regions of the poles.

In the neighborhood of the Polar Circle the sea becomes frozen, and assumes a special character. This phenomenon seems to increase as the water gets less briny, and as the rotatory movement declines in rapidity. Even in 50° of latitude pieces of floating ice are met with in the sea. These have become detached from some more northern region, and carried off by the currents which run from the poles to the equator. At 55° it is by no means rare to find the sea-shore strewn with ice. At 60° the gulfs and the inland seas are often frozen all over. At 70° the floating blocks of ice become very numerous and very large, forming sometimes regular islands as much as a half league in diameter. Finally, at 80°, there is found, as a rule, fixed ice—that is, ice which has become accumulated and bound together.

These solitary regions offer a striking spectacle.

The polar ices are tinted with the brightest hues, and seem like blocks of precious stones, forming vast plains and lofty mountains.

The fields of ice are often composed of extensive plains, perfectly level, without either fissure, hollow, or elevation. Scoresby saw one of these floating-fields upon which a carriage might have been driven for thirty-five leagues without the slightest interruption. When these masses meet, the report of the shock is like a clap of thunder.

The mountains of floating ice, as seen for the first time by the navigator who has made his way into the polar regions, present a striking spectacle. Dr. Hayes, in his voyage to the Arctic seas (1860), has conveyed to his readers the first impression produced by the sight of them. He says:

"We met our first iceberg the day before we reached the Polar Circle. Hearing the sea breaking furiously against the mass, as yet concealed by the mist, the helmsman was upon the point of crying out, 'Land ahead!' But almost immediately the formidable colossus emerged from the fog, bearing down upon us, terrible and threatening; we hastened to get out of its way. It formed an irregular pyramid, about three hundred feet wide and one hundred and fifty feet high; its summit was half hid in the mist; but the latter suddenly lifting, exposed to our gaze a dazzling peak, around which were folded light vapors. There was something very striking in the indifference of this giant, which the waves caressed in vain, while it passed on its way, deaf to their charms.

"In Davis Straits we had to pass many cruel hours; and upon one occasion I thought that our last moment had arrived. We were running against the wind, all sails bent and a heavy swell on, when the bows gave way, and all the sails fell on to the deck, nothing remaining save the chief sail, which was flapping violently against the mast; and it was only owing to a miracle of firmness on the part of the helmsman that we escaped complete shipwreck.

"For most of us Greenland was still a kind of myth; for some days we had been following the coast-line: beyond the appearance of Disco, the clouds and fog had kept it constantly hidden from our gaze. But suddenly it emerged from its mantle of mist, and stood out before us in all its splendor; its extensive valleys, its noble mountains, its abrupt and sombre rocks adding to its terrible desolation.

"In proportion as the fog and mist rolled slowly over the surface of the blue waters, the mountains of ice succeeded each other and defiled before us like the fantastic palaces in a fairy tale. Forgetting that they would come spontaneously toward the region, they seemed to us to be attracted by an invisible hand into this enchanted land."

The ice met with on the coasts of Spitzbergen and Greenland is, generally, from twenty to twenty-five feet thick, often forming immense plains, the limits of which can not be seen from the topmast of a vessel : these are called *the ice-fields.* They may be estimated as having an extent of three hundred to four hundred square leagues. An ice-field sometimes presents an entirely level surface; at others, it is rough and uneven, with, at intervals, columns twenty or thirty feet high. These columns give it a very picturesque aspect, and which are sometimes of a topaz blue tint, sometimes covered with thick snow.

The undulations of the water, the movement of the waves, or some other potent cause, break up a field of ice in a moment, and reduce it into fragments of 1000 or 2000 square feet. These fragments, becoming separated, come into collision and disperse; but sometimes they are carried off by a rapid current. In that case, if they meet a current running in the opposite direction, which is floating away large masses of ice from some other field, these mountains meet with a terrible shock.

The icebergs, lifted up out of the water, fall the one on to the other, become covered with fragments more or less voluminous, and thus compose regular mountains, with ravines and indentations, which rise from thirty to fifty feet above the water. The part out of the water is, as a rule, in regard to the portion submerged as one to four; consequently the total height of these mountains is from 130 to 200 feet. Sometimes, too, icebergs 100 to 130 feet long, which are very heavy at their two extremities, sink so deep into the water that a vessel may pass over them. But in this case the crew is exposed to the most fearful risk, as the least shock, the least cause, may disturb the equilibrium which keeps the mass submerged, and if that cause occurs, the iceberg rises suddenly and hurls the vessel into the air or, at any rate, shivers it into pieces.

In Baffin's Bay there are mountains of ice much higher than in the seas of Greenland, some having been found to measure 100 to 130 feet out of the water, which is equivalent to a total height of 660 feet. It is supposed that these fearful masses are formed upon the coasts where they shut in the valleys which abut upon the sea, and that they can become detached. In summer-time the waters flow from their summits and form immense cascades, which are sometimes overtaken by frost. This is a majestic spectacle, but it must be witnessed from a distance, as all of a sudden these columns suspended in the air will snap short and fall into the sea.

Scoresby often saw ice form upon the open sea at twenty leagues from the shore. As soon as the first embryos of the crystals become perceptible, the sea gets calm, just as if oil had been poured over its surface. These crystals soon attain three or four inches in size, and it is then that they begin to agglomerate if the cold continues, forming a sheet of ice which soon attains a thickness of from eight inches to a foot.

In these countries the density of sea-water is 1·026; when still, it freezes at 28°·4. The water which has been concentrated by the frost may attain a density of 1·104, in which case it will only congeal at 14°; and it is well known that water saturated with salt will not solidify till the temperature is less than 5°.

Fig. 55.—The last human dwelling-places. Esquimaux of the Polar Regions.

These desolate regions where mercury freezes in the open air, are nevertheless inhabited by the Esquimaux, who are the remotest inhabitants to the north, living as they do in the 79th degree of latitude. Dr. Kane visited, in 1853, two of their villages upon the Greenland coast of Smith's Straits, at 11° from the pole. These villages are called Etah and Peterovik, and the capital of the country is Upernavik, which was visited in 1861 by Dr. Hayes. An idea of the place in which these people (from whom America is descended) dwell may be gathered from Fig. 55. The huts are constructed upon landings with blocks of snow cut into the shape of domes. The entrance is by a circular and very small

Fig. 56.—Ice at the Pole.

opening, and light is admitted by means of a small window, in which a diaphanous piece of snow serves the purpose of a pane of glass.

The point nearest to the pole as yet reached is six degrees and a quarter (lat. 83° 45′), which is only about 170 leagues from it. Parry and Sir James Ross approached thus far in 1826. The ill-fated Franklin did not pass beyond 77°. Dr. Hayes navigated in the polar sea as far as 81° 40′ in the month of May, 1861.

Let us conclude this general view of climate by remarking that the last isothermal line, clearly established by observations, is that of +5°, which descends to the north of America, re-ascends to the north of Baffin's Bay, crosses the 80th degree of latitude, afterward extends to degree 70, and even to degree 65. This line forms two bends, in each of which there is recorded an increase of cold. It is not at the pole itself that the mean temperature is lowest, but on either side of it. There are thus what may be termed two poles of cold, one situated to the north of the Asiatic continent, not far from the archipelago known as New Siberia, where the mean temperature is +1°·4; the other to the north of the American continent, in the western isles of the Polar Archipelago, and its temperature appears to be −2°·2. It is probable that two analogous poles of cold exist as well in the frozen Antarctic Ocean. As to the North Pole itself, the early calculations of Plana, the mathematician, of the geometer Lambert, and of the astronomer Halley, as well as those of my regretted friend Gustave Lambert, establish conclusively the fact that the cold is much less intense there. As to our pole (taking into account refraction), the sun rises in the beginning of March, mounts slowly, skimming the horizon, and follows a spiral line which takes a greater elevation each successive day. It does not again set until the end of September. On the 21st of June it attains its greatest elevation. The maximum of heat prevails in July and August. From these calculations and the direct observations of navigators who have penetrated the nearest, it follows that the sea is not frozen at the North Pole itself.

BOOK FOURTH.

THE WIND.

CHAPTER I.

THE WIND AND ITS CAUSES: GENERAL CIRCULATION OF THE ATMOS-
PHERE — THE REGULAR AND PERIODICAL WINDS — TRADE-WINDS —
THE MONSOON — BREEZES.

WE now come to the study of the great currents of the atmosphere,
which are themselves the incessant manifestation of the sun's action
upon our planet. Without the wind the atmosphere would remain
motionless about the globe; heavy, cold, deadened, enveloping the earth
in a regular pall, never agitated by a breath of air, a receptacle for every kind of miasma—poisonous and deleterious. By its agency an immense circulation is established from one end of the world to the other,
renewing all the strata, sweeping away unhealthy exhalations, substituting for oppressive heat a refreshing coolness, or replacing the period
of frost by the warmth of spring.

What is *wind?* In this section of our work, and in the succeeding
one, which deals in clouds and rain, we take in hand the general data
of meteorology; for the currents of air on the one hand, and water on
the other, cause the varying meteorological conditions of the seasons
and of years. It is on this head particularly that we have an exact base
for our knowledge, and that we are in a position to consider the general
mechanism of this vast factory, which distributes benefits and disasters
over the earth, and among the people which inhabit it. Meteorology
will not be able to hold her own with her elder sister, Astronomy—that
is, to be precise in respect to ascertained principles, and to enable science
to announce the movements to the atmosphere, the winds, the rains, the
droughts, and the tempests, as the latter announces the movement of the
stars—until we are able to embrace, in one glance, the general circulation which is constantly going on all over the globe, and which gives
rise to divergencies which occur in different seasons and at different
places.

What is the wind?

It is neither more nor less than *a certain quantity of air set in motion
by a change in the equilibrium of the atmosphere.* The varying temperatures to which the different parts of the atmosphere are constantly ex-

posed rarely each of these parts in a different manner. When air is
heated its weight diminishes, and it has a tendency to rise; whereas
colder air becomes heavier, and flows to supply the place of the heated
air, and, in its passage toward the re-establishment of an equilibrium,
will cause a current of air which is termed *wind*, and which will continue till an equilibrium is restored.

Let us suppose, for a moment, that the atmosphere is perfectly calm
everywhere. A cloud passes over the sun, the air that is situated in
a line with its passage is rendered cooler, undergoes condensation, and
becomes denser; this air seeks an equilibrium; a primary movement
will take place in the direction of the cloud, and here we have a current
of fresh air, the tendency of which will be to occupy, as quickly as possible, the place of the hotter and more dilated air which is next to it.
Suppose that the sun, shining in a clear sky, remains motionless above
our heads. The air situated immediately underneath will become heated more rapidly than that which receives its rays obliquely. Becoming
dilated, it will rise toward the less dense aërial regions, the air which is
contiguous to it will force itself into its place, and thus another current
of air is established.

The great atmospheric currents, the winds, general and special, are
nothing else than this unceasing pursuit toward an equilibrium which
is perpetually being destroyed by the various influences of the sun.
This will be seen by applying to the entire surface of the globe the instance cited above. In what way are two contiguous parts of the atmosphere affected if they become heated in unequal proportions? Near
the equator, the sun, as its rays reach the earth in a perpendicular or
nearly vertical direction, causes a temperature which is constantly higher than at other points of the globe. It follows from this that two inferior currents must flow from the two hemispheres toward the equator.

The air, which is very heated in the equatorial zone, rises in a mass
toward the higher regions of the atmosphere. Having reached an elevation of several miles (but which we are unable to calculate exactly),
the ascending mass breaks into two, which pass away in the direction
of the two poles.

This ascensional movement thus produced gives rise to a rush of air
from the two sides of the torrid zones, and two other masses, skimming
the surface of the ground, make their way from the temperate regions
toward this line. Thus we discover all over the earth a double aërial
circuit.

Let us first take the northern circuit. A current of air, starting from the tropical regions, proceeds toward the equator. Situated in the lower regions of the atmosphere, and upon the surface of the globe, this current comes directly beneath our observation, and it constitutes the *trade-winds* of the northern hemisphere. When within a short distance of the equator, a distance which varies with the seasons, it suddenly rises, and, when it has reached a certain level, takes a directly horizontal march toward the pole, gradually descending toward the surface as its distance from the equator increases. Maury termed this kind of current the upper anti-trade-wind.

If it stopped there the current would not be complete; the trade-winds and the anti-trade-winds, connected with each other by the ascending branch of the equatorial region, are not, as yet, united on the northern side. If the earth were motionless, and the whole of its surface received light at the same time; if, moreover, its surface was universally homogeneous, the meeting of the two horizontal branches would, no doubt, take place toward the north, as it does toward the south, excepting, of course, the reversal of the direction of the movement. The upper anti-trade-wind would incline toward the ground, so as to join the trade-wind, and the circulation of the atmosphere would be almost comprised within heights of an inconsiderable elevation. Let us remark, however, that as the first origin of the movement is at the equator, the movement will be regular there, like the cause which produces it. The trade-winds and the anti-trade-winds will themselves participate of this regularity in the neighborhood of the equinoctial line; but the farther one recedes from this line the less directly will the motive force act. The descending mass will, therefore, be more diffuse, less compact, and less fixed in its quantity than the ascending mass. Its mean position will depend upon the mean activity of the equatorial draught, and upon the height to which the trade-winds reach. This height is itself dependent upon the law of the decrease in the temperature, according to the altitude. It may vary with the seasons, and has probably not been the same in all ages of the world.

The southern circuit is rather more extensive than the northern; it encroaches upon the northern hemisphere, upon the surface of the Atlantic, and in summer this encroachment is more marked than is the case in winter.

Circulation, regular as it may be, can not take place in the midst of an atmosphere always in motion like ours, without reacting upon the part which is not directly comprehended in the movement. The decrease of the temperature extends also toward the poles, and atmospheric movements are the forced consequences in these high latitudes. Two leading circumstances cause the aërial currents to travel out of the limits comprised in the above circuits, and give rise to two secondary circuits (N′ and S′); these are the rotation of the earth on its axis around the sun, and the division of land and water over the globe.

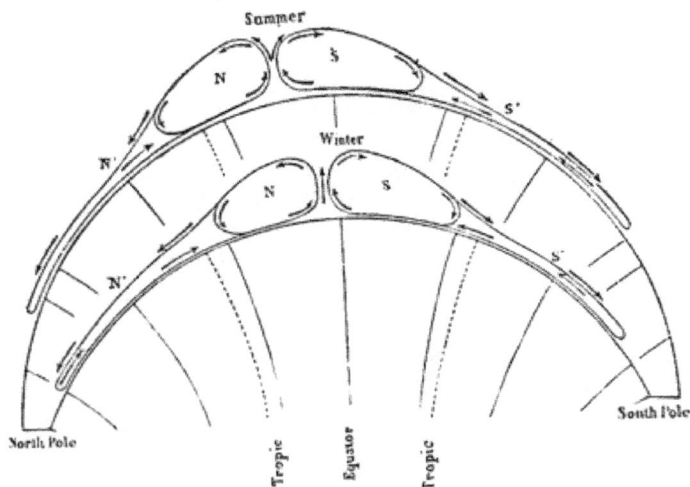

Fig. 57.—Section of the atmosphere, showing its general circulation.

The earth turns upon its axis in the direction of west to east. In virtue of this rotation, every point of it completes a revolution in the same period of twenty-four hours; but in this interval of time all parts do not traverse the same distance or move at the same rate of speed. At the equator the speed is about 416 leagues an hour; in the latitude of Paris it is 273; at degree 56 it is 231—as at Edinburgh, for instance; at the poles it is nothing.

The air which seems to us to be in repose at Paris is, in reality, moving there at the rate of 273 leagues an hour. Let us imagine this air transported to the latitude of 56° without any change in its velocity; it will continue to travel 273 leagues per hour. As each point in latitude 56° travels at 231 leagues per hour, the air will gain upon the ground, in an easterly direction, at the rate of forty-two leagues an hour! which would constitute a hurricane. The reverse would be the case if a mass of air, relatively still, in parallel 56°, were suddenly transported into parallel 49°. This air would appear to us to be traveling from east to west at the rate of forty-two leagues per hour.

In reality, these passages of air from one parallel to another always take place gradually, and, during their transition, resisting causes of various kinds tend to equalize their velocity. The lessened differences none the less continue in operation, and, as the size of the parallels of latitude diminishes the more rapidly on approaching the poles, the effects pointed out above become more and more pronounced as they occur in higher latitudes. Many tempests are derived from this cause.

The influence of the earth's rotation upon the direction of the trade-winds is as follows:

Take, first, the trade-winds of the northern circuit. We have supposed that they move from north to south toward the equator. During this movement they pass gradually by the parallels, whose diameters, and consequently whose speed, progressively increase. If their absolute velocity does not diminish, they will apparently move toward the west, and their seeming direction will be from north-east to south-west, which is, in fact, somewhere about the direction of the trade-winds in the northern hemisphere. A like result follows in the case of the southern trade-winds, which also seem to retrograde toward the west; but as these winds travel from south to north toward the equator, their apparent direction will be from the south-east toward the north-west, which is, in fact, the general direction of the trade-winds in the southern hemisphere.

When the ascending mass, having reached a certain height, divides into two horizontal masses, which form the upper or anti-trade-winds, flowing from the equator toward the poles, and, little by little, travel past parallels the speed of which is successively less and less, they soon take an easterly bend in these parallels, and their apparent direction is toward the north-east. When they have arrived at a certain distance from the neighborhood of the tropics, they descend toward the earth; then is reproduced the phenomenon noticeable in the ascending mass; the anti-trade-winds find their way with the velocity which they have acquired and their easterly tendency; the inclination of their speed in a vertical direction renders this speed less apparent, and we meet with two new regions in these latitudes, called *tropical calms*. In moving from the equator toward the North Pole, we thus encounter: 1st. The region of equatorial calms; 2d. The north-easterly trade-winds; 3d. The tropical calms; and 4th, beyond these, winds varying from south-west

18

to north-west. The same series is met with in the southern hemisphere.

In a word, we find that there are in each hemisphere two circuits which have as a common basis the ascending equatorial mass. The first, a *direct circuit*, is generally limited to the intertropical regions; the second, a *derived circuit*, is, in reality, only a prolonged arm of the first, and extends from the tropics to a varying distance from the poles. These two circuits are distinguished from each other by essential characteristics ensuing from their different positions in the atmosphere.

The direct circuit spreads upward. While the trade-winds skim the ground, the anti-trade-winds circulate in very lofty regions of the air. The distance which separates these two currents, joined to the regularity of their movements, prevents them from encroaching upon each other or influencing each other's progress. This does not hold good of the derived circuit. The prolonged arm of the anti-trade-winds there becomes superficial. It sweeps along the ground; and so, also, does the returning current. Both, therefore, are upon the same level, simply contiguous, and separated only by the action of the earth's rotation. There are points at which these currents come together; and their different qualities cause numerous, and sometimes disastrous, atmospheric disturbances. Their beds get shifted over the surface of the globe, and the succession of one after another in the same place produces sudden variations in the state of the sky. To avoid confusion, the branch of the upper anti-trade-winds which is prolonged into the derived circuit is termed the equatorial current, and the back current in the same circuit is called the *polar current*.

This general circulation of the atmosphere is influenced to a certain extent by the seasons.

At the end of our summer the regions about the North Pole have for several months had days without any nights; the temperature there has become perceptibly milder and the air rarefied. To days without nights soon succeed nights without days, accompanied by excessive cold; the air becomes contracted, and draws in a fresh supply to fill up the vacancy caused by this contraction. Each of these changes in our hemisphere corresponds with an exactly reverse change in the other hemisphere; there is, therefore, a general translation each year of the atmosphere of the northern hemisphere into the southern, and *vice versâ*.

The rush of air toward the North Pole during winter is brought about by the equatorial currents, which then acquire a very large volume. The perturbations increase there in the same proportions: it is the season of tempests. As the sun makes its way back to us, and our atmosphere becomes heated and dilates, the equatorial current slackens its speed and reaches lower latitudes. On the other hand, the polar currents become more active; but as they are diffused over the surface of Asia and of Europe, their speed is rarely very great, and summer is the calm season in our hemisphere. The atmospheric disturbances at this season never extend very far, and their local gravity is due to electrical phenomena of a special nature: it is the season of thunder-storms.

The equatorial currents take, at their polar extremities, a direction parallel to the equator, and march from west to east. Notwithstanding their variations, both in volume and intensity, it is easy to understand that they cause the atmosphere at the poles to make a continuous rotatory movement in the same direction as the earth.

For many centuries the trade-winds were an enigma, both to meteorologists and to navigators. Halley and Hadley first suggested the explanation which has been developed, and which contemporary research has modified in the course of the last century.

Between the two trade-winds there are two zones; these are the zones of equatorial calms. These calm regions occupy very different positions at the close of winter to what they do at the end of summer; they follow, in fact, but at a distance, the progress of the sun between the tropics. They never cross the equator upon the surface of the Atlantic. In February and March, months when they approach nearest to it, the north-easterly trade-winds stop at about 4° north latitude; in August and September, the months in which they are farthest away from it, the same trade-winds stop at about 11°. When a vessel sailing in the Atlantic approaches the equator, the crew begin to feel anxious, for they know that the favorable wind which has brought them thus far will gradually fail, and finally disappear altogether. The waters extend around them like a vast sheet of ice, and the ship is, so to speak, nailed to the limpid crystal. The solar rays fall vertically upon the deck.

The sun which, twice in the course of the year, pours down its rays perpendicularly upon these regions, never recedes far enough for anything like coolness to ensue. The heated atmosphere becomes so light that it is continually ascending. There evaporates also from the Atlantic and the Pacific Oceans an immense quantity of water, which becomes diffused and mixed with the heated air, and ascends with it; but as the air ascends to the lofty regions it gradually cools, sometimes very suddenly, so that a great part of the water which had accompanied it is transformed into drops. These sudden changes produce passing tempests, which are frequent in the equinoctial regions.

We have seen that, as the wind approaches the temperate zones, upon which it will descend and become converted into surface currents, the upper current encounters strata of air, the speed of which in regard to the diurnal movement is at a *minimum*. It follows that the return of the trade-winds gives rise in the temperate zones to a wind which blows from south-west in the northern hemisphere, and from north-west in the southern hemisphere. Thus, in France, the wind blows oftener from the south-west than from any other direction. At the time of the discussions upon the real movement of the earth, the followers of Copernicus adduced the trade-winds as a proof of the diurnal rotatory movement, from west to east. This was quite an illusion on their part. Carried by the movement of the globe, the observer would, had such been the case, have quitted the air of the atmosphere, which would, under those circumstances, have seemed to give rise to a wind blowing in a contrary direction, viz., from east to west. But we have seen that it is the combination of different rates of speed, on the one hand, the strata of air which are displaced by the differences of temperature in the various parts of the globe; and on the other hand, the atmospheric strata which are brought under the influence of the diurnal movement, which, in reality, produce the trade-winds. The theory of the motion of the earth does not require this pretended meteorological proof.

The existence of the upper counter-current has been ascertained directly by Captain Basil Hall, who observed that in the region of the trade-winds very high clouds are continually sailing in an opposite direction to that followed by the wind beneath. The same traveler remarked upon the summit of Teneriffe in August, 1829, a south-westerly wind; that is to say, a wind of a diametrically opposite direction to the trade-wind which was blowing upon the surface of the ground. When Humboldt ascended the same mountain in 1799, a very strong westerly wind was blowing upon the peak.

Another proof of the existence of this same counter-current of the trade-winds may be deduced from the fact of dust emitted by the volcano in St. Vincent Island falling upon Barbados.

During the evening of April 30, 1812, explosions resembling the discharge of heavy pieces of artillery were audible at Barbados; the garrison of the Château St. Anne remained under arms all night. On the following morning the horizon of the sea, to the east, was clear and well-defined, but just above it was seen a black cloud which already covered the rest of the sky, and which, soon after, spread over that part where the light of day was beginning to break. The obscurity became so intense that persons sitting in a room were unable to distinguish the window, and in the open air trees and houses, and even a white handkerchief held up at a distance of six inches before the eyes, became invisible. This phenomenon was caused by the fall of a large quantity of volcanic ashes, emitted by a volcano in the Island of St. Vincent. This new kind of rain, and the profound obscurity which accompanied it, did not entirely cease until nearly one o'clock. The trees, whose timber bends readily, bent beneath its weight, and the crash of the limbs of other trees as they snapped off short was in striking contrast to the complete calm of the atmosphere; the sugar-canes were prostrated upon the ground, and the whole island was covered with a layer of greenish ashes to a depth of one inch.

St. Vincent is fifty miles nearly due west of the Barbados, and the volcano there had shot this immense mass of ashes to the height at which the upper current was traveling—a current which was itself sufficiently strong to transport the mass.

Halley was the first to affirm the existence of the upper trade-winds as a consequence of the ordinary trade-winds. Though he advanced no direct proof of the fact, he assured himself of its truth by the almost instantaneous rotation of the wind in opposite directions, when the polar limits of the trade-winds are passed. In his opinion, as in that of all meteorologists of the present day, the equatorial south-west current which prevails in the mean latitudes of our hemisphere is, in reality, only a continuation of part of our upper trade-winds on their return journey.

The higher branch of the intertropical circuit is, at its equatorial origin, at such a height that it has been impossible to ascertain its existence with precision, even by climbing the loftiest peaks of the Cordilleras in the neighborhood of the region of calms. But, as this branch gradually descends toward the surface of the globe, in proportion as it approaches the tropics, and as, moreover, its course lies through colder regions, some few clouds appear in the air which it carries in its train. These serve as so many proofs of the direction which it takes.

The existence of trade-winds was ascertained during the first voyage made by Christopher Columbus. The regular winds, which impelled that adventurous navigator along the new route by which he expected to reach India, excited the fears of his associates, who doubted the possibility of getting back to Europe. Had Columbus, after the discovery of the New World which he alighted upon when he imagined that he had reached India, not taken pains to avoid the trade-winds, by steering to the north before he turned westward, he would assuredly never have found his way back to Spain. With his vessels both ill provided with food and defective in construction, he and his crews would have perished of hunger in the vast regions of the trade-winds. It is upon the struggle between these two currents, upon the point at which the upper current descends to the surface, and upon their reciprocal mingling, that depend the most important of atmospheric variations, the changes of temperature in the strata of air, the precipitation of aqueous vapor, and even, as Dove has shown, the varying shape and form which clouds take. The shape of the clouds, which lends so much charm to our landscape, indicates to us what is going on in the higher regions of the atmosphere. When the air is calm the clouds delineate upon the sky on a warm summer day "the projected shape" of the soil, the caloric of which radiates freely toward space.

In the great ocean and the Atlantic the trade-winds extend nearly to the tropics; but in the Indian Ocean the presence of land prevents the regular or the trade-winds from setting in; whereas in the southern hemisphere, at a certain distance from land, the south-east trade-winds prevail almost uninterruptedly. In the northern hemisphere of the Indian Ocean there prevails a south-west wind, blowing toward the peninsula of Hindoostan, to the north of India and China, from April till October; and from October to April the prevailing wind is, on the contrary, from the north-east. These are the monsoons of the Indian Ocean. This word is derived from the Malay *moussin*, which signifies season. Thus, during the summer of our hemisphere, when the sun has a north declination, it is the south-west monsoon which prevails; whereas in our winter, when the sun has a south declination, the monsoon is the north-east. These winds penetrate into the interior of continents, where they are influenced by the shape of the land. The mountain chains generally tend to attract the gaseous masses in their direction. The explanation of these periodical winds is this: In January the temperature of South Africa is at its *maximum*, that of Asia at its *minimum*. The northern portion of the Indian Ocean is hotter than the continent, but not so hot as the southern part of the same ocean at an equal latitude. We find, then, in each hemisphere, easterly winds blowing toward the hottest points. From October to April the south-east trade-winds prevail in the southern hemisphere; the north-east trade-winds are blowing in the northern hemisphere, and are termed the north-east monsoon. Between the two is the region of calms. When the sun advances toward the north, the temperature of the continent and that of the sea become more or less equalized; thus, about the period of the spring equinox, there are no prevailing winds in the northern hemisphere, but varying winds, which alternate between dead calms and hurricanes; whereas the south-east monsoon prevails throughout the year in the southern hemisphere. As the north declination of the sun increases, the temperature of Asia rises above that of the sea; whereas it declines below it in New Holland and South Africa. The relative positions of the two continents, the differences in the temperature which are most marked, and the rotatory movement of the earth, thus create a current from the south-west—a monsoon which prevails from April to October. Thus, whereas in the southern hemisphere the trade-winds from the south-east prevail throughout the year, the north-east monsoon in winter and that from the south-west in summer are met with to the north of the equator.

Thus are indicated in a brief manner the general directions of these winds. So far back as any records exist, they facilitated the communications which were then so frequent between India and Egypt. Upon the decadence of that empire these relations ceased, and the tradition of these winds was lost; for, otherwise, Nearchus would not have been so long on his voyage from the mouths of the Indus to the extreme end of the Persian Gulf.

In many places periodical winds are met with which alternate with the seasons, and which are influenced by the shape of the coast-line; thus, for instance, in Brazil there is a north-east monsoon in spring and a south-west monsoon in autumn. The Mediterranean has its monsoons, known to the ancients, who indicated their sense of dependence upon the winds by the term *etesian* winds (from ἔτος, year or season). To the south of the Mediterranean basin the vast desert of Sahara extends. Devoid of water, made up merely of sand or conglomerated pebbles, it becomes very heated under the influence of an almost vertical sun; whereas the Mediterranean preserves its ordinary temperature. Thus, in summer, the air rises above the desert of Sahara with great ra-

pidity, and floats off mostly toward the north, while underneath are
northerly winds which extend as far as Greece and Italy. In North
Africa, at Cairo and Alexandria, there are none but northerly winds.
All navigators are aware that in summer the voyage from Europe to
Africa is effected more rapidly than the return passage. Thus, if we
compare the half-duration of passages to and fro between Toulon and
Algiers, it will be found that the return passage is one-fourth longer in
the case of sailing-vessels, and one-tenth in the case of steamers. This
fact can not be attributed to the currents, which are very trifling. Besides, the north coasts of the islands of Majorca and Minorca—that of
the latter in particular—are swept by this same wind, which causes a
perceptible stunting of vegetation there. These winds prevail at Algiers, Toulon, and Marseilles. In winter, on the contrary, when the
sand radiates considerably, the air of the desert is colder than that of
the sea, and in Egypt there is a very cold south wind, though not so
strong as the summer winds.— *Kaemtz and Martin.*

To these periodical winds, to the trade-winds and the monsoons, we
may add the breezes caused upon sea-coasts by the difference between
the heat of the land and of the water. This, in the early part of the
chapter, was pointed out as produced by solar heat, like the trade-
winds.

Periodical and diurnal displacement of air takes place in mountainous regions. These consist in a breeze which creeps along the side of
the mountain at night, and in an ascending breeze during the day.
These movements of air vary according to the shape and aspect of the
mountains.

Of all the causes which are assigned to the winds, one of the most
powerful is, beyond doubt, the condensation of vapor in the atmosphere. Sometimes one inch of water will fall, in the course of an hour,
over a wide tract of country, especially in the equatorial regions.
Now, suppose this tract to be but a hundred square leagues in extent.
If the vapor necessary for the production of a depth of one inch over a
hundred square leagues were in an elastic condition in the air, and had
only 50° temperature, it would occupy a space a hundred thousand
times greater than in its liquid state; that is to say, it would occupy a
space of a hundred square leagues by 8860 feet in height. Such, therefore, would be the dimensions of a void resulting from this condensation. In reality, the vapor is not in an elastic but in a vesicular state,
although, from the very fact of its remaining suspended in the atmos-

phere, it is probably of less density than if it were in a liquid state, and its condensation into drops of rain also occasions an immense void, the filling of which must necessarily give rise to great atmospheric disturbances.

The constant circulation going on in the atmosphere renders impossible the entire consumption of any of the substances necessary to maintain the life of organized matter, such as oxygen, aqueous vapors, etc. ; and it also prevents any dangerous accumulation of deleterious matter, such as carbonic acid. The existence of animated nature is intimately connected with this circulation. These simple features do not, at first sight, seem to apply to the apparently capricious play of the weather, nor to delineate it in its true aspect or type of versatility and changeableness. The weather is not less variable, especially in our climates, as we shall presently see. We may divide the surface of the globe into two unequal parts—the regions of fixed and variable weather. The state of the air may be predicted to the limit to which the trade-winds extend, and that for several years to come. The mean zone (included between 2° and 4° N. and S. latitudes) is that where throughout the whole year great heat and calms alternate with nocturnal rain-falls and tempests. Next to them, both north and south, is another zone (4° to 10° latitude), where similar weather occurs only in summer or in winter, and trade-winds render the sky clear. There is a third zone (10° to 28° N. latitude) where, in winter as in summer, the trade-winds do not usher in the slightest moisture, where years pass without the soil being refreshed by the least drop of rain.

Finally, another zone, both north and south (from 20° to 30° latitude), which forms the limit of fixed weather; there the trade-winds cause the summer to be without rain and the winter to be mild and rainy, though the rain is never continuous. The approximate indication of the latitudes refers to the northern hemisphere and the Atlantic Ocean, the sole region where reliable observations have been collected.

We now have to consider a zone of 24° latitude, where the meeting between the polar and the equatorial currents occasions a variable climate, which only seems to us capricious and uncertain because the circumstances influencing the predominance of one of the two currents in a given locality are so complicated that we have been unable to deduce from observations a law by which these modifications can be classified. If we study the question we find, as I have said, that there are in reality but two winds in the atmosphere; that which blows from the

poles toward the equator, and that which makes its way back from the
equator to the poles. Let us now take a place situated in the region of
variable weather (the latitudes of Paris, Vienna, or London, for instance),
and further, let us admit that this place is just in the direction of the
polar current. When the north wind blows there, the cold becomes
accentuated, the sky gets clear, even if the wind, deviating slightly from
its direction, turns toward the east. The polar air which it brings with
it is, as Schleiden remarks, very dangerous for consumptive persons, by
reason of its extreme dryness and the abundance of oxygen in it. The
east wind blows until some other wind comes to take its place, and this
can only be done by the equatorial current which arrives as a southerly
wind. The immediate result produced by this meeting is to give birth
to an intermediate direction, or to the south-east wind, the hot and humid air of which, cooled by the polar current, is obliged to abandon a
part of its water in the shape of clouds, snow, or rain. The equatorial
current gradually gains the mastery, the weather clears up, becomes
warmer, and maintains itself with a southerly wind, which imperceptibly
veers to the west. There is only the polar current which can, in turn,
take its place; the fusion of these, passing to the north-west, produces
abundant atmospheric precipitation. Then we have those cold and
damp days which are so unpleasant to persons of a nervous tempera-
ment.

Strange to say, this variable zone, which one would be inclined to regard as the most unfavorable for the development of the human race,
embraces nearly all midland Asia, Europe, North America, and the
north coast of Africa, and consequently comprises the scene upon which
has been illustrated the history of humanity and of its intellectual development. Perhaps there is some secret connection between this phenomenon and the special development of the vegetable world in this region.

This sketch of the distribution of weather over the surface of the
globe is modified by many causes. The elevation of countries above
the sea-level, plains and mountains, sandy deserts and forests, cause
great disturbances in the action of these laws.

Among the influences which modify weather, one of the most important is the manner in which the sea and the land are spread over the
surface of the globe. The land, being exposed to the solar rays, is heated more rapidly than the sea, and, after a certain interval, attains a
higher temperature, which, moreover, cools again far more slowly. The

first consequence is that the hottest zone, the region of calms, is not equally extensive both to the north and to the south of the equator; but, on the contrary, occupies the largest space in the northern hemisphere.

We have seen that heat and its unequal distribution in all directions is the fundamental phenomenon around which the others, which are dependent upon it, group themselves. The moisture of the air has an intimate co-relation with this phenomenon, and the latter, together with the heat, are the causes of vegetable life. It is upon these two conditions that principally depends the distribution of plants over the globe. The animal world follows the plants, for the existence of herbivorous beings is directly connected with that of the carnivora. The first supreme principle, that which not only vivifies, but stirs up and regulates all, is the sun; its rays are the pencils with which it traces light and shadow, the burning yellow of the arid sand, and the fresh green of meadows, and even the sketch of an ethnographical map for the human race.

CHAPTER II.

THE SEA CURRENTS: METEOROLOGY OF THE OCEAN—MARITIME ROUTES —THE GULF STREAM.

WE have seen that the distribution of solar heat over the globe creates in the atmosphere a general regular circulation. In the next chapter we will prove that the irregular and variable winds are alike due to this heat, and that they are subject to laws of periodicity which science is engaged in studying. But, before having done with the great currents of the atmosphere, it is necessary that we should form some idea of the great ocean currents, also dependent on the action of the very same heat which regulates all things here below.

The sea is not motionless; neither its waters nor the atmosphere above them. A great general oscillation of the surface occurs twice a day, under the attracting influences of the moon and sun: these oscillations are the tides, whose flux and reflux alternately cover and lay bare the shores of the ocean, and give to the coast that endless variety which never fails to charm us. This movement of the waters is due to an astronomical cause, and need not be gone into here. But the sea is animated by another meteorological circulation, more complex and wider, which may almost be compared to the circulation of the blood in our veins; it is traversed by currents which, running from the equator to the poles, and *vice versâ*, thus forming a connecting link between the most distant seas, distribute heat to colder regions, exercise a cooling influence within the torrid zones, equalize the briny and chemical composition of the ocean, and form, to a certain extent, the vital circuit of the globe; like the sap which rises and falls in plants, like the blood which becomes regenerated at the heart after having carried its tribute to the farthest extremities of the organization.

These ocean currents merit our special attention, and their study will embrace at once the currents of the atmosphere which accompany and complete them, constituting the meteorology of the ocean. Both have, especially for the last thirty years, been the subject of detailed research.

Maritime travel differs *ab initio* from journeys by land, in the absence of any fixed route. For a long period, indeed, modern navigators never suspected that there existed upon the surface of the ocean numerous highways, traced by the hand of nature. The constancy of the monsoons, the periodical return of the marine breezes along the coasts of the Red Sea and in the Indian Ocean, are phenomena which our forefathers had ascertained and utilized. When the astronomer Hippalus discovered the physical fact of the return journey of the summer monsoon, he made a discovery which the Arabian sailors had for centuries been acquainted with, and which they had taken advantage of to preserve the monopoly in the trade of Ceylon spices and perfumes, which they sold as the products of Arabia. The discovery of Hippalus caused a complete revolution in the system of maritime services among the Europeans who flourished at the commencement of the Christian era. The discoveries effected by the researches of Lieutenant Maury, of the Washington Observatory, during our own day, are analogous to the above, but on a much larger scale. On account of their great intercourse with other peoples, and the geographical position of their country, which is bounded by two oceans, the Americans were more interested than any other nation in the discovery of the shortest sea-routes. To effect this purpose, it was necessary to compare with each other the thousands of routes that had been followed by thousands of navigators. This herculean task rendered it possible to deal with the whole globe as Hippalus had dealt with the short distance between Egypt and Taprobane.

The great navigators of early ages seemed to have struck out the only routes practicable, without its occurring to them to introduce the modifications which the comparative study of the data of experience might have led them to. But when the application of steam to the means of transport had proved the advantages of a rapid system of intercommunication, and the great value of time, attention naturally became turned to the discussion of better routes, and the means of deciding as to how they could be arrived at. A steam vessel, taking no account of the wind, can trace upon the sphere the shortest and the most direct line between its point of departure and its place of arrival; but with the sailing vessel, subject as it is to aërial currents which constitute its sole means of progression, the line which is shortest in point of distance often becomes the longest in respect to the time occupied in traveling along it. To find the greatest possible sum of favorable winds, without deviating more than can be avoided from the straight line, is the surest way to accomplish a quick passage. The observations taken at the surface of the seas by navigators were long allowed to remain profitless for the purposes of science and navigation. Under Maury's auspices they led, in a few years, to a knowledge of the general circulation of the atmosphere and the seas. At the same time they have been instrumental in reducing by a fourth, a third, and even a half, in some instances, the length of long voyages, and in effecting an immense saving in the cost of transport.

To awaken public interest by some practical result which would demonstrate the great importance of these new studies, he concentrated all his efforts upon one single route—that from the United States to Rio Janeiro. The data which he collected enabled him to ascertain a route far shorter and better than that followed by the great mass of navigators. The ship *Wright*, Captain Jackson, from Baltimore, was the first to steer by Maury's course. Starting from Baltimore, on the 9th of February, 1848, this vessel crossed the equator in twenty-four days, while the time occupied had previously averaged forty-one days.

This route from the United States to the equator is all the more important because it is the road of all ships sailing from North America to the southern hemisphere, whether their ultimate destination be the Pacific, the Indian Ocean, or the Atlantic. From forty-one days this passage was reduced to twenty-four days, afterward to twenty, and finally as low as eighteen. This is a gain of fifty per cent.

The passage from the States to California took, on an average, rather more than 180 days; after Maury had brought his knowledge to bear upon the subject, it was at first shortened to 135 days, and since then to 100; while one of the fleet of clippers trading there—the *Flying Fish*—cast anchor in the harbor of San Francisco on the 92d day after leaving New York.

But the most remarkable instance is furnished by the voyage to Australia. From England to Sydney, a vessel sailing under the old system used to take at least 125 days, which was the usual average. The return journey being about the same, the total length of the voyage amounted to 250 days. When Maury passed through England to attend the Congress at Brussels, he promised the British sailors and merchants that, as a recognition of the help they had afforded him, he would diminish by at least a month the voyage to Australia, and reduce the return passage to a still greater extent; or, in other words, lessen by a quarter the distance between England and its wealthy colony. A little later, when the notions with respect to this route were complete,

Maury pointed out the immense advantage that would be derived by making the voyage to Australia a regular circumnavigation of the globe—that is, doubling the Cape of Good Hope on the outward voyage, and Cape Horn on the return passage. The total length of these two voyages would, he said, occupy 130 days, or even less, in place of 250, as were taken before. This prediction has been fulfilled, and even exceeded, the saving of time being equivalent to fifty per cent.

Let us see what is the economy from a pecuniary point of view. The price of freight to Australia is about one shilling per ton per day. Taking the average tonnage of the vessels upon this line as being only 500 (they are 700 in reality), and assuming a reduction of only thirty days in the passage, it will result that each ship will have realized in each passage a saving of 15,000 shillings. If we take Maury's calculations, and put the number of vessels of all flags that ply annually between the North Atlantic ports and Australia at 1800, there will be a clear gain of twenty-five millions of shillings at the expiration of a twelvemonth.

For English commerce alone, in the Indian Ocean, the annual economy is nearly £500,000. Taking all passages effected by ships of various nations, this discovery must effect an annual saving of four millions of pounds sterling.

The greater the distance to be accomplished, the greater is the advantage in deviating from the straight line to seek a region where continuous breezes will impel the vessel at the greatest speed. Thus, generally speaking, if one is sailing from east to west, it is in the intertropical region that the speed is greatest; whereas, in order to sail very rapidly from west to east, it is necessary to go beyond the tropics, either north or south.

Each day's delay in the arrival of a merchant-vessel beyond the fixed date or the average of passages is not only a more or less considerable cause of annoyance to the passengers, whose health, and even life, may be depending upon their speedy arrival; it is also a cause of loss to the shipper and the merchant. The expenses of a large vessel vary, as Admiral Fitzroy pointed out—including wages, provisions, material, a full cargo, and an average number of passengers—from £50 to £200 a day; moreover, to these immediate expenses must be added the diminution in the annual earnings of the vessel which are consequent upon the forced delay in its next departure. The evils incident upon a long passage are, therefore, complex in nature, affecting the interest of the ship-per and of the public at large.

The progress realized by the "Sailing Directions" in shipping indus-try is, consequently, equivalent to that effected by the adjunction of a new motive power. Thus a ship which, sailing in the ancient track, would have been at sea for one hundred days, now follows the new course, and reaches its destination in half the time, and is thus, so to speak, supplied with a traction-engine powerful enough to double its speed. These fortunate results have been universally accepted. In a conference held at Brussels in 1853, the United States, France, England, Russia, Sweden, Norway, Denmark, Holland, Belgium, and Portugal agreed upon a uniform plan of meteorological observations at sea, and this plan was soon adopted by Prussia, Austria, Spain, Italy, and Bra-zil. Since then, all the trans-oceanic vessels belonging to these powers have become floating observatories, which register night and day all the incidents of navigation calculated to secure a complete knowledge of the movements of the atmosphere and the sea.

Thanks to these researches and to the development in late years of meteorological observations, I am enabled to give, in the previous and following chapter, a general sketch of the *distribution of winds over the surface of the globe.*

Let us now consider the circulation of water, also due to the influence of solar heat.

It is well known that the seas are divided, first into three great oceans, viz., the Atlantic, which separates Europe and Africa from America; the Pacific, which covers half of the globe between the two Americas upon the one hand, and upon the other Eastern Asia and New Holland, with the Archipelago between; and thirdly, the small ocean known as the *Indian Ocean,* which is almost entirely upon the south of the equa-tor between Africa, Asia, and New Holland.

If the two great oceans be divided into two parts, that to the north and that to the south of the equator, and if the polar seas be taken into account, we shall have altogether seven divisions in which the move-ment of the hot or cold waters, their flow from the equator toward the poles, and their return to the point from which they started, can be studied. It is to this movement that are due, throughout the sea, cur-rents of hot and currents of cold water, the majestic and steady changes of which, and the more or less varying temperature of which, give rise to effects of a far more important nature in the economy of climates than might be supposed by those whose only knowledge of the globe is derived from ordinary maps.

Let us analyze and weigh these important currents, taking as an example the circuit formed by the waters of the Atlantic to the north as being best known to us, and which is continually being traversed by vessels coming and going between Europe and North or Central America.

In the equatorial regions, the waters of the ocean are impelled toward the west by an incessant movement which, in the Atlantic, carries them toward tropical America. This vast current, 30° in width, twenty of which are to the north and ten to the south, breaks against the shores of the New World. In accordance with the shape of America, the eastern point of which is a long way below the equator, the greater part of these waters make their way toward the *Gulf of Mexico*, the bends of which it follows, and finally makes its way out again by the extreme point of Florida, running along the coast of the United States from south to north. This gulf, situated in the torrid zone, is on all sides surrounded by lofty mountains, which shut in the solar rays as within a vast funnel, and store up therein the heat of a burning climate. It is from this focus that the equatorial current starts. It runs across the Straits of Florida, and produces an impetuous stream, nearly 1000 feet deep and fourteen leagues wide, running at the speed of five miles an hour. Its waters, which are warm and very salt, are of the color of indigo blue, and differ from their greenish borders formed by the waves of the sea. This mass creates in its passage a great agitation, and thus follows its course without becoming confounded with the ocean. Shut in between two liquid walls, the waters of the Gulf Stream form a moving vault which glides along the sea, carrying off to a great distance all objects which get drifted into it. "In the greatest droughts it never fails, in the greatest floods it never runs over. Nowhere in the world does there exist so majestic a current. It is more rapid than the Amazon, more impetuous than the Mississippi; and the collective waters of these two streams would not equal the thousandth fraction of the volume of water which it displaces."— *Maury.*

By means of the thermometer the navigator can follow the great liquid vein. The instrument, plunged alternately into its edges and its mid-stream, shows a difference of 27° of temperature.

Powerful and rapid the Gulf Stream runs northward, following the coast of the United States, as far as the Banks of Newfoundland. There

19

it encounters the tremendous shock of a polar current, upon which are
floating enormous icebergs, veritable mountains of ice, the force of
which is such that one of them, weighing more than twenty billions of
tons, carried the vessel commanded by Lieutenant de Haven more than
three hundred leagues southward. The Gulf Stream, whose waters are
lukewarm, dissolves the floating ice. The icebergs melt, and the earth,
and even the fragments of rock which they contained, are swallowed up
by the waters.

Upon reaching the neighborhood of Europe, it sends a great part of
its waters in the direction of the Polar Sea, along the coasts of Ireland,
Scotland, and Norway; the remainder turns off to the south, opposite
the west coast of Spain, and regains the great tropical current off the
centre of Africa. After having effected their junction with this current, of which they are, so to speak, the source, its waters make their
way westward, to reach once more the coasts of Mexico and the United
States, and to traverse, for the second time, the space which separates
the United States from Europe, thus forming a continuous circuit from
Africa to Mexico, returning to the point from which they started by the
route given. The bottles which sailors throw into the sea, with a mention of the day and the spot where they were confided to the ocean,
have shown us that this voyage of from 13,000 to 19,000 miles is accomplished in three years and a half. The winds have about the same
direction as the waters, that is to say, that between the tropics the easterly trade-winds prevail, driving the atmosphere from Africa to America, just as the tropical current conveys the water thither. Between the
United States and Europe, just as this current causes the sea to flow
eastward, so also do the counter-currents of the trade-winds blow toward
Europe, whence it happens that the passage from the United States to
Europe is effected more rapidly than the return journey, for in this latter case the wind and the current are against the vessel. It is well
known that when Christopher Columbus ventured to give himself up to
the west winds, he got as low down as Africa to take advantage of the
easterly winds which, according to his calculation, would lead him to
China. As the late M. Babinet remarks, it is difficult to understand
how, at this epoch, when geographical knowledge was sufficiently advanced to permit of the globe's dimensions and the distance from India
and China being pretty accurately known, any one could have expected
to reach the eastern coasts of China after a navigation equal to three
or four times the distance between the Old and the New World. If

America had not been in existence, he would have perished a hundred times over before he could have reached China.

Before passing to the other maritime circuits analogous to these of the North Atlantic, let us consider carefully the circumstances by which it is characterized.

The tropical waters, in their journey from the coasts of Africa to those of America, pass beneath the rays of a zenithal sun, and are continually being heated until they reach the Gulf of Mexico; they then flow by way of the Straits of Bahama, where they form a rapid current of hot water, which re-ascends to the east of the United States, toward the Banks of Newfoundland. There the current turns eastward on its way to Europe, but still preserves the high temperature due to its tropical origin, and this is one of the most powerful agencies of nature for increasing the temperature of our globe—viz., the conveyance, by means of these waters, of the heat which the sun sheds between the tropics toward the northern regions. In proportion as this current advances, it parts its heat, which it distributes into the atmosphere and over the seas which it traverses; then it returns, leaving Spain and the north of Africa to its left, to resume its place in the tropical current, and again to receive heat, which it will transfer as before to European latitudes.

It is by means of the winds that the heat of the sea communicates itself to the main-land. We shall see presently that in Europe the prevailing winds of the globe are westerly, inclining to south-west. It is seen at once that these currents of air, having a current of hot water for basis, will share its temperature, and pass over Europe and be much warmer than if the sea, deprived of this warm current, had only the same degree of warmth as is due to latitude. To demonstrate this assertion, we have only to compare the climates and temperatures of American cities with those of France and England which are in the same latitudes.

None of the masses of water which move from place to place in the seas merit such close attention as that of the Gulf Stream; none are of greater importance in regard to the commerce of nations, nor exercise a greater influence upon climate. It is to the Gulf Stream that the Britannic isles, France, and neighboring countries, owe, in a great measure, their mild temperature, their agricultural wealth, and, moreover, a very large part of their material and moral strength. Its history is almost that of the whole of the North Atlantic, so great is the hydrological and climacteric influence of this current of the seas.

Owing to the rotatory motion of the globe, and probably also to the general direction of the coasts, the current follows without intermission a north-easterly course, and comes in contact with none of the advanced points of the continent. Beyond New York and Cape Cod it bends farther eastward, and, ceasing to run parallel with the American coast, turns off into the mid-Atlantic, toward the shores of Western Europe. As Maury says: "If an enormous cannon could fire a ball from the Strait of Bahama to the North Pole, the projectile would follow almost exactly in the curve or course of the Gulf Stream, and, deviating gradually as it went, would reach Europe traveling eastward."

From the 43d to the 47th degree of north latitude, in the neighborhood of the Banks of Newfoundland, the Gulf Stream, traveling from the south-west, encounters upon the surface of the sea the polar current. The line of demarkation between these oceanic streams is never absolutely the same, and varies with the seasons. In winter—that is, from September to March—the cold current drives the Gulf Stream toward the south; for, during this season, all the circulatory system of the Atlantic — winds, rain, and currents — veer toward the southern hemisphere, above which the sun is situated. In summer—that is, from March to September—the Gulf Stream regains the preponderance, and repels the polar current farther north. After having come in collision with the waters of the Gulf Stream, those of the arctic current cease, in a great measure, to flow upon the surface, and sink by reason of their being cold, and consequently heavy. It is easy to trace the direction of this counter-current, which is exactly opposite to that of the Gulf Stream, by the mountains of ice which the mild temperature of lower latitudes fails to melt, and which float in a south-easterly direction, until they meet the superficial current, which they cleave like the prow of a vessel. Farther south, it is only by sounding that the existence can be ascertained of this hidden current, the cold waters of which serve as a bed to the warm stream proceeding from the Gulf of Mexico. It descends lower and lower until it reaches the Straits of the Bahama Islands, where the thermometer indicates it at a depth of 1300 feet.— *Reclus.*

We have the pendant of the Gulf Stream in the Pacific Ocean, in the shape of a warm current which follows the coasts of China and Japan, and which has long been known to Japanese geographers by the name of Kuro-Siwo (the Black Stream)—a name which originated, no doubt, in the dark hue of its waters. In the southern seas the currents are not so well known to us, and are, in fact, much less numerous. It is, moreover, probable that the marine streams are not isolated currents, but several portions of one net-work, distinct veins in a comprehensive system of circulation.

The quantity of heat which the Gulf current carries northward forms a very considerable part of the caloric which is stored up in the waters of the torrid zone. The total heat of the current would suffice, if it were concentrated upon a single point, to melt mountains of iron and to form a stream of metal as voluminous as the Mississippi; it would, further, suffice to raise from winter to summer temperature the whole column of air which lies over France and Great Britain.

Notwithstanding the march of the sun, it is, upon an average, as warm in Ireland at 52° N. latitude as it is in the United States at 38° N. latitude, or a place more than 1000 miles nearer the equator.

The Gulf current, which carries the tropical heat to the temperate regions of Europe, often serves, too, as a highway for the hurricanes; hence the names of Weather-breeder and Storm-king, which have been given to the Gulf Stream. The movements of the atmospheric ocean and those of the ocean of waters are so completely parallel that we are tempted to view them as one and the same phenomenon both in the currents aërial and marine. Thus the Gulf Stream seems to be for the winds what it in reality is for the waters—the great intermediary between the two worlds. It transmits to the seas of Northern Europe the saline matters of the Gulf of the Antilles; it carries with it the tropical heat for the benefit of the temperate regions; it marks the route followed by the torrents of electricity proceeding from the storms in the Antilles. It is, in fact, the great serpent of the Scandinavian poets, which displays its immense ring along the ocean, and which, by the motion of its head, either causes a mild breeze to blow or emits the raging hurricane. While, in the North Atlantic, the equatorial current, which falls into the Gulf of Mexico, returns from whence it came, traversing high latitudes, another part of this current, much less voluminous, after having touched Cape St. Roch, which forms the eastern extremity of Southern America, descends along the eastern coast of that same continent, and then, crossing the Atlantic from west to east, returns toward Lower Africa, running along its western shores and rejoining, by the south, the great tropical current, just as the Gulf Stream meets it northward. Down to the quantity even of water which it contains, this current bears a marked resemblance to the circuit which occupies the north of this ocean. The portion which runs off beyond the tropics, and which returns from west to east, from South America to South Africa, is also a current of hot water, like the Gulf Stream between the United States and Europe. The comparison of the masses of water which each of these circuits separately conveys shows how much better the north is provided with hot waters than the south. It is not too much to say that the north circuit forms a current five or six times more abundant than the south circuit. If we now consider the Pacific Ocean, there also we find tropical waters which flow on to the shores of New Holland, the Northern Archipelago, and Lower Asia. Most of these waters re-ascend northward in vast currents of lukewarm water which give to High California and to Oregon climates very similar to those of Europe.

The North and South Atlantic, the North and South Pacific, and the Indian Ocean, each contain a current, that of the former ocean being the most voluminous. The Arctic seas, north and south, also appear to be traversed by a current running eastward, round the Pole.—*Babinet.*

The circulation of the sea is completed by submarine currents. There must exist one of these, conveying the waters of the Mediterranean into the Atlantic. Its existence is, in a way, demonstrated by a calculation which shows that the quantity of salt water in the upper current of the Straits of Gibraltar is 2900 cubic miles per year, the quantity of soft water contributed by the rivers 240 cubic miles, and that which is lost by evaporation 480 cubic miles. Thus there would be an annual excess of 2660 cubic miles, if the equilibrium were not re-established by a submarine current. This hypothesis seems to have received confirmation by a very curious fact.

Toward the close of the seventeenth century a Dutch brig, pursued by a French corsair, the *Phœnix*, was overtaken between Tangier and Tarifa, and disabled by a single cannonade. Instead of sinking at once, the brig, which had a cargo of oil and alcohol, floated beneath the surface of the waters, and did not finally go to the bottom for two or three days, after having been carried twelve miles toward Tangier from the point at which she first disappeared. It was evidently carried this distance by an under-current, in an opposite direction to that of the surface-current. This fact, in conjunction with some recent experiments, confirms the opinion which admits the existence of a current issuing from the Straits of Gibraltar. Lieutenant Maury also considered it certain that there is a submarine counter-current to the south of Cape

Horn, which carries the overflow of the Atlantic into the Pacific. As a matter of fact, the Atlantic is continually being fed by very large rivers; whereas the Pacific, into which debouches no important stream, must, on the other hand, lose an immense body of water, owing to the evaporation which takes place from its surface.

Certain lower currents have been ascertained by weighting a piece of wood and plunging it into the water, keeping hold of it at the same time with a piece of string, so as to let it sink to any depth which may be desired. At the other end of the line is attached an empty barrel strong enough to support the apparatus, and then the whole is set free. The sailors who tried this experiment for the first time were astonished to find this little barrel traveling in an opposite direction to the wind and the sea at the rate of a knot or more per hour. The crew were even inclined to look upon it as a supernatural phenomenon. The speed of the barrel was evidently equal to the difference in speed between the upper and the lower currents.

In 1773, Captain Deslandes cast anchor in the Gulf of Guinea; a strong current running into this bay prevented him from going farther south. Deslandes then noticed that there was an under counter-current at the depth of eighty feet, and he adopted an ingenious plan for availing himself of it. A machine, with considerable surface, was let down to the depth of this submarine current. This was hurried along with so much force that it towed the vessel at the rate of one and a half miles per hour.

In the Antilles seas a vessel is sometimes brought to a halt even in the middle of a current.

In the Sound there has long been known to exist both an upper and an under current.

The mean temperature of the surface of the sea differs but little from that of the air, so long as warm currents do not add their influence. In the tropics it appears that the surface of the water is rather warmer than that of surrounding air.

An examination of the temperature at the surface at various depths gives the following results:

1st. In the tropics the temperature *diminishes* with depth.

2d. In the polar seas it *augments* with increased depth.

3d. In the temperate seas, included between 30° and 70° latitude, the temperature decreases in a smaller degree as the latitude gets higher, and beyond degree 70 begins to increase.

There exists, then, a zone the temperature of which is almost stationary, from its surface down to a great depth.

It is scarcely possible to doubt that the currents caused by the difference in pressure which strata of the same level are subject to, at the equator and toward the poles, contribute materially to this distribution of heat. It seems certain that there is, as a rule, a surface current which carries the warm waters of the tropics toward the polar seas, and an under-current which takes back from the poles to the equator the frigid water of the polar regions; but the direction and intensity of these currents are modified by a number of causes, which depend upon the depth of the sea basins, their shape, and the influence of winds and tides. In very deep water there is a uniform temperature of 39°, which corresponds, as physical science has proved, to the maximum of the density of water. This temperature exists at the equator at a depth of 7200 feet. In the polar regions, where the water is colder upon the surface, this same temperature is met with at a depth of 4600 feet. The isothermal lines of 39° form the demarkation between the zones where the surface of the sea-water is colder, and those where it is hotter than the stratum which marks 39°.

Lastly, the quantity of salt in the waters of the ocean differs according to the points of the globe, and is unquestionably an important element in the density, and, consequently, in the actual formation, of maritime currents.

CHAPTER III.

THE VARIABLE WINDS: THE WIND IN OUR CLIMATES — MEAN DIRECTIONS IN EUROPE AND IN FRANCE—RELATIVE FREQUENCY OF THE DIFFERENT WINDS—RISE OF THE WINDS ACCORDING TO THE TIMES AND PLACES—MONTHLY AND DIURNAL VARIATION IN INTENSITY.

HAVING observed the regular and periodical currents of the atmosphere and the seas, let us now consider the *irregular* winds which blow over our climates. These latter are only apparently irregular, for in nature there is no such thing as chance, and each molecule of air that changes position is obeying laws as absolute as those which regulate the worlds of space. We will endeavor to throw some light upon the multitude of winds which succeed each other in our regions, and to ascertain the causes which set them in motion.

Beyond the changing limits within which blow the trade-winds and the periodical breezes of the two hemispheres, the temperate zones are the seat of variable winds. Europe, for instance, is entirely subject to that régime, and the masses of air float off sometimes in one direction, sometimes in another. Now and then one kind of wind will prevail for weeks together; sometimes, on the other hand, the wind will blow from two or three different points of the compass in as many hours; sometimes, again, the air remains calm, and there is not a breath of wind to agitate even the foliage of the poplar-tree. Thus the instrument used to indicate the direction of the winds in our climates, the weather-cock, has long been taken to signify inconstancy.

Nevertheless, even inconstancy has a cause, and is often more apparent than real. The winds in our climates, which seem to us so capricious and variable, leave behind them a trace of the laws which they follow.

We have seen that the *upper* trade-winds, which travel from the equator to the pole, modify their primitive direction from north to south in our hemisphere, and veer gradually to the south-west as they reach higher latitudes. They lose at the same time both in velocity and heat, and gradually come nearer to the ground. About 30° latitude they are almost on a level with the surface. This *south-west* wind, in fact, prevails throughout Europe. Thus, amidst the variety of winds, we already find that there is one which is regular, since it is no other than the upper trade-wind which has descended thus far, and which occupies the largest place in the meteorology of our climates.

We have seen that the great oceanic current, the Gulf Stream, reaches the coasts of Europe from the south-west. The air circulates in the same direction, and increases still further the inflection of the upper trade-winds, or, to speak more correctly, it is the same equatorial aërial and maritime current turned off in a south-west direction by the rotation of the earth.

To ascertain precisely the direction of the winds, it is necessary to keep an account of the time during which each wind has prevailed, taking a supposititious total upon which the calculation is based. Thus, for instance, let us suppose that the south-west wind has been blowing for a little more than ninety days of the year; it would be put down that it had prevailed for a quarter of the whole time. If the total 1000 be taken to signify this time, 250 would be placed to the account of the south-west wind (that is, if it had been blowing for ninety-one days and seven hours, which is exactly a quarter of a year). In the same way all directions indicated by the vane would be similarly put down, and thus we should obtain a comparative table giving the average result for a long series of years.

This plan has been adopted in Europe for many years, and the following table will show the result of the observations made. It indicates a decided preponderance of a south-westerly wind over the European continent and even North America:

RELATIVE FREQUENCY OF DIFFERENT WINDS.

	N.	N.E.	E.	S.E.	S.	S.W.	W.	N.W.	Mean Direction of the Wind.	Mean Force of the Wind.
France........	126	140	84	76	117	192	155	110	S. 88° W.	133
England......	82	111	99	81	111	225	171	120	S. 66° W.	198
Germany.....	84	98	119	87	97	185	198	131	S. 76° W.	177
Denmark......	65	98	100	129	62	198	161	156	S. 62° W.	170
Sweden	102	104	80	110	128	210	159	106	S. 50° W.	200
Russia.........	99	191	84	130	98	143	166	192	N. 87° W.	167
N. America..	96	116	49	108	123	197	101	210	S. 86° W.	182

It will be seen that the *south-west* is the prevailing wind. By adding up the numbers set down, as they run horizontally, the total will be found to be 1000; thus in France the south-west wind blows $\frac{192}{1000}$, or nearly a fifth of the whole time. The proportion is greater still in England. By adding together the west and the south, it will be seen that the continuance of wind from this quarter amounts to nearly one-half of the prevailing winds; $\frac{146}{100}$ in France, and $\frac{51}{100}$ in England. The careful observations taken at Brussels and in various parts of Belgium since 1830 show a like preponderance to exist there. The prevailing wind is, indeed, just S. 45° W. In Russia there is a greater variety, owing to its distance from the ocean.

Thus we are under the benign influence of the equatorial current. But, if the return trade-winds reach so far, and even to the pole, the lower polar current, which conveys the cold air from north to south, and forms in the tropics the north-easterly trade-winds, must also have its influence upon our regions. It must pass us somewhere on its way from the pole to the equator; and if the air which travels from the equator to the pole did not return, there would cease to be any atmosphere at all in the tropics. Now, let us study for a moment the preceding table of the relative frequency of the winds. The maximum is to the south-west, as is shown by the figures underlined, whence the totals become smaller and gradually swell again, giving a second maximum in the shape of a N.E. wind. That is our polar current. The N.E. wind forms the $\frac{11}{100}$ of the winds in France, $\frac{11}{100}$ of those in England, and $\frac{12}{100}$ in Russia.

There exist, therefore, in our hemisphere *two general directions* of winds. Now it is the equatorial, now the polar current which predominates. The first is warm and moist, the latter cold and dry. Each has an opposite influence upon the productions of the soil, and the state of the crops depends in a great measure upon the epoch and continuity of their prevalence.

The S.W., W., and S. winds on the one hand, the N.E. and N. winds on the other, constitute the general *primitive* winds to which our regions are subject. All the other directions of the wind are due to these two currents, and for the following reasons:

If the two currents are blowing in proximity to each other, each occupying a certain extent, as they are proceeding in opposite directions, there must exist, about the limit which separates them, whirlwinds and circular blasts engendered by the action of the two currents of air. These circular blasts will revolve from N.E. to S.W. at the tangent of the polar current, from S.W. to N.E. at the tangent of the equatorial current.

As an instant's reflection will show, this is a simple horizontal movement like that of a grinding-stone. Each point in the circumference of
this grinding-stone will have its own direction, since we are supposing
that this mass revolves in its entirety. It would be, in fact, a zone of
variable winds which would be liable to change its place under the influence of the two great currents from which it springs, and which themselves vary in position, extent, and intensity. Here we have one cause
for the change of wind which is almost constant (since the two currents
are always in existence), and which must be multiplied to a vast extent.
There is a second and not less important cause.

There is a constant difference of temperature in the various regions
of the same country. In one place there is water, in another land; here
deserts, there forests; at one point low and sultry plains, at another
bleak table-lands. These differences of temperature modify our two
currents on their passage through them. A cloudy sky is favorable to
the progress of the one, and arrests the march of the other. Thus partial winds spring out, like lateral branches, from the trunks of the two
great trees which are lying prostrate.

A third cause must be superadded to the above—the protuberances
upon the land. The general currents which pass over a chain of mountains do not blow with the same regularity that they do in the plains.
In fact, the winds must be all the more unequal in their successive
blasts in proportion as the surface over which they sweep is uneven.
The same aërial surface which moves over the waters with the uniformity of a vast river, loses the regularity of its movement when it is interrupted in its course by the protuberances of the soil. At the foot of the
Swiss mountains, and especially around Geneva, where the ground is
very uneven, the alterations in the force of the wind are so great that
the anemometer sometimes shows a variation in intensity from one to
three. In the lofty ravines of the Alps it often happens, even in the
midst of the fiercest storms, that the atmosphere is at intervals perfectly
calm. Even in the countries which are not very hilly, and in the plains
studded with houses and plantations, the wind does not blow with the
regularity of the trade-winds at sea, but advances with a succession of
blasts, each of which represents a victory of the atmospheric current
over some obstacle upon the plain.

At the level of the ground the wind is always intermittent, whereas
in the heights of the air it almost always proceeds with the regular and
majestic motion of a river.

Thus laws regulate these minor changes as well as the general move-ment of circulation. We may now consider whether there is any law as to the succession of winds.

Let us revert to the first cause of change dealt with above. As a rule, our hemisphere is divided into large oblique bands composed of masses of air running in an inverse direction, some toward the poles, others toward the equator. These bands shift their position around the globe, so that at one moment the polar wind, at another moment the tropical wind, will prevail in the same place; but there is always a compensating balance between these atmospheric currents, and the wind, which is neutralized or repelled in one part of the hemisphere, is soon felt at some other point. As long as the struggle between the two masses of air animated by opposing movements continues, the vicissi-tudes of the conflict and the general preponderance of one of the winds cause a temporary modification in the march of the air, and make the vane turn toward the different points of the horizon. It is from the en-counter of the two regular winds that chiefly arises the apparent irregu-larity of the whole atmospheric system.

Although the struggle between the two aërial streams is continually going on at one point or another, nevertheless they are not of equal force, and one of them always obtains the mastery after a more or less prolonged period of resistance. This wind, which proves the superior in impulse, is the back current that has come down from a great eleva-tion, and reaches the level of the ground outside the zone of the trade-winds. The atmospheric currents coming from the equator naturally incline toward the east, whence it results that, in the northern hemi-sphere, the majority of the winds are from the west.

Many centuries ago the savants ascertained that in the northern hemi-sphere the normal succession of the winds is from south-west to north-east by west and north, and from north-east to south-west by east and south. This is a rotatory movement analogous to that which the sun seems to describe in the sky when, after rising in the east, it travels westward, developing its vast curve around the zenith. Aristotle, in his "Meteorology," wrote, more than two thousand years ago, "When a wind ceases to give place to another, the direction of which is next in order to that of the former, the change always takes place with the sun." Since the time of the great Greek naturalist, several authors enumerated by Dove have re-affirmed this fact of the regular rotation of the winds, which was, indeed, known to sailors in the earliest ages. Dove was the first who collected the scattered proofs of this generally accredited theory, and transformed the primitive hypothesis into a scientific certainty. It no longer admits of any doubt that, in the northern hemisphere, the winds generally succeed each other in the following order: S.W., W., N.W., N., N.E., E., S.E., S., S.W.

In the southern hemisphere the normal rotation of the aërial currents is exactly the opposite. Thus, as E. Reclus remarks, the procession of the winds in each of the two hemispheres coincides with the apparent march of the sun, which, so far as Europe is concerned, describes its daily course to the south of the zenith, and, in Australia, passes to the north of it. Such is the regular order which Dove termed the law of gyration, but which is generally called after the name of its discoverer.

I have noticed in my aërial travels a gyratory deviation, which shows that the wind can not extend in a straight line when it spreads over a great area, but inclines in the direction indicated by the above theory.

Immersed in the atmospheric current which bears him along, the aëronaut is placed in the most favorable position imaginable, both for ascertaining the continuous direction of the current and for measuring its speed. Upon each occasion I took care to trace accurately on a map of France or Europe the aërial line taken by the balloon, which is done with extreme ease when the sky is clear, and which may always be obtained even with a cloudy sky, either by availing one's self of the momentary breaks or by descending every now and then below the clouds.

The balloon marks so accurately the direction and speed of the current, that the first sensation in navigating the air is that of being completely at a stand-still. It is a peculiar and always surprising impression experienced when, traveling along with the velocity of the wind, one feels neither the slightest breath of air nor the least movement, even when hurriedly carried off into space by the most violent tempest. I never felt but once any thing like a breeze. This was on the 15th of April, 1868, and then only for a few minutes. This I attribute to the fact that the balloon, which was traveling at the rate of thirty-four miles an hour, had reached a region where the air was shifting its position less rapidly. One capital fact is brought to light by the aërial lines which I have traced, and that is, that these routes all incline in the same direction, by virtue of a general gyratory deviation.

The actual direction of a wind is the most easily observed of its characteristics. To ascertain it, we suppose the horizon to be divided into four equal sections by two diameters perpendicular to one another, one

running from south to north, the other from east to west. The points
at which the diameters intersect the horizon are called the four cardinal
points. But they would not of themselves suffice, for it is necessary to
have a number of intermediate directions. These are indicated by other diameters, which divide the horizon into sixteen equal parts; and
thus we obtain the indications of the wind at as many different directions, called, starting from N. round by E., N.N.E.; N.E.; E.N.E.; E.;
E.S.E.; S.E.; S.S.E.; S.; S.S.W.; S.W.; W.S.W.; W.; W.N.W.;
N.W.; N.N.W.; N.

When the points of the compass are known, and objects are affected
by the movement of the air, it is easy to ascertain the direction of the
wind; but often recourse is had to an instrument which is no doubt
the oldest of those used in meteorology, viz., the weather-vane. This
simple apparatus consists of a metal plate, generally of tin or zinc, cut
into a figure of some kind, and turning upon a rod, to which is attached a horizontal cross with the letters N., S., W., E., at its extremities.
The weather-vane is placed upon the highest part of a building, and in
by-gone days no house of moderate size was deemed complete without
it. Exposed to the weather, it becomes corroded, and ceases to follow
implicitly the impulsion of the winds. Sometimes the rod gets out
of order, and the vane inclines to one side. Its indications are not
worth consideration unless they are verified from time to time, and the
vane is situated beyond the influence of obstacles which obstruct the
free passage of the wind. It is not a rare occurrence for the atmosphere to be influenced by several different currents, one superposed
upon the other. In this case, the principal current — that which, so
to speak, governs the weather—is generally placed at a considerable
height, even when it is not the highest of all; it is discovered by the
motion of the clouds. This is the best and surest indication of the
direction of the wind.

As the mass or density of the air only varies within very restricted
limits, the force of the wind depends almost entirely upon its speed,
and varies as the square of its velocity, or very nearly. The terms
"force of the wind" and "velocity of the wind" are therefore almost
identical. To measure the speed, an apparatus, called an *anemometer*,
is used. One of those most frequently in use is that by Dr. Robinson,
of Armagh. This instrument is composed of a vertical axis support-
ing four horizontal radii of the same length, crossing at right angles,
and at the extremities of which are four hollow half-spheres.

A moment's reflection will suffice to make it clear that the wind is always pressing against two concave and two convex half-spheres. As it has more power over the former than over the latter, it causes a rotatory motion, and the number of revolutions which the half-spheres make is proportional to the velocity of the wind. The number three represents with approximate accuracy the relation which exists between the horizontal movement of the air and the horizontal movement of the half-spheres. Thus, by measuring the circumference of the circle which the centre of one of the demi-spheres describes, and by multiplying half its length by three, we obtain the distance traveled by the wind for each revolution of the apparatus.

The monthly averages of each wind referred to eight points of the compass, as found from sixty years' observations at the Observatory at Paris (1806–1866), are as follows:

PROPORTION UPON 10,000 WINDS.

The N.. 1039
" N.W.. 1084
" W.. 1782
" S.W.. 1935
" S.. 1476
" S.E.. 799
" E.. 694
" N.E.. 1191

These numbers show the dominant winds to be S.W. and S.

The monthly averages of the winds at London show a prevalence of south-westerly winds to an even more marked extent than in Paris. The result of observations taken for twenty consecutive years at the Greenwich Observatory, which I have received from Mr. Glaisher, the director of the meteorological service there, gives the following averages of the relative frequency of each wind (see Fig. 58):

The N.   wind blows on an average for 41 days.
N.E.   "   "   "   "   48   "
E.   "   "   "   "   22   "
S.E.   "   "   "   "   20   "
S.   "   "   "   "   34   "
S.W.   "   "   "   "   101   "
W.   "   "   "   "   38   "
N.W.   "   "   "   "   24   "
       Days of complete calm   34   "
                                       ___
                                       365   "

The average direction of the winds at Brussels gives the same result (see Fig. 59), and we have already remarked the predominance of the equatorial current in the study of the general mass of observations taken throughout Europe.

It seems certain that the wind is propagated not only by *impulsion*, but by *aspiration*. This second mode deserves attention because it furnishes important data as to the cause of the movement. Franklin appears to have been the first to observe this fact. He mentions in one of his letters that, when attempting to watch an eclipse of the moon at Philadelphia, he was prevented from doing so by a hurricane from the north-

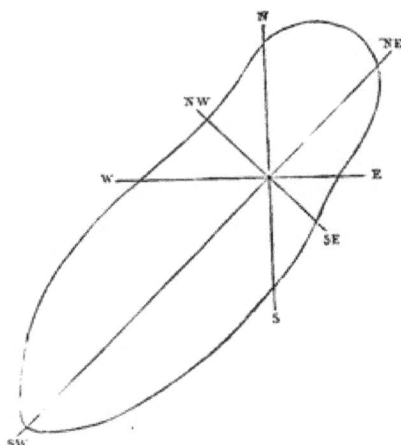

Fig. 58.—Average annual prevalence of the different winds at London.

east, which took place at about seven in the evening, and was followed, as is usually the case, by clouds which obscured the whole sky. He learned to his surprise, some time afterward, that at Boston, which

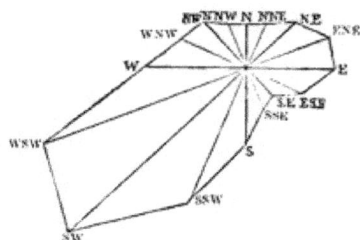

Fig. 59.—Average annual prevalence of the different winds at Brussels.

is about 400 miles to the north-east of Philadelphia, the storm had not commenced until 11 P.M., long after the first phases of the eclipse had been observed; and, by a comparison of the various accounts collected in different colonies, Franklin remarked that, according as the place was farther north, the later was the hour at which this north-easterly tempest occurred there, and that thus the wind was blowing in one direction and was advancing progressively in another.

Since that time a great number of tempests have been remarked, which presented this peculiarity in respect to their direction. Nevertheless, in nearly every case, the wind advances in the direction toward which it is blowing.

The terrible storm from the south-west, which occurred on November

ber 29, 1836, passed over London at 10 A.M., the Hague at 1 P.M., Amsterdam at 1·30 P.M., Emden at 4 P.M., Hamburg at 6 P.M., and Stettin at 9·30 P.M. It traveled, therefore, in the same direction as that in which it was blowing, and took ten hours to reach Stettin from London.

The following is a general sketch of the prevailing distribution of wind over the surface of the globe:

Suppose a ship to start from the Arctic Polar Circle for the equator, to cross it, and proceed onward to the Southern Arctic Circle, it will meet with the following succession of winds:

1st. At the outset, it navigates in the region of south-westerly winds or of the northern anti-trade-winds, so called because they blow in an opposite direction to the trade-winds of their hemisphere.

2d. After having crossed the parallel of latitude 50°, and until it reaches that of 35°, it encounters the zone of partially western winds, in which south-west predominates, and in which the north-easterly current also prevails over the other winds.

3d. Between N. latitudes 40° and 46° there is a region where the winds are very variable, and where there are calms. The winds blow, in the course of the year, in equal proportions from the four quarters during three months.

4th. To the west winds, which have predominated thus far, succeeds the calm region of the Tropic of Cancer, then that of the trade-winds which conduct the vessel to the latitude of 10° north, where it reaches the zone of equatorial calm, which is only 5° in breadth.

5th. From 5° north to 30° south the south-easterly trade-winds prevail.

6th. Then succeeds the calm zone of the Tropic of Capricorn, analogous to that of the Tropic of Cancer.

7th. From S. latitude 35° to 40° there prevail, as a rule, westerly winds, which sometimes veer to N.W. and to S.W.

8th. Lastly, the vessel reaches at S. latitude 40° the southerly anti-trade-winds, which have a north-westerly direction, and prevail, as far as observations in the direction of the Southern Pole have extended.

If we now consider the *intensity* of the wind, we notice that its variation, apparently so irregular, is dependent, like every thing else, upon the movements of the earth, in the seasons and in the days. Twenty years' comparisons made at Brussels show that the wind is less intense during the longest days than during the shortest, as in June the indi-

catious of intensity are 0·832, and in December 1·227. The month
of September, however, seems to be an exception, for it gives the
minimum, averaging only 0·804; but this month is, in many respects, an exceptional one in our
climates.

It is, moreover, remarkable that
during the six months when the
sun is below the equator the force
of the wind is above the average
of the year; whereas, on the contrary, its force is generally below
the average during each of the
other six months.

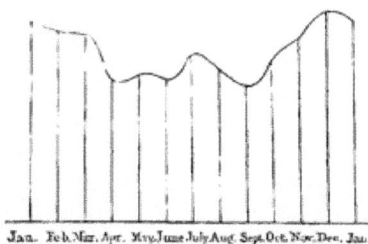

Jan. Feb. Mar. Apr. May. June July Aug. Sept. Oct. Nov. Dec. Jan.

Fig. 60.—Monthly intensity of the winds.

The intensity of the wind varies, too, according to the time of day.
The anemometer at the Brussels Observatory, which registers the wind
every five minutes, shows that this diurnal variation in the intensity of the winds extends from
an average of 0·15 (midnight to
4 A.M.) to 0·21 (10 A.M.), 0·26
(noon), 0·29 (2 P.M.), 0·28 (4
P.M.), and 0·23 (6 P.M.) This is
shown by Fig. 61.

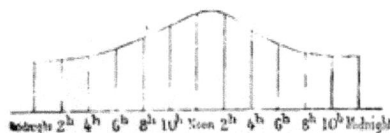

Midnight 2ʰ 4ʰ 6ʰ 8ʰ 10ʰ Noon 2ʰ 4ʰ 6ʰ 8ʰ 10ʰ Midnight

Fig. 61.—Diurnal intensity of the winds.

Thus the wind is almost twice as strong at 2 P.M. as in the middle
of the night.

The time will arrive when the march of the variable winds in our
climates will be ascertained, just as the general circulation of the trade-
winds and the monsoons in the tropical regions has long been made
known. The day will come, too, when observations of the upper winds
will have revealed to the meteorologist the route which they follow,
just as observations of the planets have discovered to the astronomer
the orbits they describe. Then we shall be able to tell the daily and
yearly direction of the atmospheric wave which passes over our heads.

The currents, the laws of which we have been studying, play a great
part in nature. They favor the growth of flowers by causing the
branches of the plants to oscillate, and blowing the seeds a long distance. They renovate the air in cities, and render northern climates
milder by supplying them with heat from the south. Without wind
rain would be unknown in the interior of continents, which would be
transformed into arid deserts. Without wind the earth would be al-

most uninhabitable, and whole districts would become centres of contagion—vast cemeteries, in fact. We have seen the deleterious effects of air when confined. Man acts as a deadly poison to man, as typhus fever and plagues clearly demonstrate. The winds alone can avert these calamities, by blowing away the emanations, by disseminating them in the regions of space, and substituting for vitiated air a fresh, salubrious atmosphere. Moreover, it is the same with air as with water; motion alone keeps it pure, whether because it has a principle of life unknown to us, or because animalculæ, or vegetable and animal débris, becoming decomposed when at rest, spread their deleterious principles throughout a motionless atmosphere.

The winds not only bring life upon their blast, they may also transmit death to countries where the yellow fever, the plague, or cholera prevails.

A distance of twenty leagues does not protect Rome from the deadly air which has blown over the Pontine marshes. In Paris the west wind blows for seventy days in a year; place an *Agro Romano* in the Mayenne, the Sarthe, or Touraine, and the population of Paris would be decimated by intermittent fever.*

It has been mentioned that in all latitudes similar to those of Europe, and even rather more southerly, the prevailing wind is west, which conveys to Europe the warm air of the Atlantic, and endows it with that unique climate which admits of the cultivation of barley and other cereals as far as the North Cape; whereas in Greenland, which is deprived of these balmy breezes, it never thaws, although this latter country is in about the same latitude as the north of Scotland. The city of Boston, in the United States, is in the same latitude as the olive-growing districts of Spain. Nevertheless, during the winter there, the small lakes in the neighborhood are sometimes frozen a yard deep. The five great American lakes (which are, in truth, inland seas) freeze over, and are traversed by temporary railroads. What a striking contrast between the climate which produces this ice and that where the olive-oil and wine afford an easy subsistence to the indolent cultivators about Bordeaux and in Spain! Yet the intelligent activity of the inhabitant of the United States has transformed even this ice into a prof-

* There are at times strange variations in the Bills of Mortality which can be due to no other cause than the wind. Thus, for instance, on July 26th, 1871, half of the inhabitants of Paris were attacked by a mild form of cholerine. There had been no other perturbation than a heavy gale of wind which raged all the previous night.

itable crop, which is exported to India and the tropical regions, fetch-
ing a higher price than that obtained for the olives of the Asturias.

Toward the centre of France there exists the most exquisite climate
of the whole world, so that if a locality be selected somewhere about
the east of the meridian of Paris, it will possess a more favorable climate than any other place in the same latitude.

Let us now consider the influence which the wind has on climatol-
ogy. The winds have a dominating influence upon the distribution
of temperature, as they effect in different countries, according to their
positions in respect to the four cardinal points, permanent modifications in the climate which these countries would otherwise have. The
régime of the winds leads to a régime of temperature which is indis-
solubly connected with it. The currents of the atmosphere bring with
them the temperature of countries whence they come. Every one may
have noticed that the north wind is generally cold, and the south wind
generally warm. But it would be commonplace to be satisfied with
these vague indications, and the rôle of science is to analyze facts. Consequently, for many years past, the temperatures which the thermometer denotes for the directions of the wind have been carefully compared,
and one of the first results was to show that in France the winds blowing from the south-east and the south cause an increase of 5° or 7° in
the temperature over those which blow from the opposite direction. A
comparison of the mean corresponding temperatures of the different
winds throughout the various cities of Europe has made it evident that
the influence of the wind varies according to places, as may be seen by
the appended table:

INFLUENCE OF THE WINDS UPON TEMPERATURE.

Stations.	N.	N.E.	E.	S.E.	S.	S.W.	W.	N.W.	Differ-ences.
	Deg.	Deg.	Deg.	Deg.	Deg.	Deg.	Deg.	Deg.	Deg.
Paris......................	52·2	52·7	55·8	59·2	59·4	58·5	56·1	53·4	7·2
Carlsruhe	50·9	47·5	50·9	55·6	54·5	51·6	54·3	52·2	8·1
London..................	45·9	46·6	49·3	51·1	52·5	51·4	50·4	47·7	6·6
Dublin...................	45·3	46·6	48·2	49·3	50·9	50·7	48·0	45·5	5·6
Hamburg................	46·4	45·7	47·1	49·1	50·0	50·2	48·6	47·1	4·5
Zeeken (Silesia)........	42·3	43·5	45·7	46·8	49·3	49·1	46·8	44·4	7·0
Arys (Prussia)..........	39·4	39·9	38·1	46·2	43·7	43·5	44·6	46·6	8·5
Reikiawick (Iceland)..	35·1	35·8	41·2	45·0	46·6	38·5	45·9	45·7	11·5
Moscow..................	34·2	34·5	38·3	39·2	42·8	42·3	41·7	37·9	8·6

Thus the mean difference between the influence of the warm and of
the cold winds reaches 7°·2 in Paris, and as much as 11°·5 in Iceland.
There are often differences even more marked.

The coldest wind is nearly always that which blows from a direction between north and east. The warmest wind is nearly always from S.S.W. The farther one passes inland the nearer it approaches to the west.

The preceding fact is a confirmation of the meteorological truth that no phenomenon stands alone: all act and react upon each other. No sooner does the S.W. wind begin to blow than it takes effect upon the temperature, not only by its warmth, but by the vapor which it brings, and the condition of the sky which is the consequence. In winter, the moist west winds are remarkably warm, because they cover the sky with clouds, and thus prevent loss of heat by terrestrial radiation.

The winds affect not only temperature, but also atmospheric pressure.

When the north and north-easterly winds are blowing, the barometer rises; it falls when the wind is from the S. or the S.W.

The following is the result of a great many years' observations in the principal cities of Europe, and it shows very clearly the influence of the wind upon the reading of the barometer:

INFLUENCE OF THE DIFFERENT WINDS UPON THE BAROMETER.

Winds.	Paris.	London.	Copen-hagen.	Berlin.	Halle.	Vienna.	Stock-holm.	St. Pe-tersburg.	Moscow.
	Inches.	Inches.	Inches.	Inches.	Inches.	Inches.	Inches.	Inches.	Inches.
N........	29·892	29·850	30·108	29·902	29·753	29·536	29·843	29·919	29·283
N.E.....	29·938	29·953	30·136	29·921	29·761	29·511	29·910	30·028	29·358
E........	29·830	29·891	30·101	29·902	29·709	29·584	29·819	30·001	29·288
S.E.....	29·699	29·809	29·898	29·749	29·629	29·461	29·726	30·030	29·219
S........	29·672	29·728	29·920	29·592	29·571	29·451	29·682	29·918	29·171
S.W.....	29·675	29·755	29·890	29·658	29·619	29·465	29·669	29·949	29·164
W........	29·776	29·847	29·992	29·764	29·619	29·584	29·782	29·911	29·204
N.W....	29·866	29·856	30·086	29·836	29·711	29·520	29·811	29·898	29·228
Mean ...	29·798	29·858	30·035	29·780	29·694	29·478	29·803	29·963	29·257

The general result of these researches is that the barometer rises highest with the wind between north and east—that is to say, when the current is coldest; and that its minimum elevation is when the wind is anywhere between south and west, the points from which its current blows the warmest. Analogous conclusions have been obtained in other countries. Thus, upon the eastern coasts of the United States and China, the barometer is generally highest when the wind is in the north-west—the coldest which prevails in those regions—and, as a rule, lowest when it is in the south-east, the temperature being at its maximum when the wind is in this direction.

The fact of the reading of the barometer increasing with cold winds, and decreasing when the winds are warm, is one that has been made evident wherever observations upon the point have been taken.

It may be generally stated, so far as our hemisphere is concerned, that *the barometer reaches its maximum when the winds blow from the north and the interior of continents, and its minimum when they come from the equator or the sea.*

In Europe the most rain-bringing winds are those between south and west, and the driest those between north and east: this is the reason why it rains oftener when the barometer is low than when it stands high.

Just as the winds, according to the direction whence they come, influence the temperature and the pressure of the air, the reading of the thermometer and the barometer, so do they affect *humidity*, announcing, bringing on, and keeping off rain. Daily experience tells us that the air has not always the same degree of moisture irrespective of the direction of the wind. When the farmer desires to harvest his hay or corn, when the laundress puts out her linen to dry, their task is accomplished far more rapidly with an easterly than with a westerly wind. Certain dyeing operations can only be attempted with the wind in the east. Instructive as these observations may be, they can not, however, provide us with rigorous and unchanging laws.

The air always contains, in addition to the gases of which it is composed, a certain quantity of vapor of water, and this element plays a principal part in the absorption and distribution of heat over the surface of the globe.

It would be to the highest degree important to be able to ascertain numerically the quantity of vapor which exists in the several regions of the globe. The life of plants and of animals, the nature of the landscape, are dependent upon this element as well as upon temperature: the dryness and the humidity of the air have the greatest influence upon the development of disease. What we do know is that the air above all the seas is saturated with vapor of water.

The farther inland, the drier the air becomes: nevertheless, after long-continued rain, it is at times saturated with moisture overland, because soft water vaporizes more readily than salt water. But, generally speaking, the quantity of vapor of water contained in the air varies according to the country; and there are regions—the deserts of Africa and Asia and the steppes of Siberia, for instance—where there is not the slightest evaporation from the soil, and where the air is dry in the extreme. The winds which come from the sea bring humid air; those which blow from the land bring dry air.

The quantity of vapor with which the air may be laden varies, according to temperature, in the following proportions:

At 14°, a cubic foot of air is saturated with water by the weight of one grain.						
30°,	"	"	"	"	"	two grains.
41°,	"	"	"	"	"	three grains.
49°,	"	"	"	"	"	four grains.
56°,	"	"	"	"	"	five grains.
66°,	"	"	"	"	"	seven grains.
80°,	"	"	"	"	"	eleven grains.
88°,	"	"	"	"	"	fourteen grains.
100°,	"	"	"	"	"	twenty grains.

At 212°, the air is capable of absorbing a quantity of vapor of water equal to its own volume; the tension of the water becomes equal to that of the air; it boils; and the pressure of the vapor is equal to one atmosphere.

Thus the hotter the air the more it can contain of water in a state of invisible vapor. Let us suppose a cubic foot of air to be saturated with vapor at 100°: it contains twenty grains. Now, if a current of cold air sets in and reduces it to 30°, as it can only now contain two grains, it is obliged to part with about eighteen grains of water. This condensation would lead to diurnal rains if cold currents were to encounter daily saturated masses of air.

The quantity of vapor is at its minimum when the wind is blowing between N. and N.E.; it increases when the wind is in the E., the S.E., and the S., and attains its maximum when the vane points to S. and S.W., diminishing again when the breeze is from the W. and the N.W. The cause of these differences is very simple. Before reaching us, the west winds pass over the Atlantic, and are loaded with vapor, whereas those which blow from the east come from the interior of Europe and Asia. These vapors resolve themselves into rain when the west winds reach France; but this water is vaporized almost immediately, and the result is that these winds continue to be more charged with vapor than those which come from the east. The W.S.W. wind, blowing both from the sea and from warmer countries, is capable of containing a larger quantity of vapor of water than the west wind, which is colder. This is not the case in regard to *relative* humidity.

Thus, although with a north wind the air may contain a much smaller proportion of vapor of water than when the wind is south, it is far more humid, because of its low temperature. The seasons again modify this general rule. The following is the influence of the wind for each season, complete saturation being represented by 1000:

RELATIVE HUMIDITY ACCORDING TO THE DIRECTION OF THE WINDS DURING THE FOUR SEASONS.

Winds.	Winter.	Spring.	Summer.	Autumn.
N.	895	750	676	787
N.E.	912	723	674	826
E.	926	669	613	757
S.E.	855	714	663	792
S.	830	703	674	762
S.W.	819	703	699	786
W.	809	717	714	806
N.W.	832	734	688	827

The contrast here shown between winter and summer is striking. Although in these two seasons the proportion of vapor is less with an easterly than with a westerly wind, nevertheless the low temperature of these winds in winter re-establishes the equilibrium, and in this season the east wind is the most humid and the west the driest. In summer it is just the contrary; it is when either of these winds begins to blow that the contrast is the most striking. If, for instance, in winter the westerly winds have prevailed for some time, the sky being clear, and there suddenly springs up an east or a north-east breeze, then the sky becomes cloudy, and the lower regions of the atmosphere become filled with mist. But if the wind continues to blow, then the sky becomes clear again, although the air remains moist. If the reverse takes place—that is to say, if the sky is overcast, the wind being in the east, and if it suddenly veers round to the south—the sky becomes clear and the atmosphere dry, the reason being that the heated air dissolves the vapor of water and becomes further removed from the point of saturation. It is only when this wind has prevailed for several days and collected a large quantity of vapor that the atmosphere again becomes humid.

We will now consider the force and velocity of the wind. It is at times very gentle, and at others extremely powerful. No other element is so capricious and so changeable; none so capable of soft caresses or of wild rage. The scale of its variations is so extensive that it is difficult to give a very exact account of its range, from the breeze which scarcely raises a ripple on the surface of a lake to the hurricane which uproots trees and throws down buildings. The following table will give an idea of the different degrees of velocity which it acquires:

TABLE OF THE VELOCITY OF WIND.

	Velocity per Second nearly in Feet.	Approximate Velocity per Hour in Miles.
Scarcely perceptible wind	$1\frac{1}{2}$	1
Perceptible wind	3	2
Light breeze	6	4
Moderate wind	17	12
Good breeze	25	17
Fresh wind (swelling sails)	32	22
Wind that causes windmills to revolve	50	35
Good sea-breeze	65	45
Strong breeze	75	50
Very fresh (reefing top-sails)	90	60
Violent wind	117	80
Tempest	147	100
Hurricane that blows down buildings	160	110
Maximum speed of a cyclone's rotation	220	150
Maximum of the rotation and of translation as well	260	180

It is not known to what degree of speed masses of air borne off by cyclones may attain, for it is in the upper regions of the atmosphere, where there is but a feeble resistance to aërial currents, that the wind of the tempest must be most rapid. Therefore it is not enough to ascertain the rate of speed of the moleculæ of air near the level of the ground in order to form an idea of the rapidity at which the atmospheric mass moves when hurried along by the tempest. I have remarked in my aërial travels that the speed of air generally increases in proportion to the height.* The balloon which, during the siege of

* [On March 31, 1863, the balloon left the Crystal Palace, Sydenham, at 4 hrs. 16 min. P.M., and fell at Barking, in Essex, a point fifteen miles from the place of ascent, at 6 hrs. 30 min. P.M. Leaving out of the calculation all motion of the balloon, excepting the distance between the places of ascent and descent, its hourly velocity was seven miles ; the horizontal movement of the air at Greenwich, as shown by Robinson's anemometer, was five miles per hour.

On April 18, 1863, the balloon left the Crystal Palace at 1 hr. 16 min. P.M., and descended at Newhaven at 2 hrs. 46 min. The distance is about forty-five miles passed over in an hour and a half, or at the rate of thirty miles per hour. Robinson's anemometer had registered less than two miles per hour.

On June 26, 1863, the balloon left Wolverton at 1 hr. 2 min. P.M., and fell at Littleport at 2 hrs. 28 min. P.M. The distance between these two places is sixty miles ; the velocity was therefore forty-two miles per hour. The anemometer at Greenwich registered ten miles per hour.

On July 11, 1863, the balloon left the Crystal Palace at 4 hrs. 53 min. P.M., and fell at Goodwood at 8 hrs. 50 min. P.M., having traveled seventy miles, or at the rate of eighteen miles per hour. The anemometer at Greenwich registered less than two miles per hour.

On July 21, 1863, the balloon left the Crystal Palace at 4 hrs. 52 min. P.M., and fell near

Paris, traveled from that city to Christiania accomplished the distance (nearly 1000 miles) in fifteen hours, or at the rate of $66\frac{1}{4}$ miles per hour; and this, although there was but little wind on the ground. The balloon sent up from Paris at the coronation of Napoleon, in 1804 (at 11 P.M.), carried the news of the Pope's submission to the emperor direct to Rome, reaching that city at seven the next morning, having done the 800 miles at an average hourly speed of 100 miles! These facts serve to give us an idea of the speed of the cyclone at a certain height above the ground, when even along the earth, which is covered with points of resistance to it, its rapidity is as much as 100 miles in the hour, and upon the ocean 150 to 170 miles.

As to the pressure exercised by the aërial current which moves at so great a rate, it is indeed formidable. In a notice upon the construction of light-houses, Fresnel calculated that the highest wind-pressure was sixty pounds on a square foot, but it is very probable that in many

Waltham Abbey, having traveled about twenty-five miles in fifty-three minutes, or at the rate of twenty-nine miles per hour. The horizontal movement of the air by Robinson's anemometer was at the rate of ten miles per hour.

On September 29, 1864, the balloon left Wolverhampton at 7 hrs. 43 min., and fell at Sleaford, a point ninety-five miles from the place of ascent, at 10 hrs. 30 min. A.M. During this time the horizontal movement of the air was thirty-three miles, as registered at Wrottesley Observatory.

On October 9, 1864, the balloon left the Crystal Palace at 4 hrs. 29 min. P.M., and descended at Pirton Grange, a point thirty-five miles from the place of ascent, at 6 hrs. 30 min. P.M. Robinson's anemometer during this time registered eight miles at the Royal Observatory, Greenwich, as the horizontal movement of the air.

On January 12, 1865, the balloon left the Royal Arsenal, Woolwich, at 2 hrs. 8 min. P.M., and descended at Lakenheath, a point seventy miles from the place of ascent, at 4 hrs. 19 min. P.M. At the Royal Observatory, by Robinson's anemometer, during this time the motion of the air was six miles only.

On April 6, 1865, the balloon left the Royal Arsenal, Woolwich, at 4 hrs. 8 min. P.M. Its correct path is not known, as it entered several different currents of air, the earth being invisible, owing to the mist; it descended at Sevenoaks, in Kent, at 5 hrs. 17 min. P.M. Five miles was registered during this time by Robinson's anemometer at the Royal Observatory, Greenwich.

On June 13, 1865, the balloon left the Crystal Palace at 7 hrs. 0 min. P.M., and descended at East Horndon, a point twenty miles from the place of ascent, at 8 hrs. 15 min. P.M. Robinson's anemometer during this time registered seventeen miles at the Royal Observatory, Greenwich.

On August 29, 1865, the balloon left the Crystal Palace at 4 hrs. 6 min. P.M., and descended at Weybridge at 5 hrs. 30 min. P.M., a point thirteen miles from the place of ascent. During this time fifteen miles was registered by Robinson's anemometer at the Royal Observatory, Greenwich.—ED.]

cases this is exceeded. Leaving out of the question the effects of strong cyclones in the tropics, several cases have occurred in the temperate zones where the pressure exercised by the wind in a very limited space was much above the calculations of meteorologists. To cite only one instance, the tempest which occurred on the 27th February, 1860, and which blew from the west in the plains of Narbonne, was so violent as to blow trains off the rails on the line between Salees and Rivesaltes. The pressure must have been at least eighty pounds to the square foot.

It has been calculated that, approximately, the mechanical force of the wind is in proportion to the surface of the object exposed to it, and in direct ratio to the square of the velocity, and that for a velocity of a yard per second, for each square yard, the effect produced is about a quarter of a pound. With strong winds, the velocity of which is twenty yards per second, there is a pressure of ten pounds per square foot; when, as in hurricanes, the speed is forty yards, the pressure becomes quadrupled. This renders it easy to understand how trees are uprooted and houses blown down.

The extreme smallness of the molecules of air is often more than compensated by the rapidity of their motion, so that they are capable of producing effects which appear incredible, but which are in conformity with the laws of mechanics.

To give a correct idea of these effects, I may anticipate the chapter upon Cyclones, and cite a few of the great disasters caused by certain hurricanes. At Guadaloupe, on July 25th, 1825, solidly-constructed houses were demolished, and a new building belonging to the State had one wing completely blown down.

The wind had imparted such a rate of speed to the tiles, that many of them penetrated through thick doors.

A piece of deal, thirty-nine inches long, ten inches wide, and nearly one inch thick, moved through the air so rapidly, that it went right through a palm-tree, eighteen inches in diameter.

A piece of wood about eight inches wide, and four or five yards long, projected by the wind along a hard road, was driven a yard deep into the ground.

A large iron railing, in front of the Governor's Palace, was shattered to pieces.

Three twenty-four pounders were blown from one end of a battery to the other.

In 1823, a hurricane, about half a mile in diameter, passed close by

Calcutta, killed in the space of four hours 215 persons and wounded 223, blew down 1239 fishermen's huts, and drove a piece of bamboo through a wall five feet in thickness: the blast of the air must have had a force equal to that of a six-pounder cannon.

At St. Thomas, in 1837, the fortress which protects the entrance into the harbor was demolished as if by bombardment. Fragments of rock were projected from a depth of thirty to forty feet, and hurled on the shore. In other places, strong houses, torn up from their foundations, were swept along the ground before the wind. On the banks of the Ganges, the Antilles coast, and at Charlestown, several vessels were carried from the sea some distance inland. In 1681, an Antigua vessel was carried out of the water to a point ten feet above the highest known tide. In 1825, the vessels which were in the harbor of Basseterre disappeared, and one of the captains, who had escaped, said that his ship was lifted by the hurricane out of the sea, and was, so to speak, "shipwrecked in the air." A quantity of the débris from Guadaloupe was carried to Montserrat, over an arm of the sea fifty miles wide. In the tempest which blew across the English Channel on January 11th, 1866, stones weighing from four to six hundred pounds were hurled over the Breakwater at Cherbourg to a height of more than eight yards. Admiral Le Noury states that the sea dashed against the fort, which is 185 feet above the level of the shore.

The only difficulty in explaining these phenomena is to discover how the air can attain in the atmosphere so prodigious a velocity; for, granting that velocity, the most extraordinary chemical action becomes the necessary consequence. It is gas in motion which drives the cannon-ball from the gun, and which hurls into the air vast masses of rock when a mine explodes. An oak plank, nearly an inch thick, may be pierced by a candle fired out of a gun; the force of the projectile being only due in this case to its velocity.

CHAPTER IV.

RESPECTING CERTAIN SPECIAL WINDS: THE BISE — THE BORA — THE GALLEGO — THE MISTRAL — THE HARMATTAN — THE SIMOOM — THE KHAMSEEN—THE SIROCCO—THE SOLANO.

HAVING considered the theory and the action of the general winds (both those that are regular and irregular) which blow over the surface of the globe, we must now turn our attention to special winds which characterize certain countries, and to atmospheric movements which at times traverse oceans and continents with the rapidity of a bird of prey, and which seem to form an exception to the system of organized laws by which nature is regulated. Scientific analysis has shown that these phenomena are obedient, like every thing else in the universe, to definite and fixed laws.

In France, the temperate climate which we enjoy precludes the intense atmospheric phenomena which occur in less favored regions. Among the winds, properly so called, which differ slightly in their character from most of the general winds, may first be cited the *bise*, or north wind, which is very cold, and occasionally very violent. In the east of France it is much dreaded, for it comes nearly in a straight line from the North Sea; and having traversed Holland and Belgium when those countries are covered with snow, it becomes even colder during its passage. At Istria and in Dalmatia the bise is known as the *bora*, and it is so strong that it sometimes blows over a horse and cart. In Spain, this same north wind—which is sometimes a north-east wind in that region—is designated the *gallego*.

In the south of France, the cold and violent south-west wind which has passed over the snows of the Alps and the Pyrenees, and which is known as the *mistral*, deserves particular notice.

Its cause was long unknown. It was attributed to a sudden coldness of the wind that passed over the Alps and the Pyrenees. M. Marié-Davy, in several notes published in the *Bulletin de l'Observatoire* in June, 1864, proved that the cause of this wind is not local, and that the movements which give rise to it pass eastward like whirlwinds. Kaemtz, in a communication to the Institute in July, 1865, shows, by means of a list of barometrical pressures in France, Spain, and Italy, before, during, and after the passage of the mistral, that it is a regular tempest, coming from a great distance, and that it is not due to a sudden fall in the temperature of the wind while passing over the mountains.

It is remarkable that, in proportion as meteorology advances, we learn not to look for the causes of most phenomena in the localities where they occur, but to general preponderating causes to which the local circumstances are subordinate.

Whenever the mistral blows, there is an excess of atmospheric pressure to the west of the Gulf of Lyons. Whatever may be the origin of this pressure, it always is an accompaniment of the mistral.

For the mistral to occur, no matter in what season, there must always be a combination of circumstances which are always identically alike. Whether there be fine or bad weather in the south-west of Europe, there is always an excess of pressure to the west of the Cévennes.

The violence of the wind is due to the form of the Pyrenean isthmus. As soon as the general direction of the atmospheric movement veers slightly from west to north, the central plateau and the main body of the Alps cause an inclination of the current toward the Gulf of Lyons. This current, compressed between the Alps and the Pyrenees in the direction of length, and by the Cévennes in a vertical direction, constitutes a rapid upon the coast of Languedoc, and this is one of the causes of the excess of pressure upon the north-east slope of the Cévennes, and of the diminution in pressure upon the Mediterranean, where the wind maintains a velocity no longer commensurate with the width of the channel. Hence also arises the violence of the north wind in the valley of the Rhône between the spurs of the Alps and those of the central plateau.

The mistral is the driest of the winds in these regions, because it has been rendered dry in its passage over the Cévennes. It is, indeed, pluvious or moist upon the north-western slope of those mountains. The winds from the east or south regions bring rain with them, because they are sea-winds upon the coasts and upon the south-western slope of the Cévennes; they are dry upon the opposite slope.

The high temperature of the interior of Africa is the cause of the extraordinary winds which are met with on the coasts of Guinea and Barbary, in Egypt, Arabia, Syria, the steppes of Southern Russia, and even in Italy. These winds, the names of which are harmattan, simoom, and khamseen, have unusual accompaniments, some details of which it may be interesting to mention.   They are remarkably hot and dry, and are attended by whirlwinds of dust.

The name *harmattan* is given to a wind which blows three or four times each season from the interior of Africa toward the Atlantic, in the space comprised between Cape Verd (lat. 14° 44′ N.) and Cape Lopez, on the African coast near the equator.   The harmattan is generally noticed in the months of December, January, and February.   Its direction is from E.S.E. to N.N.E., and its ordinary duration is one or two days—sometimes five or six.   This wind is only moderately strong. A peculiar kind of mist, so thick as to shut out all but a few red rays of the sun at noon, always rises when the harmattan begins to blow.   The particles of which this mist is composed alighting upon turf, the leaves of trees, and the skin of the negroes, make every thing white.   Their nature is not known, and all that has been ascertained respecting them is, that they are carried but a very little way out to sea: at a league from the shore, for instance, the mist is very slight, and at three leagues there is no trace of it, though the harmattan may be blowing with its full force.

The extreme dryness of the harmattan is one of its most marked characteristics.   If it continues any length of time, the branches of the orange and lemon trees become parched and begin to die; the covers of books, not even excepting those which are wrapped up in linen and inclosed in a case, become bent as if they had been laid before a fire.   The panels of doors and windows, the furniture of rooms, crack, and often snap.   The effect of this wind upon the human body is not less pronounced.   The eyes and the lips dry up and smart.   If the continuance of the harmattan be for four or five consecutive days, the cuticle of the hands and the face begins to peel, and it is necessary to anoint the body with grease.

All this would lead one to suppose that the harmattan must be very unhealthy; but, so far from this being the case, it is the opposite.   Intermittent fever, for instance, is radically cured by the first breath of the harmattan.   Persons who have become weakened by the excessive blood-letting practiced in those countries at once recover their strength; remittent and epidemic fevers also disappear as if by enchantment.   So salutary, in fact, is the influence of this wind while it lasts, that it is said to be impossible to communicate infection even artificially, for it appears that vaccine virus will not act during its continuance.

Its asserted poisonous properties are therefore pure invention, and may possibly have been circulated by the Arabs to deter travelers from penetrating into what they consider their kingdom.

Kaemtz says: "At every epoch the Arab of the desert, poor and of nomad habits, has detested the inhabitant of towns who leads a steady and orderly life. Thus, when the merchant is compelled to cross the desert, the Bedouin exacts an enormous price for protecting him. . . . . In the eyes of the inhabitant of a town the desert was the theatre of scenes horrible beyond description. All the marvelous stories of adventure found a ready audience, just as in our days the Turks have the most grotesque and unreal ideas concerning Europe. The dwellers in the desert took care not to destroy these fancies, but rather to confirm them, and the few merchants who knew the exact truth kept it to themselves in order to maintain a monopoly of commerce. It is in this way that visionary ideas maintain their sway."

The Arab writers teem with falsehoods concerning the desert, and the European travelers have outrivaled them. The Mussulman believes he is acting meritoriously when he deceives the unbelievers and keeps them away from the desert. L. Burckhardt, of Bâle, was the first who supplied reliable information concerning the phenomena of the desert, and especially touching the winds which prevail there. He thus reduced to their true value the fabulous stories of his predecessors, Beauchamp, Bruce, and Niebuhr.

Burckhardt relates that this wind of the desert surprised him between Siout and Esné. He says, "When it rose I was alone, mounted upon my dromedary, and far away from any houses or trees. I endeavored to protect my face by covering it with a handkerchief. In the mean while, the dromedary, into whose eyes the sand was driving, became alarmed, and began to gallop, causing me to fall off. I remained flat upon the earth, for I could not see ten yards in front of me, and I covered myself with my clothes as well as I could until the wind became less violent. I then went in search of the dromedary, which I found some distance off, lying with his head against a bush to protect it from the sand." Malcolm and Morier, who crossed the Persian deserts, Ker-Porter, who visited that which lies to the east of the Euphrates, agree with Burckhardt that when they were exposed to the simoom they felt no ill effects from it beyond the momentary inconvenience.

"It is not only in the sandy deserts of Africa and Asia that the hot winds are to be dreaded, but in nearly all continental countries near the tropics. In India they are known as the 'devil winds.' They frequently occur during the dry season, and scatter terror and desolation through country and town. Without being absolutely poisonous, it may be admitted that winds whose speed is so formidable, laden with grains of sand, and the temperature of which is as much as 104°, may exercise an unhealthy influence upon the regions through which they pass, and be especially dangerous for Europeans who do not know how to protect themselves."

About the time of the equinox the tempests in the desert become terrible. Every one has heard of the burning wind, the *simoom*—a wind which, in Arabia, means poison. This formidable wind blows also in Egypt, where it is called khamseen (fifty), because it lasts that number of days, five-and-twenty before, and five-and-twenty after the spring equinox.

The simoom is preceded by a black spot which rises in the horizon. This spot grows rapidly larger. A murky veil obscures the sky, gusts of sand darken the sun and dry up all verdure. As soon as it begins to blow, birds fly off affrighted; the dromedary seeks a bush to protect him from the sand; the Arab covers his face, rubs his body with grease or wet mud, and lies on the ground or hides himself behind a tree until it is over. The simoom is the most dangerous of the accidents to which a caravan crossing the desert can be exposed, and to it is attributed the destruction of the 50,000 men sent by Cambyses to destroy the Temple of Jupiter Ammon.

In 1805 the simoom buried in the sand a whole caravan, causing the death of 2000 men and 1800 camels. More than once French generals have feared that columns of troops which they have been obliged to send into the desert have been overtaken and destroyed by the simoom.

The impalpable dust which the air carries along in thick clouds enters the nostrils, the eyes, the mouth, and the lungs, and causes asphyxia. When it does not absolutely kill, the rapid evaporation from the surface of the body dries up the skin, inflames the throat, makes the breathing rapid, and produces violent thirst. The terrible blast of the simoom dries up the sap of trees in its passage, and causes the water contained in the skins carried by the camels to evaporate. The caravan is then a prey to horrible thirst, which sets the blood on fire, and the route which they follow is strewed with the whitened bones of men and animals who perish for want of water.

Thomas William Atkinson was a witness, in 1850, of the fierce hurricanes which swoop down upon the steppes of Mongolia. He says,

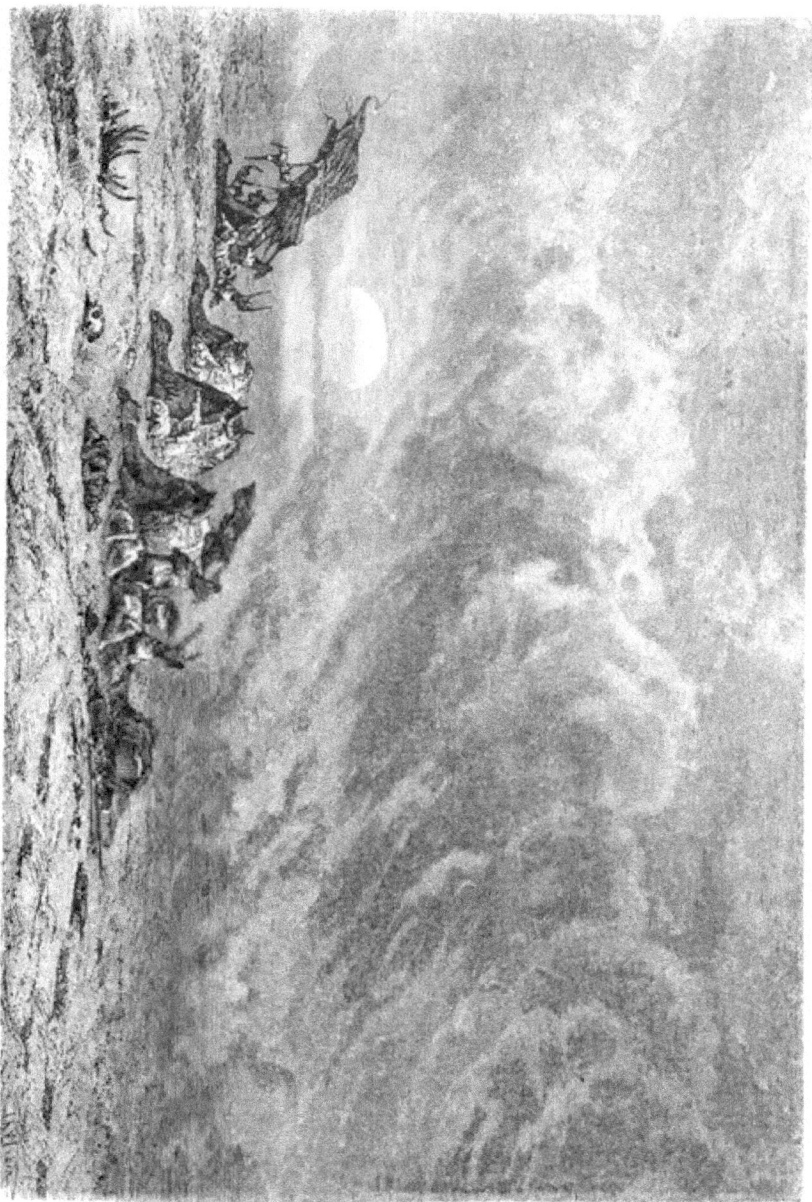

Fig. 62.—The Simoon.

"A solemn silence prevails in these vast and arid plains, which are deserted alike by men, quadrupeds, and birds. People talk of the solitude of forests. I have often ridden through their sombre alleys for days together: the soughing of the wind, the cracking of the branches, and the murmur of the leaves are to be heard there; sometimes, too, the crash of one of the giants of the forest as it falls to the ground wakes the distant echoes and startles from their lairs the tenants of the glades, making the birds utter a cry of terror. This is not solitude: trees and leaves have a language which man recognizes from a great distance; but in these arid deserts there is no sound to break the death-like silence which prevails.

"The sand was raised into circular terraces, some from fifteen to twenty feet high, and they extended as far as the eye could reach into the desert. Seen from one of the knolls they presented the singular appearance of an immense necropolis, over which were dotted countless tumuli.

"While I was taking a sketch of this, I was witness of the formation of a hurricane above the level of the water. It was coming from the north direct upon us. The Cossacks and Tchuck-a-boi went to place their horses in security behind the reeds and bushes, leaving two of their companions with me. The tempest came on at a fearful rate, driving enormous waves into the air, and striking down all vegetation in its path. A long white wave came moving along the lake, and when it was within half a verst its roar became audible. The men begged me to move away, and we rejoined the rest of the troop behind the bushes. I had scarcely reached there when the hurricane burst forth, bending the bushes to the ground. When it reached the sand it began to revolve in a circle, lifting whole mounds of sand into the air, and causing others to spring up where there had been none previous to the storm. This tempest lasted a quarter of an hour, when the atmosphere became as calm as it had been before.

"It is very dangerous to be overtaken in the plain by one of these typhoons. I have since seen them swoop down from the mountains or rise from the hollow of some deep ravine, in the shape of a black and compact mass, with a diameter of as much as 1000 yards or more, and rushing along the steppe with the speed of a race-horse. All animals, whether tame or savage, take flight before it, for once enveloped in its sphere of action they must infallibly perish. The wild horses gallop off in terror before the storm which pursues them."

In Europe we have the sirocco (Italy), and the solano (Spain), both of which have a very enervating effect upon those exposed to them.

Brydone, who was at Palermo on July 8th, 1770, during a sirocco, writes, "I opened my door at eight in the morning without suspecting there was any change in the temperature, when all at once I felt a burning impression upon my face like the air from a hot oven. I closed my door, exclaiming to Fullarton that all the atmosphere was on fire." At this moment the thermometer, in the open air, marked 111°.

An army surgeon who accompanied the French troops in a march between Oran and Tlemcen, in the desert, gives the following account of a sirocco: "It was toward the end of July, 1846. Several soldiers had succumbed to the heat. The sirocco assailed our little column. Under the influence of this dry, heavy, and enervating air, the breathing became difficult; the lips and the nostrils, cracked by the burning dust driven up by the wind, were both dry and painful, and the throat, as it were, became contracted. The face was burned by gusts of heat, sometimes followed by tremor, and a fainting away which resembled syncope. The perspiration ran off in streams, and the water, which was eagerly swallowed, did not quench the thirst, but increased the stomachic pains and the difficulty of breathing. To walk would have been impossible; we felt half suffocated under cover of the tents, and in the open air the burning breeze caused a choking sensation. . . . . But for the water, our column must have perished."

CHAPTER V.

THE POWER OF THE AIR: THE HURRICANE — THE CYCLONE — THE
TEMPEST.

THE two great general currents which have been adverted to, the
one moving from the equator to the poles, and the other from the poles
to the equator, come into collision with each other in the equatorial
zone. Various causes counterbalance the periodical action of the solar
rays, and place obstacles in the way of the ordinary progress of air.
The diversity in the temperature of continents and seas causes a variation in the normal direction and intensity of the currents. The state
of the sky in the tropics, according as it remains clear or cloudy for
any length of time together, condenses the heat as in a focus of absorption, or disseminates it over vast tracts of country. The undulations
of the soil, the high chains of mountains and their temperature, the less
lofty plateaux, and even the valleys themselves, cause in one place the
storing up and the repose of large masses of air, in another their distribution in different directions, while in other cases this same unevenness of the ground forces the currents back right and left, causing them
to eddy like the waters of a stream, or to rush furiously past the obstacles in their way. The blasts of the air as they meet, either join
forces or oppose each other, increasing or destroying their mutual
power. It is in this way that strong winds, hurricanes, and tempests
arise. These atmospheric contentions, which sometimes attain gigantic
proportions, create a great disturbance in the course of nature. They
have been studied by sailors and meteorologists, and the principal laws
which seem to regulate them have been ascertained. Redfield and
Reid, Professor Dove, of Berlin, and Admiral Fitzroy, have, after great
labor, succeeded in forming a theory of the tempests which explains
the most violent of the movements in the atmosphere, and their researches will be useful in considering this subject here.

One of the chief observations made is, that hurricanes do not proceed
in straight lines, but follow a curve, turning horizontally upon their
own axes by a rapid rotatory movement.

This characteristic movement of horizontal rotation has earned for these gigantic whirlwinds the name of *cyclone* (κύκλος, circle).   They are the general hurricanes, which are not local tempests resulting from the deviation of the wind, owing to the configuration of the soil or the meeting of several ordinary currents, but extend over several hundred square leagues, and travel a distance of many thousand miles.

The cyclones are vast whirlwinds, of various size in diameter, in which the force of the wind increases from all the points of the circumference to the centre, where a calm prevails.   In this centre, however, the sea remains rough.   There is no cloud in this calm region ; the sun shines brightly, the stars appear, and fine weather seems to have returned, when in reality it is surrounded on all sides by a vast belt of fierce hurricanes.

All around this central calm the rotatory movement is of the same force, and this force is the greatest.   Consequently, on passing to this central region, a ship passes from a violent tempest into a complete calm, and, on crossing this calm space, passes on the opposite side into a violent tempest again.   But in this latter case the hurricane blows in the opposite direction to what it did before entering the calm, since the movement of cyclones is circular.

The first central zone, which constitutes in reality a hurricane, and during the passage of which occur all the disasters, is generally from 100 to 120 leagues in diameter, whatever may be the extreme limits which the phenomenon reaches, for its power is not in proportion to its extent.

The rotatory speed of the hurricanes varies very much : it is that which constitutes chiefly the violence of the whirlwind, and which causes it to be, in regard to the places against which it blows and the vessels it assails, either a hurricane, a gust of wind, or a simple gale.   In violent storms, it is estimated that the molecula· of air turn around its centre with a rotatory speed of sixty leagues an hour—a rapidity which explains the ravages and disasters produced by the passage of this terrible wind.

The cyclone generally begins between the latitudes of 5° and 10°. It moves, in our hemisphere, in a north-westerly direction, continuing thus until it reaches a particular latitude, when it turns toward the north-east, and thus forms a parabola, the two branches of which diverge farther and farther.

The difference in the density of the different atmospheric strata which are encountered in its passage, the rotatory movement itself.

must impart an oscillating movement to the cyclone, so that, instead of describing a regular parabola, the course of the cyclone is rather spiral, infolding itself around the parabola. Ships that happen to be in the midst of it are exposed to its oscillating action: hence those terrible gales which are succeeded by a more or less complete calm; hence those dramatic situations in which the ship in distress sees the wind veer round to all the points of the compass in a short space of time.

The sudden and dangerous variations of the wind, which were formerly considered as essential to hurricanes, typhoons, tornadoes, etc., can not, and in fact do not, occur save in the immediate path of the centre of the cyclone. The cyclone contains in itself the germ of its own early destruction: in proportion as it advances it is approaching nearer to regions which are colder than those whence it started; the vapor which it contains becomes condensed into torrential rain; the electricity issues from it in large currents; the equilibrium which existed becomes destroyed; and the centrifugal force, being no longer counterbalanced, permits it to extend to an immense size. It then loses in force what it gains in volume; at its starting-point it does not measure more than a few leagues, but when, having lost its equilibrium, it collapses—as generally occurs between the latitudes of 40° and 45°—it extends over hundreds of miles.

The more rapid the escape of electricity the quicker will the meteor collapse: thus it sometimes happens that a cyclone terminates its course before reaching these high latitudes, and without describing the second branch of its parabola, which therefore remains incomplete.

Between latitudes 5° and 10°, and longitudes 45° and 60°, when a cyclone is near its starting-point, it has been ascertained that the rate of revolution is inconsiderable, varying from one to five miles an hour, increasing as the hurricane advances westward.

In latitudes 35° to 45°, and in longitudes from 50° to 30°, the rate of revolution varies from six to twelve miles an hour. In the higher latitudes it is greater, and has been known to be as much as twenty miles per hour.

The greatest rate of *revolution* ever registered is that of a cyclone which reached the Banks of Newfoundland from the Antilles in August, 1853, when the speed was thirty-one miles per hour. This velocity gradually increased to ninety; and without affecting the speed of rotation, which was sixty leagues an hour. Thus the wind is capable of traveling along the surface of the sea at a speed of seventy-five leagues an hour, perhaps more.

The origin of cyclones, so far as can be judged by the comparisons that have been made, is probably due to the encounter of two currents of air moving in opposite directions. The place of meeting forms a neutral point, where the air receives a rotatory movement from the collision of the two currents. It is like an eddy in a stream, and a moment's reflection will enable the reader to form an idea of it.

These immense whirlwinds come into existence to the south as well as to the north of the equator. The astronomer Poey, Director of the Observatory at Havana, has ascertained, by a laborious research into the hurricanes which have raged in the West Indies since the discovery of America (1493) to the present day, that, out of 365 grand cyclones, more than two-thirds have occurred between the months of August and October, that is, during the period when the heated shores of South America are beginning to attract toward them the colder and denser air of North America. In the Indian Ocean cyclones are most frequent when the change occurs in the direction of the monsoons and at the end of summer. In the list of hurricanes in the southern hemisphere, drawn up by Piddington and completed by Bridet, there is not a single mention of a hurricane in the months of July or August: more than three-fifths took place in the first three months of the year. It is at the epoch of change of seasons that the powerful masses of air, loaded with electricity, enter into a struggle for the mastery, and give rise, by their meeting, to those great eddies which develop themselves in a spiral shape over sea and land. At the same time, the whirlwind never extends very high into the aërial ocean. According to Bridet, the average height of hurricanes in the Indian Ocean is less than 10,000 feet; Redfield puts it at no more than 6000. As a rule, the stratum of air which revolves in this way is not nearly so thick; sometimes it is so shallow that the crew of a vessel which is exposed to a cyclone see above their heads a clear sky. Above the cyclone the storm-winds follow their regular course.

The analysis of cyclones is especially due to Redfield. America is a country peculiarly well adapted for observing these phenomena, as the hurricanes which run along the shores of the United States pass, during their progress through the tropics, over the West India Islands, where their remarkable character has earned them the appellation of "West India hurricanes."

In regard to the cyclones which occur in Central Europe, it is rarely possible to ascertain through what part of the tropics they have passed, and this is a sufficient proof that the wider the extent of our observations, the less likely shall we be to form incorrect ideas of these phenomena of nature.

The meteorologist Dove proved, in his work upon the "Law of Tempests" (Paris edition, p. 173), that a cyclone movement occurs whenever any obstacle stands in the way of the regular change in the direction of the wind (which is due to the rotation of the earth), and therefore prevents the regular rotation of the vane to some given point. He says:

"The hurricanes in the West Indies generally commence at the inner limit of the zone of trade-winds, in the region of calms, where the air rises and becomes disseminated in the upper strata of the atmosphere, and in a direction contrary to that of the trade-wind. This renders it probable that the primary cause of cyclones is the intrusion of a part of this upper current into that below.

"Let us also imagine that the air which rises over Asia and Africa flows laterally into the upper strata of the atmosphere—a fact which is made evident by the sand which falls in the North Atlantic, and which rises to a great height, for on the Peak of Teneriffe the sun is sometimes obscured by it. A similar current must have a tendency to oppose the free passage of the upper anti-current of the trade-winds, and must force it back into the lower current, or the direct trade-wind. The point at which this intrusion takes place must advance with as great rapidity as the oblique upper current which causes it. The interposition of a current traveling from E. to W. with another traveling from S.W. to N.E. must necessarily create a rotatory movement in a direction the opposite to that followed by the hands of a watch. According to that, the cyclone, which advances from S.W. toward N.E. in the lower trade-winds, represents the point of contact of two other currents which in their higher layers advance in directions perpendicular to each other. That is the origin of the rotatory movement, and the ulterior progress of the cyclone will necessarily be based upon the same principles. The cyclone being thus considered as the result of the meeting of currents at different points, one after the other, may therefore preserve its diameter unchanged for a considerable period, and it may even diminish in size, though it ordinarily increases.

"It is, moreover, quite clear that, if the above explanation be correct, a cyclone which turns in the same direction may originate by the interposition of some mechanical obstacle in the route of the current, as it travels toward the high latitudes of the north—an obstacle which compels this current to assume a more southerly course (that of a south wind) upon its eastern than upon its western edge, where it always remains nearly due west. This was what happened, to cite one instance, during a cyclone in the Bay of Bengal on the 3d, 4th, and 5th of June, 1839."

The name of cyclone is therefore, in a certain measure, the geometrical designation of the more ancient term *hurricane*, like the *tornadoes* which are seen on the coasts of Africa, like the *typhoons* of the Chinese seas; the great tempests that occur in these regions are of the same kind as the cyclones in the Atlantic. Dampier, the prince of navigators, describes the approach of the typhoon with that accuracy which renders all his works so reliable. We read in his "Voyages" (vol. ii., p. 26):

"The typhoons are a special kind of violent tempests which blow along the coast of Tonquin and the neighboring shores in the months of July, August, and September. They generally occur about the period of full moon, and are, as a rule, preceded by very fine weather, light breezes, and a clear sky. These light breezes are the ordinary trade-winds, which blow from the S.W. at this season, and which veer to about N. or N.E. Before the tempest begins, a thick cloud forms in the N.E.; it is very black near the horizon, copper-colored toward the summit, and gradually lighter in color toward its outside edge, which is perfectly white. The aspect of this cloud is very strange, and it appears sometimes twelve hours before the storm breaks. When it begins to move very rapidly, the wind breaks out at once, augmenting in force with great suddenness, and blowing with great violence for about twelve hours. It is often accompanied by thunder and lightning, and thick rain. When the wind begins to diminish, it drops very suddenly for about an hour, after which it recommences to blow from the S.W. for about the same period as it did from the N.E., rain falling as before."

The course followed by the centre divides the hurricane into two equal parts, but into parts which differ from each other. In the one the movement of rotation and that of translation have the same direction; in the other, on the contrary, the direction of translation and that of rotation are different. It follows that at an equal distance from the centre there is much more wind in the first hemicycle than in the second; hence the name of *dangerous hemicycle* is given to the one, and that of *manageable hemicycle* to the other.

In the northern hemisphere the cyclone turns from right to left—that is to say, that an observer placed in the centre of the whirlwind would see the wind pass before him from right to left. The dangerous hemicycle will be to his right if he follows the same route as the centre of the hurricane, and the manageable hemicycle to his left.

In the southern hemicycle, on the contrary, the hurricane turns from left to right; the dangerous hemicycle is to the left, and the manageable hemicycle to the right of the line through which the centre passes if he follows the same direction as the hurricane.

The direction of the wind observed at a given point of the cyclone is very near to a tangent drawn to the concentric circle upon the circumference of which it is placed. Consequently, it is always nearly perpendicular to the radius drawn from this point to the centre of the concentric circle or cyclone. Now, the law of gyration indicates that if one faces the wind the centre will be to the right in the northern hemisphere, and to the left in the southern, but always at right angles to the direction of the wind.

It is upon this latter fact, which numerous observations place beyond a doubt, that all the theories as to the means of avoiding the centre of a cyclone, by moving away from the line which it takes, are based. The nearer the centre, the more violent the wind, and the more sudden its variations. The sea is also roughest at the centre, being subject, at very short intervals, to violent gusts of wind from all directions — and this after having been under the influence of comparatively regular winds which have had time to cause a heavy swell, and to impart to the water a direction different from that of the wind. Hence arise the short and chopping waves which assail a vessel on all sides at once.

It is easy, however, to avoid the place over which the centre of a cyclone passes.

Let us suppose that the centre of a cyclone is coming toward a vessel. It will pass over this vessel, or to the right or to its left. If it is about to pass over, its direction in respect to the ship will not change; but then the direction of the wind, which is always perpendicular, will not change either, and the crew will find the wind increase in violence without changing direction.

If the centre pass to the right of the vessel, it will shift slightly toward the right. Its direction will vary from left to right; but that of the wind, which is connected with the first, will vary in the same direction—that is, from left to right.

The exact opposite will take place if the centre pass to the left of the vessel.

Thus, if the wind increases without changing its direction, the vessel will be upon the line along which the centre passes; if the wind veers from left to right, it will be to the left of this line; if the wind changes from right to left, it will be to the right of the same line.

From these laws regulating cyclones, we may gather that the worst position in which a vessel can be is that which leads to the centre of the hurricane, and, to avoid this, all efforts should be directed.

The premonitory signs of the cyclone are: Some days before the hurricane, both at sunrise and sunset, the clouds assume a reddish and orange hue, which becomes reflected upon the sea; and it is this which renders them so brilliant and splendid, and which inspires with such sentiments of admiration those who do not dream of the imminent danger which they foreshadow.

As the cyclone approaches, this reddish tint gets deeper in color; then a black and deep band extends across the sky; the edges of the *cumulus* are of a copper hue, imparting to the sea and the land an analogous glitter which makes the atmosphere look as if it were on fire; the sea-birds fly rapidly inland to seek shelter from the fury of the tempest which they have an instinct is coming, thus hoping to escape the death which would overtake them at sea.

But of all the premonitory signs of the tempest, the surest and the easiest to interpret is the movement of the mercury of the barometer.

As the pressure of the air gradually diminishes from the circumference to the centre of the whirlwind, the approach of the phenomenon is always made evident by a fall of the barometer. This same symptom characterizes the tempests in our temperate regions, which are in reality, so to speak, the continuations of the oceanic cyclone.

The barometer begins to fall twelve, twenty-four, and even forty-eight hours before the arrival of the cyclone.

An oppressive calm, accompanied by a suffocating air, prevails for four-and-twenty hours; Nature seems to be collecting all her strength to accomplish the work of devastation.

Whatever may be the course taken by the hurricane, the point nearest to its centre is known when the barometer reading ceases to decrease. Then, for a space of two or three hours, the reading of the barometer will rise and fall every half-hour, without making any decided movement up or down, this being a certain sign of proximity to the centre, that the heaviest blasts have been felt, and that the violence of the storm will gradually abate.

The total decrease of the barometer is proportionately greater as the central rarefaction is more complete, and this rarefaction, chiefly caused by centrifugal force, augments in ratio to the increase of the rotatory movement, which causes hurricanes to be so violent. The barometer, therefore, declines in proportion as the violence of the wind becomes more intense, and the most disastrous hurricanes are those which influence it to the greatest degree.

The rarefaction of the atmosphere at the centre of cyclones is clearly proved by the following table, taken from the register of a barometer during the hurricane that passed over St. Thomas on the 2d of August, 1837, when the central calm occurred at 8 P.M.:

			Inches.					Inches.	
August 2,	6	A.M.	29·922		August 2,	7·50 P.M.		28·032	Calm.
"	2	P.M.	29·764		"	8·20 "		28·032	
"	3·20	"	29·646		"	8·22 "		28·386	
"	4·45	"	29·480		"	8·38 "		28·583	
"	5·45	"	29·292	Hurricane to the N.W.	"	8·50 "		28·780	Hurricane to the S.E.
"	6·30	"	29·134		"	9 "		28·938	
"	6·35	"	28·898		"	9·25 "		29·213	
"	7	"	28·780		"	9·50 "		29·410	
"	7·10	"	28·583		"	11 "		29·607	
"	7·22	"	28·268		August 3,	2 A.M.		29·725	
"	7·35	"	28·111		"	9 "		29·922	

Variation, 0·89 inch!

These large perturbations of the air are perhaps, next to great volcanic eruptions, the most fearful phenomena that take place upon the globe, and, as Reclus remarks in his work upon the "Earth," we can not be astonished that, in Hindoo mythology, Rudra, the chief of winds and storms, should have become, under the name of Siva, the god of destruction and death. For some days before the outbreak of a hurricane, Nature, desolate and gloomy, seems to foresee a disaster. The small white clouds which travel in the air with the anti-trade-winds are concealed by a yellowish vapor; the stars are surrounded by halos with a vague iris, and by heavy banks of clouds which, about evening, are beautifully tinted with purple and gold. The air is suffocating, as if it issued from the mouth of a furnace. The cyclone, which is already revolving in the upper regions, gradually descends. Jagged remnants of reddish or black clouds are borne furiously along by the tempest, which plunges rapidly through space, and the column of mercury descends in the barometer. Soon an obscure mass becomes visible in the stormy part of the sky, and, increasing in size, gradually covers the firmament with a veil of darkness and a blood-red glitter. This is the cyclone, which is swooping down upon the earth, and a terrible silence succeeds the moaning of the sea and of the skies.

In the early part of the cyclone a strange, dull sound is sometimes heard, "with a noise like that of the wind in very old houses during winter nights."—*Piddington.* The gusts which rend the air during the time the cyclone continues are said to create a noise like that of the roaring of wild beasts, a tumult of countless voices, and cries of terror. At the points where the centre passes, a formidable sound like the discharge of artillery, an incessant rolling of thunder (the voice of the hurricane, as it in fact is), is heard above all others.

The progress of the winds meets with a certain degree of resistance upon land, but the destruction caused is none the less terrible. Buildings which lie in their path are overturned; the waters of a stream are driven back toward their source; isolated trees are torn up by their roots; forests are bent down as if they formed but one compact mass, and their branches and leaves are scattered; the grass is swept off the ground. In the track of the hurricane fly countless débris like the flotsam carried along by a stream. Generally speaking, the action of electricity is superadded to the violence of the air in motion, and helps to augment the ravages of the tempest: sometimes flashes of lightning are so rapid that they descend like a sheet of flame; the clouds, and even drops of rain, emit light; the electric tension is so great that, according to Reid, sparks have been seen to fly from the body of a negro. A whole forest in the Island of St. Vincent was killed without the trunk of a single tree being blown down. In Europe, too, upon the shores of Lake Constance, many trees were skinned of their bark, though they still remained upright in the ground.

The most terrible cyclone of modern times is probably that which occurred on October 10, 1780, which has been specially called the Great Hurricane, and which seems to have embodied all the horrible scenes that attend a phenomenon of this kind. Starting from Barbados, where trees and houses were all blown down, it ingulfed an English fleet anchored before St. Lucie, and then ravaged the whole of that island, where six thousand persons were buried beneath the ruins. From thence it traveled to Martinique, overtook a French transport fleet, and sunk forty ships conveying four thousand soldiers. "The vessels *disappeared;*" such is the laconic language in which the governor reported this disaster. Farther north, St. Domingo, St. Vincent, St. Eustache, and Porto Rico were also devastated, and most of the vessels that were sailing in the track of the cyclone were lost, with all on board. Beyond Porto Rico the tempest turned north-east toward Bermuda, and though its violence gradually decreased, it nevertheless sunk several English vessels. This hurricane was quite as destructive inland. Nine thousand persons perished in Martinique, and a thousand at St. Pierre, where not a single house was left upstanding, for the sea rose to a height of twenty-five feet, and 150 houses that were built along the shores were ingulfed. At Port Royal, the cathedral, seven churches, and 1400 houses were blown down; 1600 sick and wounded were buried beneath the ruins of the hospital. At St. Eustache, seven vessels were dashed to pieces against the rocks; and of the nineteen which lifted their anchors and sailed to sea, only one returned. At St. Lucie the strongest buildings were torn up from their foundations; a cannon was hurled to a distance of more than thirty yards, and men as well as animals were lifted off their feet and carried several yards. The sea rose so high that it destroyed the fort, and drove a vessel against the hospital with such force as to stave in the walls of that building. Of the 600 houses at Kingstown, in the Island of St. Vincent, fourteen alone remained intact, and the French frigate *Junon* was lost.

In the Leeward Islands, the inhabitants of the Government Palace took refuge in the centre of the building during the height of the storm, thinking that the immense thickness of the walls (nearly a yard) and their circular shape would preserve them from the fury of the wind. At half-past eleven they were obliged to repair to the cellar, as the wind had penetrated everywhere and lifted off the roof. The water rising there to the height of more than a yard, they were driven into the battery and protected themselves behind cannons, some of which were driven from their places by the force of the wind. The hurricane was so violent that, seconded by the sea, it carried a twelve-pounder a distance of more than 400 feet. (This cannon was, it must be supposed, upon its carriage, which had wheels.) By the light of day the country looked as it does in midwinter; there was not a single leaf, or even a branch, remaining upon the trees. Human passions are quelled in presence of such a war of the elements. When the *Laurier* and the *Andro-*

22

mède were lost at Martinique, the Marquis de Bouillé set at liberty the five-and-twenty English sailors who had survived the shipwreck, writing to the Governor of St. Lucie that he was unwilling to retain prisoners men who had fallen into his hands during a disaster to which every one was liable.

The last memorable tempest is that of March 3, 1869, when the three-masted vessel *La Lérida*, of Nantes, was lost off Le Havre, on her way from Haïti. On March 2, at 10 A.M., this vessel, which for two hours had been struggling against a fearful sea, approached the jetty, where a tremendous current, the force of which was further increased by the north-west wind, raised up an insuperable barrier. The vessel soon felt the first shock of the current which, two hours later, would have had little effect. Hitherto it had managed to sail with the wind blowing aft, and this manœuvre, diminishing its speed, left it almost at the mercy of the hostile elements. A feeling of despair came over the spectators, most of whom were seamen. They saw that this movement had gravely compromised the chances of the vessel's escape. The captain tried another effort. He endeavored to luff, so as to run his ship into the mouth of the Seine, but this was attempted too late. One last chance remained—the two anchors were cast out, but they did not grip! Still, for a moment there seemed room for hope; the anchors had caught, but the heavy sea snapped the chains. It was all over in less time than it takes to write these lines: the *Lérida*, at the mercy of the waves, ran against the angle of the bastion, which stove in its poop and bulwarks. The only thing that remained was to endeavor to save the crew. Fortunately the ship was near enough to land to admit of ropes being thrown to them, and all were rescued save two, who, losing their presence of mind, clung to a rope that was not strong enough to bear their weight. The captain, who had staid on the vessel last of all, had scarcely left when she went down.

I may finally add that in the torrid zone, and in climates where temperature is high, hurricanes are numerous, and extremely violent; in our temperate climates they are at once rarer and less violent; and in the polar regions, the great atmospheric disturbances, which occur very frequently, are limited to winds which, though tempestuous, do not constitute a hurricane.

CHAPTER VI.

TROMBES, WHIRLWINDS, OR WATER-SPOUTS.

AMONG the chief phenomena which disturb the apparently regular order and harmony of Nature, scattering terror and desolation in their paths, there is one remarkable for its peculiar and colossal form, for the forces which it seemingly obeys, for the unknown and apparently contradictory laws which it follows, and for the disasters which it causes. These disasters are themselves accompanied by such strange circumstances, that their origin can not be confounded with the other phenomena which prove so fatal to man. This meteor, fortunately rare in this part of the world, is designated in France by the general term *trombe*.

Previous to Peltier's explanation of this peculiar atmospheric phenomenon, it was imperfectly known. We are now able to describe with precision its nature and its character; and we know that a *trombe* is a column of air which generally turns rapidly upon its own axis, and which revolves comparatively slowly, for, as a rule, a person can keep up with it at a walking pace. This whirling column of air is both caused and set in motion by electricity. The sometimes violent wind which its movement produces, and which acts so disastrously, as we shall presently see, is not the result of atmospheric currents upon a large scale, as with the cyclones, but is confined to very limited dimensions. The *trombes* are often only a few yards in diameter, but their force is very great. They sweep the soil over which they pass, destroying trees and houses so completely that sometimes nothing remains upright in the track along which they have passed. This phenomenon generally has its origin as follows:

By virtue of considerable electric tension, the lower surface of a stormy cloud descends toward the earth in the shape of a cylinder, or rather of a cone, like a great speaking-tube, the top of which is lost in the clouds, while the orifice is relatively close to the surface. This reversed cone may be more or less developed, more or less different in shape, according to the special condition of the clouds or the locality. That which is always present is a connecting link of vapor between the clouds and the earth.

Beneath the cloudy column there is a great agitation upon the sea or upon the ground. Sailors compare it to a boiling process which would emit vapor and streams of liquid sheaves. Upon land the dust of the roads and light substances form an analogous kind of smoke. In a short space of time the lower whirlwind rises sufficiently high, and the upper column descends low enough to admit of their joining and being fused into one and the same column, which is thicker at its higher than at its lower part, and which is often transparent like a tube, within which vapor can be seen rising and falling.

When the centre of waters raised over the sea is more compact, it appears like a pillar placed to sustain the descending column. There proceeds from this column a noise which varies considerably, from what seems like the hissing of a serpent to the noise of heavy wagons being driven over stony roads. This noise is much more pronounced on land than at sea.

The germs of destruction seem to be embodied in this singular formation. The *trombe* advances slowly, to all appearances, blowing violently, writhing convulsively, leaving its mark upon all the productions of nature and humanity, and rending into atoms all that oppose its advance. The disasters caused by this formidable agent show that its pressure is sometimes as much as eighty to one hundred pounds to the square foot. Flocks of cattle, men, and even rivers, are lifted to an immense height. The roofs of houses are carried into the air; walls are leveled by the sudden violence of an irresistible pressure. To judge of the force of this strange phenomenon, let us consider some of its most remarkable effects.

Take, for instance, two *trombes* which were observed to the south of Paris, May 16th, 1806, from one to two P.M., and which are particularly good instances of these phenomena. Peltier copies them from Professor Debrun. They may be termed *the Paris trombes.* "The first began about one o'clock, and seemed to be at least twelve feet wide at its base near the cloud, like that of a cone turned upside down. It then became successively fifteen, twenty, and forty feet long. The lower it descended the more pointed became its conical form, for, when it first left the cloud, it formed a perfect cone. Gradually increasing in length and decreasing in breadth, it finally became no bigger than a man's arm.

"This whirlwind traveled very slowly toward the south, then west and south-west, and seemed to be suspended over the last houses of the Faubourg St. Jacques, then above the plain of Montrouge and Montsouris. It was of a gray and white color like ordinary clouds, and stood out very clearly against the background of the darker clouds.

"What struck me most was that it formed a long tube, partially *semi-transparent*, gradually making several curves and inflections, something like a long flexible piece of gut, in which I saw *vapors mounting* with an undulating movement, like smoke which might ascend a stove-pipe in glass. The most curious fact was, that the ascent of the vapors was much more marked and active toward the lower part, which was then about 300 or 400 feet above the ground.

"As the cloud which formed the head of the column advanced, the main mass described a curve and followed it, becoming elongated by 1500 or 1600 fathoms, and remaining attached to it. But when the column became extremely long, and consequently very slight in volume, and when it formed an angle of 20° or 25° with the horizon, then the main body of the column began to curl off (or become detached). This whirlwind, when its inflection was most pronounced, seemed to have its head over Châtillon and its tail over Arcueil; but while the head of the column was moving forward, I remarked that the lower part seemed to be attracted by the valley of Arcueil, and that it had great difficulty in emerging from it.

"It lasted for more than three-quarters of an hour, and went off to a point at last. Its upper part seemed to me to work its way back into the cloud whence it started, though, as it was then at a great distance to the S.S.W. of Paris, and very small in volume, I could not affirm this positively.

"About twenty minutes after the formation of this whirlwind I saw a second, which did not, indeed, present so many marked peculiarities as the first, but which was far more majestic in appearance. It was produced by a cloud, not nearly so high in the air as that which caused the first, and it was visible above the Rue du Faubourg St. Jacques and the Observatory. It was of a grayish hue, and was traversed from top to bottom by a tube as luminous as the moon. I saw the vapors rising and falling in the lower part of it very distinctly. At short intervals the body of this whirlwind lengthened and shortened, and sometimes rapidly. It passed before the first, and seemed not to be more than from 1600 to 2000 paces to the north; but the first, just before it disappeared, traveled much more rapidly southward. It followed about the same direction as the first, and its lower part curved slightly toward the west.

"There was a thunder-clap from a cloud not very far from the whirlwinds, especially from the second; they did not seem to be in any way affected by it. We judged, from the loudness of the report, that the lightning had struck the ground. Drops of rain as large as a man's thumb fell at the point where I was standing, followed by hailstones as big as nuts.

"The second whirlwind gradually made its way back to the cloud out of which it had proceeded, and by which it was rapidly re-absorbed. It had not lasted altogether more than five-and-twenty minutes."

These whirlwinds were, as will have been gathered, harmless. They do not seem to have touched the ground; but there is no doubt they would have proved more dangerous to any balloon which might have approached them.

We now come to *trombes* of another kind, the passage of which along the surface of the ground leaves unmistakable traces of their power.

"At 1·30 P.M. of the 6th of July, 1822, in the plain of Assonval, six leagues distant from St. Omer and Boulogne, the clouds, coming from different points of the horizon, suddenly effected a junction, and covered with one mass the whole sky. Directly afterward, a thick vapor, with the bluish hue of burning sulphur, was seen to descend from this cloud. It formed a reversed cone, the base of which touched the cloud. The lower part of the cone, which reached to the ground, soon formed an oblong mass of about thirty feet detached from the cloud, revolving very rapidly.

"As it rose, it emitted a sound like that caused by the bursting of a large shell, leaving an indentation upon the ground about twenty-five to thirty feet in circumference, and to a depth of three or four feet in the middle. When at about a hundred yards from its point of departure, and moving in an easterly direction, the whirlwind blew down a barn, and shook a solidly-constructed house with the force of an earthquake. On its way, it rooted up a group of very large trees, which were found lying in many different directions, showing that the whirlwind was revolving while advancing. Others had their topmost branches torn off, and some of these were found hanging to the tops of other trees sixty or seventy feet from the ground.

"The whirlwind then went a distance of two leagues without touching the soil, tearing off large branches of trees which it scattered right and left: reaching the corner of a wood, it carried off the tops of some large oaks which were blown over the village of Vendôme, situated at the foot of the hill to the east of the forest.

"Globes of sulphurous vapor were from time to time emitted from the centre of this whirlwind, and the noise which it made was like that of a heavy carriage driven rapidly over paving-stones. Each time that a globe of fire or vapor was emitted there was an explosion like that of a gun, the wind, which was very violent, adding a wild shriek. After having torn up the soil and every thing which resisted it, the whirlwind rose into the air and went on to a distance of a league and a half, where it recommenced its ravages.

"Thence it reached the valleys of Witernestre and Lambre. In the first of these villages, composed of forty houses, only eight were left intact; and it was noticed that the gables and walls of the houses were blown in all directions—showing that the wind had blown from every quarter.

"The disasters which it caused at Lambre were not less extensive. Several persons remarked the circular progress of the trombe, its sulphurous hue, and the focus of flaming fire which issued with the sparks of bituminous vapor. The trees around the church were broken and uprooted, the curé's house carried away, and eighteen others, mostly of brick, were snapped off at their foundations, with the curious phenomenon of the walls falling outward."

The following whirlwind was not less remarkable: At 3 P.M. on the 26th of August, 1823, after some calm and warm weather, a whirlwind appeared at Rouvier (Eure et Loir). It was preceded by a black cloud from the S.W., followed by others of a yellowish hue, with intermittent thunder and hail. Apparently touching the cloud at its summit, and with its base on a level with the ground, it threw down every thing in its passage, hurling the soil and trees to a great distance. The whirlwind was of a dark yellow color—due, no doubt, to the dust and other substances which it carried off. The leaves of the hedges and trees which were not blown off were dried as if by fire. In the hamlet of Marchefroid, where it continued only a minute, it destroyed fifty-three houses. The inhabitants heard no thunder, nor did much hail fall. A child of three years of age was killed. A deep wound was found in its neck, but it was impossible to tell what body had caused it. In the valley of St. Ouen, the meteor destroyed a range of trees extending 800 feet, and then moved toward Mantes, extending over a width of from

forty to fifty fathoms. Whole houses were swept away, and in the direction followed by the whirlwind branches of trees were found scattered on all sides. Trees were snapped off at a height of four, six, and
ten feet from the ground in the valley—a fact which would lead one to
suppose that the tempest there did not reach quite so low as the ground.
In one instance the destruction was very regular. The four walls of a
garden, built of solid stone, were blown down, each wall in a straight
line, and as if the stones had been placed there for constructing a wall.
The body of a three-horse wagon loaded with grain was blown off the
carriage and carried on to the top of a building, the roof of which it
stove in. Pieces of the wood-work of the wagon were found upon the
other side of the building. The grain had disappeared, and the horses,
though uninjured, had been entirely stripped of their harness.

The following case is equally remarkable: On the 26th of August,
1826, the neighborhood of Carcassonne was visited by an enormous
column of fire which, sweeping along the surface of the soil, destroyed
every thing that lay in its passage. A young man was carried off by
it into the air and hurled head foremost against a rock. Fourteen
sheep were taken off their legs and asphyxiated. This column of air
and fire overturned walls, displaced enormous rocks, uprooted the largest
trees, and did great damage to a very solidly-constructed country house.
The air, wherever it passed, was impregnated with sulphur.

Among the whirlwinds which have left traces of great destruction
behind them must be cited that of Monville on the 19th of August,
1845. The valley in question, which is so attractive a point in the
railway journey between Rouen and Dieppe, was visited at about 1
P.M., the weather being hot and oppressive, by a whirlwind of a very
remarkable kind. The large mills existing at Monville were suddenly
enveloped and blown down. The factory, in which hundreds of wom-
en were at work, fell in, amidst a sudden discharge of electricity, and
they were buried beneath its ruins. Some of them who escaped death
were unable to understand what had happened, and believed that the
end of the world had arrived. Men were hurled over hedges; others
were cut to pieces by the machinery which was whirled about in the
air; others, without being actually hurt, were so terrified that they died
from the effects of the fright in the course of a few days. Whole rooms
and walls were turned upside down, so as to be no longer recognizable.
At other points the buildings were literally pulverized, and their site
swept clean. Planks measuring a yard long, five inches wide, and near-

Fig. 63.—Whirlwind.

ly half an inch thick, archives and papers, were carried to distances of fifteen to twenty-five miles, almost to Dieppe. Trees situated in the track of the storm were blown down and dried up. The extent of the ground thus devastated was as much as nine miles, increasing from 100 yards in width near the Seine, at Canteleu, to 300 yards about Monville, and decreasing again to thirty yards at Clères. The barometer fell suddenly from 29·92 to 27·75 inches.

This sudden dilatation of the air necessarily upset the equilibrium of the atmosphere in the immediate neighborhood. An inhabitant of Havre informed me that on the day this catastrophe occurred he saw a vessel which was three leagues off the shore enveloped in a tempest, although the sea just outside Havre was relatively calm.

The catastrophe of Monville is remembered in Normandy, just as a terrible shipwreck is handed down in the recollection of a sea-port town. Fortunately whirlwinds do not often assume such immense proportions, or do not occur at the spot where large masses of people are congregated. Several others, equally violent, perhaps, have not found any element of resistance in their path. That which occurred in the neighborhood of Trèves, in 1829, was in the form of a chimney hanging from a cloud, and emitting jets of flame and vapor. It soon changed to the shape of a serpent, undulating above the land, and leaving a track from ten to eighteen paces broad, along a distance of 2000 yards, where even the grass, plants, and vegetables growing upon the ground were swept away. There was, however, no loss of life nor destruction of houses. That which devastated Chatenay, near Paris, in June, 1839, burned up the trees that lay within its circumference, and uprooted those which were upon its line of passage. The former, in fact, were found with the side which was exposed to the storm completely scorched and burned, whereas the opposite side remained green and fresh. Thousands of large trees were blown down and lay all one way, like wheat-sheaves. An apple-tree was carried over 200 yards on to a group of oaks and elms. Houses were gutted inside without being blown down. Several roofs were carried off as if they were kites. An inside wall was cut into five nearly equal parts of eight yards each; the first, the third, and the fifth were laid in one direction; the second and the fourth in an exactly opposite direction. Several rows of slates had their fixings torn out, without being themselves displaced. In a whirlwind which raged over the village of Aubepierre, in the Haute-Marne, on the 30th of April, the slates on the roof of a wash-house were turned completely

upside down, each rank being reversed as if by the hand of a work-
man.

In the sandy regions of the African and Asian deserts, the traveler

Fig. 64.—Sand whirlwind.

sometimes encounters gigantic whirlwinds of sand which rise from the
earth to the clouds, twisting convulsively, and emitting a sound like the
hissing of a serpent. This is the phenomenon represented in Fig. 64,

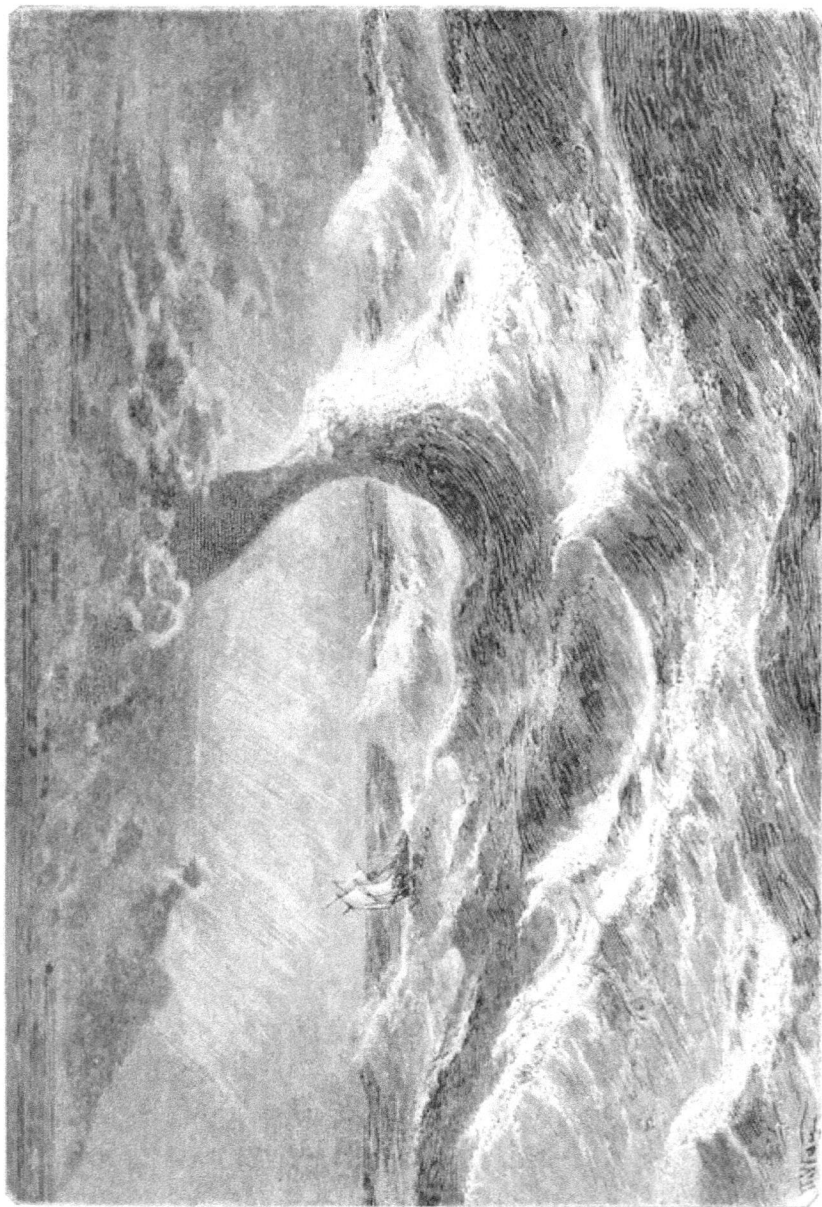

Fig. 65.—Water-spout at Sea.

and it is taken from the travels of T. W. Atkinson on the frontier-land of Russia and China.

The water-spouts that occur upon the water differ only from the whirlwinds of the air in respect to their situation. In place of dust, leaves, and other solid substances drawn up by the whirling column, they are composed of water, generally in the form of very condensed vapor, but sometimes in a liquid state, which becomes mixed with the air of the water-spout. Peltier cites many instances, extending over every degree of latitude. I can find no case in which they have been proved to have swallowed up a vessel. Generally the base of the column is severed by discharging a cannon into it. Upon one occasion, however (in the Ionian Sea, on October 29, 1832), it appears that a ship was caught in a water-spout and tossed up and down, to the great alarm of the passengers, who were situated "like a person at the bottom of a well who is looking up into the air."

The cloud which is attracted may descend near enough to the ground to raise up masses of water as well as floating substances; the heaviest of these will become detached from the mass one by one by reason of their specific weight, but the smaller bodies may be carried a great distance, and may fall all together. This is one cause of showers of frogs and fish.

BOOK FIFTH.

WATER—CLOUDS—RAIN.

CHAPTER I.

THE WATER UPON THE SURFACE OF THE EARTH AND IN THE ATMOS-
PHERE: THE SEA—VOLUME AND WEIGHT OF THE WATER THROUGH-
OUT THE GLOBE—PERPETUAL CIRCULATION—VAPOR OF WATER IN
THE ATMOSPHERE — ITS VARIATIONS ACCORDING TO THE HEIGHT,
THE LOCALITY, AND THE WEATHER — THE HYGROMETER — DEW —
WHITE FROST.

THE globe to which the force of attraction attaches us is nearly 7958
miles in diameter, and is therefore about 25,000 miles in circumference.
It is a sphere, the cubic volume of which is about 264,000,000,000
cubic miles. If it consisted entirely of water, it would weigh about
1080 trillions of tons, inasmuch as water weighs about $62\frac{1}{2}$ pounds
per cubic foot. But, as the earth is more than five times (5·44) as
heavy as water, the weight of the globe is about 5880 trillions of
tons. The atmosphere which envelops our planet weighs, as I have
said, scarcely the millionth part of the weight of the entire earth (the
$\frac{1}{1,110,000}$ part). Water occupies a not less important part in the terrestrial system than air. The mean depth of the oceans is about two
and a half miles, taking into account the uneven nature of their beds,
the level of which, owing to the shores, table-lands, valleys, and mountains, varies from a few yards to six miles.

Embodied in one mass, the water of the sea (taking its average depth
at two and a half miles) would form a sphere 900 miles in diameter.
Spread over the whole spherical surface of the globe—supposing the
surface to be perfectly even—it would cover it to the depth of nearly
two miles. The density of sea-water is rather above that of soft water:
its entire mass would weigh less than the $\frac{1}{4000}$ part of the weight of the
earth.

The maximum depth of the ocean is about six miles, and the part of
the atmosphere in which we can breathe is of about the same extent.
It is within this limited zone of twelve miles that all the phenomena of
life take place, from the submarine forests and strange animals which
inhabit the lowest depths, to the plants which vegetate upon the surface
where man has his being, to the various kinds of animals which live in

the open sky, to the condor which soars above the limit of perpetual snow. This zone of life is very limited, when compared to the size of the earth, which itself appears so small in relation to the planetary system.

To form an idea of the immense difference, we have only to examine an equatorial section of the globe. Even if the sinuosities are increased fifty-fold, the terrestrial rind is almost a complete circle. The continents and islands are but the summits of table-lands and mountains, the lower parts of which are submerged.

This water covers nearly three-fourths of the earth, in the state which corresponds to the mean temperature of its surface—that is to say, in a liquid state. Its currents constitute, as we have seen, the grand arterial circulation of the planet. Not content with thus prevailing in its ordinary state, it reigns in a *solid* state, in the silent regions of the poles and upon the ice-bound sides of inaccessible mountains; and, in a *gaseous* state, it reigns with more absolute sovereignty in the atmosphere, the life of which it regulates, and in which it in turn promulgates abundance and dearth, the gladness of fine weather, and the gloom caused by sombre skies.

This water is not motionless, either in the depth of the oceanic basin, in the solid ice, or in the atmosphere. Thanks to the always active power of the sun, to the aërial currents, the water rises vertically from the bed of the sea to its surface, becomes vaporized at all temperatures, ascends in the shape of invisible vapor through the ocean of the air, becomes condensed into clouds, travels across continents, falls again in the shape of rain, filters through the surface of the soil, passes along the strata of impermeable clay, springs up as a source or fountain-head, descends by the streamlet into the river, and falls from the river back into the sea again.

Every source, every streamlet, every river, every stream, has its origin in rain. Even the mineral waters are produced by the same cause, and their heat is merely due to the profound depths from which these meteoric waters have been brought up; and they, moreover, continue to ascend through the interstices of the rocks, afterward returning to the level of their primitive reservoir, like a siphon. The sun, as it evaporates the sea-water, leaves behind it the salt, which is not volatile. That is why rain-water is soft, and that of a running stream also. The salt never leaves the sea, and its quantity is such that it would cover the whole surface of the globe to a depth of ten yards.

Just as the blue color of the sky is due to the vapor of water, so also is the color of water itself, taken in a mass, blue; its shades vary from that hue to green, according to the action of light. The vapor of water mixed with the air is of the highest importance in the distribution of temperatures; both its formation, as well as its movement from place to place, represent a formidable force which is permanently in action. The air can contain more vapor of water in proportion as it is heated. A given diminution of temperature brings it to its saturating limit, without in any way adding to the quantity of vapor which it contains. To ascertain the quantity of vapor of water mixed with the air at a given moment, a thermometer, for instance, suspended in the air might be made gradually colder until it indicated the limit of saturation—that is, until its bulb was covered with condensed vapor, or dew. By ascertaining what quantity of vapor of water corresponds to this thermometrical degree of saturation, we should learn the real quantity suspended in the air at the moment of the experiment.

The instruments for measuring the moisture of the air have received the name of *hygrometer* (ύγρός, moist, μετρον, measure). That in most general use is formed of two thermometers exactly alike,* and placed side by side. The bulb of one is enveloped by a piece of muslin, which is kept constantly moist. The moistened thermometer has a lower temperature in consequence of evaporation proceeding from the moist muslin. The difference of reading between the two thermometers is, therefore, dependent upon the more or less moisture in the air. The hygrometrical state of the atmosphere is not the same from the top to the bottom, like the proportions of oxygen and nitrogen. As a rule, it increases, beginning from the surface of the ground up to a certain height, where there exists a zone of maximum moisture; above that point it decreases. I will not venture to trace a diagram of this variation in the moisture according to the height, as I have done in regard to the decrease of atmospheric pressure and of the temperature, for the observations which I have made upon this head are wanting in number and in precision. Those taken by Glaisher are much more precise, and have been made by different hygrometers. They show that, generally speaking, the moisture increases from the surface of the soil to a height of 3500 feet, and that after that it diminishes, there being, however, spaces which represent moist strata of air of varying thickness.

The observations taken upon mountains confirm the increase which

---

* See "Glaisher's Hygrometrical Tables" for description and use of dry and wet bulb thermometers.—ED.

was first noticed to be in proportion to elevation. Kaemtz has ascertained the degree of humidity to be a mean of 84 upon Mount Righi as against 74 at Zürich beneath it. Bravais and Martins registered 76 upon the summit of the Faulhorn, when at Milan there was only 63. At heights exceeding 3300 feet the moisture decreases, in spite of the special augmentations due here and there to currents which lie one over the other.

Upon the surface of the ground, the relative moisture of the air varies according to the time of day, in inverse ratio to the temperature. The warmer the air, the drier it will be; the colder it is, the more readily will it be saturated with moisture. In our temperate regions, the hygrometrical state of the air augments, with little fluctuation, toward sunrise during the minimum of temperature; afterward falls, until about 2 P.M., at the maximum of heat; and rises again toward evening and at night. Twenty years of daily repeated observations (1843–1863), taken at Brussels by the aid of the Saussure hygrometer and dry and wet bulb thermometers, have furnished M. Quételet with the information that the mean degree of humidity at noon is as follows:

January	87	July	67
February	84	August	68
March	73	September	74
April	66	October	80
May	65	November	85
June	64	December	89

Where complete saturation is represented by 100.

We see that the maximum of relative humidity occurs in December, and the minimum in June. This invisible atmospheric moisture, the presence of which is only revealed by aid of delicate instruments, confers upon the landscape all the variety with which it is endowed—the emerald green of the Irish pastures, the blue sky of the Mediterranean, the splendor of tropical vegetation—and it becomes visible in the shape of dew as soon as a diminution of temperature brings it to the point of saturation. If it is the air itself which becomes colder, it is made opaque by the passage of the vapor in a liquid state, and hence arises fog. If it be a solid body which is thus rendered cold, the moisture becomes condensed upon its surface, and the result is dew.

Dew does not come down from the sky, as is still taught in the French primary schools. Its production is in no degree assimilated to that of rain. It *is formed* at the spot where it is seen. If small portions of grass, cotton, or other fibrous substance be exposed to the sky on a fine night, it is found that, after a certain time, their temperatures are fifteen, eighteen, and even twenty degrees below that of the circumambient atmosphere.

In places where the sunlight does not penetrate, and whence a large extent of sky can be seen, this difference between the temperature of the grass, cotton, wool, etc., and the atmosphere is noticeable between 3 and 4 P.M.—that is to say, as soon as the temperature diminishes; in the morning it continues for several hours after sunrise.

The observations of Wells, continued by Arago,* have proved that on a clear night the grass of a meadow may be ten to twenty degrees colder than the air; if the weather becomes cloudy, the grass at once increases several degrees in temperature, without any increase in that of the atmosphere.

This diminution of heat is due to nocturnal radiation. When there is nothing to prevent the heat of a body from becoming dispersed, it gradually becomes irradiated and lost. The transparent air does not suffice to prevent this loss of heat. But a cloud, a wooden screen, a sheet of paper, a little smoke even, will answer the purpose. Without obstacles of some kind, the substance becomes colder according to its power of radiation, which is itself dependent upon the nature of the substance (it is, for instance, very great in the case of glass, and very trifling with metals); and when the temperature of the body thus exposed has reached that of the point of saturation, the atmospheric moisture is deposited upon it, taking at first the shape of spheroidal drops; then, when these drops are sufficiently weighty and near together, they extend like a shallow pool of water over the surface of the substance.

Dew is never abundant except when the nights are calm and bright. A little dew may be seen when the nights are cloudy, if there be no wind, or even with wind if the weather is bright; but there is never any sign of it when there is wind, and the sky is cloudy as well. The circumstances which favor an abundant deposition of dew more generally occur in spring-time and in autumn—the latter especially—than in summer. It must be remembered, too, in addition to the above fact, that the differences between the temperature of day and night are never greater than they are in spring and autumn.

The phenomenon of the deposit of dew upon a dense and smooth

* [And by the Editor. See "Phil. Trans.," part ii., 1847, for paper on "Radiation at Night from the Earth, and several Substances placed on or near it."]

substance — upon a sheet of glass, for instance — resembles that seen
when a pane of glass is exposed to a current of vapor of water warmer
than itself: first, a light and uniform layer of moisture dims the surface; then are formed irregular and flat drops, which run together after they have acquired a certain volume, and flow in all directions.

This may be seen whenever any substance which has been rendered
cold by exposure to a low temperature is taken into a warm room; the
substance at once becomes covered with moisture. In the same way
glass placed in a room where a large number of persons are dining is
at once dimmed by the thick stratum of dew which the invisible vapor
mixed with the surrounding air deposits. The glasses of a pair of
spectacles which have been exposed to cold air will often be found obscured in the same way.

If, during a frost, the windows of a room in which a large company
has been dining are suddenly thrown open, a cloud forms instantaneously in the path of the cold air, and the ceiling is made damp by a
long stain of condensed vapor.

Dew is a phenomenon of importance, not only because of the absolute
quantity which any one point of the globe receives, but because of the
extent of ground over which it may be deposited. It is mostly in
tropical regions that its effects upon vegetation are the most marked
and the most favorable. When the air, nearly saturated with vapor at
the temperature of 86°, contains more than thirteen grains of water to
the cubic foot, the water falls abundantly during the declining temperature of night; it makes the leaves drip, and in the morning grass is
as wet as if there had been heavy rain. The dew is known to deposit
in greater or in lesser quantities, but it has not been found possible to
measure it, because it does not fall like rain; its appearance depends
upon the radiating power of the body which it moistens, for it is only
deposited upon substances which are colder than the surrounding air,
and in increased quantity according as the difference of temperature is
greater. Plowed land, fallow, forests, rocks, and sand vary much in respect to the dew which deposits upon them; and, more than that, the
leaves of all plants do not possess an equal radiating power, and the
intensity of their diminution of temperature, influencing the deposit of
dew which ensues upon it, is dependent upon their distance from the
ground, their color, the smoothness or the ruggedness of their epidermis.
The dew alights upon the leaves of mangel-wurzel, while the tops of
potatoes in an adjoining field will hardly be moist.

M. Boussingault has endeavored to measure these quantities of dew. After certain nights, when the dew had fallen abundantly, he used to repair to the meadows on the banks of the Saüer before sunrise; there, by aid of a sponge, he soaked up the water over forty-three square feet of glass, and this he placed in a bottle and weighed. In some instances it was found to exceed two pounds in weight.

Dew and mist contain about the same proportions of ammonia and nitric acid; both, moreover, have a great analogy to rain when it begins to fall—when it is, so to speak, in process of washing the air. It is, in fact, in the first part of a shower of rain after a season of long drought, that there is present the greatest amount of carbonic acid, carbonate and nitrate of ammonia, organic matters, and dust of every kind. If a close examination be made of the substances which air contains in infinitesimally small quantities, it is in the mist, the dew, the first drops of rain, the first flakes of snow and hail, that we must look for them.

White frost, which is so fatal to vegetation in spring, and which has given such a bad reputation to the harvest-moon, is, in reality, the dew frozen by the same cause as that which led to its formation—nocturnal radiation.

In 1871, A. Wilson, having followed the movements of a thermometer during a winter night when the weather alternated constantly between clear and foggy, found that it always rose about a degree at the same moment that the atmosphere clouded over, and fell to the point at which it had previously stood when the mist cleared off. His son, Patrick Wilson, asserts that the instantaneous effect of a thermometer hung up in the open air is to cause an elevation of $5°$. The researches of Pictet, undertaken in 1777, and published in 1792, coincide nearly with the above.

It is a curious circumstance, which was discovered by Pictet, that, when the nights are still and clear, the temperature of the air, instead of diminishing the higher from the ground, shows, on the contrary, a progressive rate of increase, at least up to a certain height. A thermometer nine feet above the soil marked throughout the night $4\frac{1}{2}°$ Fahr. less than an exactly similar instrument which was attached to the summit of a pole fifty feet high. About two hours after sunrise, and two hours before sunset, the two instruments were exactly the same. Toward noon the thermometer nearest the ground was often $4\frac{1}{2}°$ higher than the other. When the sky was covered with clouds, the two instruments corresponded exactly, both by day and night.

These observations have been confirmed. Wells having placed at the four corners of a square four small pegs, which stood perpendicularly four inches above the surface of a meadow, spread over them horizontally a fine cambric handkerchief, and during five nights compared the temperature of the small square of grass covered by it with the surrounding portion which remained fully exposed to the air. The turf that was protected from radiation by the handkerchief was at times 11° warmer than the other. While the latter was completely frozen, the temperature of the turf protected from the air was several degrees above 32°. With the sky very cloudy, a screen of cambric, matting, or any other substance, produces scarcely any effect.

Mr. Glaisher finds, after three years' consecutive observations at Greenwich, that the temperature of the air twenty-two feet above the ground is higher than that at four feet at every hour of the night and day during the months of November, December, January, and February; that it is higher at night and in the evening in May, June, and July; and that it is also higher during night-time and in the afternoon in March, April, August, September, and October. At an elevation of fifty feet the temperature is also higher during the night throughout the whole year. With the sky cloudy, the temperature remains the same.

In June of 1871 the attention of the Academy of Sciences was directed to the subject of late frosts by M. Ste. Claire-Deville and M. Elie de Beaumont. The immediate instance in hand was the frost which occurred on the 18th of May (Ascension-day) and extended to the vines and their crops around Paris and in the centre of France. As I myself had seen a vine which had been frozen in the Haute-Marne, I showed by a few comparisons that this disastrous frost extended over quite one-half of France at the same moment. It would certainly be most desirable to find some means for protecting crops during the critical period which follows the blossoming, as many severe losses would thus be prevented.

CHAPTER II.

THE CLOUDS: WHAT A CLOUD IS—THE MANNER OF ITS FORMATION—
MIST—OBSERVATIONS TAKEN FROM A BALLOON AND FROM MOUNT-
AINS — DIFFERENT KINDS OF CLOUDS — THEIR SHAPES — THEIR
HEIGHTS.

THE *invisible* vapor of water spread throughout the atmosphere, the
distribution and variations in which I have just pointed out, becomes
visible when a decline in the temperature or an addition of moisture
brings it to the point of saturation. Suppose, for instance, that a certain quantity of air at eighty-six degrees contains 478 grains of vapor
of water, this air will be quite transparent. If by some cause or other
this air descends to seventy-seven degrees, or receives an accession of
moisture, it will become opaque. A diminution of nine degrees of
heat will cause 108 grains of vapor of water to be condensed and to
become visible. That is what a cloud really is: vapor of water which
the air, being saturated, is no longer able to absorb, and which becomes
separated from it by passing into the state of small vesicles.

This passage from the gaseous to the liquid state takes place indifferently at all elevations. When it occurs at the level of the soil, it is
termed mist. But there is no essential difference between a cloud and
mist. Traveling through the clouds in a balloon, meeting no resistance, the air is simply more or less opaque, more or less cold, more or
less damp, just as is the case upon the surface of the ground, according
to the diversity of the mists. This is also the case with the clouds
when one is enveloped in them upon the summits of a mountain.

Though there is *no essential difference* between mists and clouds, there
is, however, one in fact, viz., that a mist is vapor of water passing from
the visible to the invisible state ; whereas a cloud is a grouping of visible vapors in some given shape. The first is *motionless,* the second is
endowed with movement. Let us consider the mist first.

Seen through a glass, mist is composed of small and opaque bodies.
A closer study shows that these small bodies are composed of water,
obeying the laws of universal gravitation. The molecules of water are
grouped together in the form of spherules. Are these spherules full

or hollow? Such is the question upon which meteorologists are divided. The opinion already given by Halley that these spherules are hollow, and that the water is but an envelope, seems the best founded.

Take a cupful of some dark-colored liquid, such as coffee or China ink, dissolved in water; warm it, and place it in the sun's rays: if the air be still, the vapor will ascend and soon disappear; if looked at through the glass, it will be seen that globules are rising. The smallest run rapidly over the surface of the magnifying-glass, the others fall back on to the liquid mass. Saussure adds that the small vesicles which rise differ so much from those which fall back that it is impossible to doubt that the first are hollow.

The way in which they act when exposed to the light is also favorable to this supposition, for they do not scintillate like the full drops when they are exposed to a bright light. Every one must have remarked that soap-bubbles are generally very brilliant in color. The same must also have been noticed with bubbles from other viscous substances, and it is the easier to observe them because they continue a longer time. These colors rise from the division of the incident rays into two parts. Some of the rays are reflected by the outside surface: others penetrate through, and are reflected by, the inner surface. The envelope of the sphere must be thin, to admit of this taking place. Kratzenstein having examined in the sun and through a magnifying-glass the vesicles which ascend out of hot water, observed upon their surface colored rings like those of soap-bubbles; and not only was he convinced that their structure is analogous to that of soap-bubbles, but he was further successful in calculating the thickness of their envelope.

De Saussure and Kratzenstein attempted to measure by aid of the microscope the diameter of the vesicles which compose the vapor of water. But it is difficult to arrive at any positive result, for it is the vesicles rising from mist, and not those from hot water, which it is necessary to measure. Fortunately, some of the optical phenomena which occur, when the sun shines through clouds, furnish us with a means of arriving at this result.

Kaemtz has taken a great number of measurements in Central Germany and Switzerland; he has ascertained that upon an average the diameter of the vesicles of mist is about ·00087 of an inch, and that it varies in the different seasons as follows:

DIAMETER OF THE VESICLES OF THE MIST.

	Inch.		Inch.
January	0·0106	July	0·0066
February	0·0138	August	0·0055
March	0·0079	September	0·0087
April	0·0075	October	0·0079
May	0·0059	November	0·0095
June	0·0071	December	0·0134

It will be seen that there is an almost regular progression from winter till summer; the anomalies arise from the small number of observations that have been taken. Thus in winter, when the air is very moist, the diameter of the vesicles is twice as great as in summer, when the air is dry; but this diameter also varies in the course of a single month. It attains its *minimum* when the weather is very fine; it increases when there are signs of rain; and before the fall it varies considerably in the same cloud, which probably contains a large number of drops of water mixed with vesicular vapor.

Autumn, like spring, is the season of abundant dew. The cooling process to which the ground is subject, when the nights are clear, and the moisture of the air nearer precipitation than in summer, causes the atmospheric water to be deposited upon terrestrial objects which have diminished in temperature, just as in a crowded room the moisture of the heated air affects the glass brought in from outside. The steam of hot dishes, the breath of the persons present, the combustion of the lights, make the air of the dining-room hot and moist, and cause water to trickle down the vases containing ice. In autumn, the nocturnal coldness of the ground often communicates itself to the stratum of air immediately above, and hence arise the low fogs which are soon dissipated by the sun's rays. If the ground be uneven, the cold air of fogs descends into the valleys, and seems, to any one standing upon an eminence, a white *sea* perfectly level. As a child, I have often watched before sunrise, from the ramparts of Langres, the ocean of grayish vapors that extend through the valley of the Marne, and the waves of which reached to within a few feet of where I was standing. The height of the ramparts at Langres is near 1500 feet above the level of the sea. In winter the view sometimes extends at sunrise so far beyond the mist in the plain, that the white outline of Mont Blanc is discernible with the naked eye.

To witness a spectacle of this kind at its best, it is necessary to be

upon the top of a lofty mountain, whence the view embraces a vast
horizon, and at sunrise after a day when the clouds have obscured the
sky of the country below. The clouds, disturbed in a thousand ways
by the rays of the sun and the light winds which are the natural consequence, are not very level during the day-time. But at night the
equilibrium and the level are restored, and a sea of aërial vapors extends far as the eye can reach beneath the feet of the observer. The
elevated summits of the isolated mountains around him break here and
there through the nebulous ocean, above which soars from time to time
an eagle in quest of its prey. Standing in the valley, in the midst of
the mist, the sun's rays, as they play through the foliage, delineate brilliant beams of light, the *ensemble* of which forms what is called a *glory*
not more than a few yards above the head of the spectator. This *glory*, which emanates from the tree immersed in the fog, recalls to mind
Moses's burning bush.

Sometimes only the surface of rivers is covered with fog, because
water emits vapor which becomes condensed in the air which lies over
them, and which becomes cold after sunset. The air takes almost instantaneously the temperature of the bodies to which it is in contiguity. During a calm and clear night, the portion of the atmosphere
which lies over water will be warmer than that above dry land.

In calm weather, where water is abundant, the lower strata of the
atmosphere become laden with the extreme amount of moisture compatible with their temperatures. I have already stated that the moisture which the air contains when it is saturated is of a fixed quantity,
which varies according to its temperature. If saturated air becomes
cold by contact with a solid body, it deposits upon the surface of that
body a portion of its moisture; but when the cooling process takes
place in the very midst of the gaseous mass, the moisture that is set
free passes off in small floating vesicles, which affect its transparency;
it is these vesicles which constitute clouds and mists.

Let us suppose that some circumstance—a small declivity of the soil,
for instance, a slight puff of wind—causes a fusion to take place at
night between the air that lies over a river or sea, and that which is
above the land: the latter, which is colder, diminishes the temperature
of the former; the former also loses a part of the humidity which it
contained, and which did not at first cause any alteration in its diaphanous condition. But as this moisture gradually resolves itself into a
state of vesicular vapor, the air becomes thick; when the number of

floating vesicles becomes very large, a heavy fog comes on. The distribution of fog throughout the year corresponds with that of humidity and temperature. Fogs are much more numerous in winter than in summer. The Brussels Observatory, which has recorded them with great care, gives the following as the number of days on which there have been fogs for the last thirty years (1833–1863) :

January	259	August	76
February	168	September	159
March	138	October	228
April	62	November	276
May	71	December	315
June	42	Total	1822
July	28		

Under certain circumstances the fog is very thick, and is bounded by a plane surface like a sheet of water, rising slowly in the still air, and enveloping all surrounding objects with a cold and damp embrace. M. Raynal, whose vessel was wrecked off Auckland Island in 1864, was witness of a curious instance of fog, which he relates in this way : Having, on the 9th of August, climbed one of the mountains in the island, he was making his way down again with one of his companions, following a narrow path between two precipices. "I was unable," he says, "to move a step, for we could not see where to put our feet. We passed at least an hour in this way, absolutely motionless, and holding each other by the hand, while the cold began to benumb our limbs. Fortunately a breeze sprang up, and dividing the fog into two parts, gradually carried it away."

But it is in the frozen latitudes that the fogs are thickest. At Spitzbergen, says M. Martins, the mists are almost continuous, and so thick that it is impossible to make out objects which are a few paces off. These damp, cold, and piercing mists often wet as much as rain. Thunder-storms are unknown in these regions, even during summer. Toward autumn the fogs increase, rain changes into snow. Fig. 67, illustrative of an incident during the scientific voyage to which I have referred, gives an idea of these immense and perpetual fogs.

In countries where the soil is damp and hot, and the air damp and cold, thick and frequently recurring fogs must be expected ; this is the case in England, the shores of which are surrounded by seas with a high temperature. It is the same with the polar seas and Newfoundland, where the Gulf Stream, which comes from the south, has a higher temperature than that of the air.

Fig. 66.—Intense fog in one of the islands of the Antipodes.

In London the fogs are at times dense. Every year the journals record that it has been found necessary to light gas in the middle of the day, both in the streets and houses. Very heavy fogs* also occur in

* There are at times *dry fogs*. They have no connection with the hygrometrical states I am now discussing. They are generally due to the smoke of burning prairies, and may ex-

Paris and Amsterdam, the sky, at a short distance from these cities, being at the same moment perfectly clear.

Fig. 67.—Intense fog in the Spitzbergen Mountains.

Thick fogs emit, too, a noxious odor when they become impregnated

tend over a vast distance. The smoke of the heath in Holland sometimes reaches as far as Austria, hundreds of leagues off. The smoke of volcanoes also extends very far, that from Honolulu having been seen in 1868 at a distance of 200 miles from the mouth of the volcano. In 1865 the smoke from a great fire at Limoges covered the sky seventy-five miles off. The most intense dry fog known is one that occurred in 1783.

24

with the different exhalations which may find their way into the lower strata of the atmosphere. Ammonia may often be discovered. It is not rare to find it accompanied by a smell of peat in Belgium and the north of France. During the cold and damp fogs of the month of October, 1871, in Paris, the smell of petroleum was several times perceptible.

If a chain of mountains be looked at from a distance, it will often be seen that a cloud hangs over each peak, but that the intervals between them are clear. This state of things may last for hours and even days, but this absence of motion is only apparent, for there is frequently a strong wind blowing over these summits, which condenses the vapor as it ascends the flanks of the mountains. As soon as it disappears from the summits, the wind also vanishes. In Alpine passes, the formation, the movements, and the disappearance of clouds form a spectacle of very varied beauty.

The clouds which ascend the mountain side of a day-time, by virtue of the diurnal ascending currents, often dissolve when they reach the summits under the influence of an upper wind, which is comparatively dry and warm. It is of an evening especially that this is the most noticeable, and the phenomenon generally occurs upon the ridges and summits of the passes which lead to them. The fog then seems to make its way in the direction from which the wind is blowing, yet, notwithstanding the surface by which it is bounded, remains stationary.

Very often, sombre clouds, passing rapidly over the St. Gothard Hospice, are precipitated in vast masses into the deep gorge of Lake Tremola. It might be fancied that all Lombardy would be obscured by a thick fog, but before it has issued from Lake Tremola, the warm ascending currents dissolve it.

Let us now consider the clouds in themselves, their formation, and the manner in which they are suspended in space.

We saw, in the previous chapter, that the moisture of the air increases up to a certain height until it reaches a zone of *maximum humidity*, the elevation of which varies according to the seasons and hours, and above which the air is drier and drier. This zone was seen by De Saussure in his Alpine travels, and by Commander Rozet both in the Alps and the Pyrenees. It is a blue, transparent vapor, which it is difficult to distinguish when one is immersed in it, but the upper surface of which is easily made out when situated beyond it. This surface is always horizontal, like that of the sea. From a great height upon some peak of the Alps or the Pyrenees, the topmost limit of this atmosphere of vapor is clearly delineated on the horizon by a bluish line, like that which bounds the horizon of the sea. Its height varies according to the season and hour; it has been found to vary between 3500 and 13,000 feet. Its temperature never falls below 32°.

It is upon this surface of the atmosphere of vapor that clouds are formed, and on which they seem to repose. On the 15th of July, 1867, I rose to a height of 5000 or 6000 feet before sunrise, and for once I was present at the formation of clouds in the workshop of Nature. It was above the Rhine plain, between Cologne and Aix-la-Chapelle. The atmosphere had remained pure, when small white flakes began to appear in the zone of maximum moisture. These gradually ran together, became grouped in large numbers, and dissolved with as much rapidity as they had formed. The small white clouds, agglomerated together, formed *cumuli*. This formation of clouds was proceeding several hundred yards below us. As the sun rose, the moisture on the balloon evaporated, and we gradually ascended to a height of 7900 feet. It was the same with the clouds, which indeed rose rather more rapidly than the balloon, and finally surrounded and surmounted it.

The clouds are generally carried along by the wind, following its course and being relatively motionless in the current with which they float. The measurement of their speed gives, indeed, the measurement of the velocity of the upper wind. But this rule is not without exceptions. There are, however, *clouds which do not progress*, even when they are traversed by a more or less powerful wind, which it would be thought must take them along with it.

When traveling in company with M. Eugène Godard in a balloon, while we were over the forest of Villers-Cotterets, I was much surprised to see for more than twenty minutes a small cloud which might have been about 200 yards in length and 150 in breadth, suspended *motionless* about eighty yards above the trees. As we approached, we noticed five or six smaller, which were disseminated and also motionless, notwithstanding the air was moving at the rate of eight yards per second, and we were curious to ascertain what invisible anchor retained these small clouds. When we were above them, we found that the principal was suspended over a piece of water, and that the others were over the course of a stream, from which arose a current of humid air, the invisible moisture of which, reaching its saturating point, became visible in its passage through the cool wind that prevailed above the wood.

Kaemtz witnessed an analogous occurrence near Wiesbaden after heavy rain. He says, "The clouds dividing, the sun burst forth, and I saw a column of mist that continued to ascend from the same point. I hastened thither, and found a newly-mown meadow surrounded by pasture-lands, the high grass of which, being less heated than the bare surface of the mown meadow, gave rise to a less active evaporation." In Switzerland the phenomenon occurs on a smaller scale. While it is fine upon the Faulhorn, the Swiss lakes are often covered with fogs of very different densities. The same meteorologist has observed that the fogs over lakes Zug, Zürich, and Neuchâtel were very thick, while those which rested over lakes Thun and Brienz were merely light vapor. This phenomenon has occurred too often to be attributed to chance. Lake Zug is rather deep, and its tributaries do not descend directly from the regions of perpetual snow. Its temperature must be higher than that of Lake Brienz, into which the Aar empties itself immediately after having descended from the Grimsel glaciers. With the temperature the same, the first would become more readily involved in fog than the second.

I must now explain the causes which lead to the suspension of clouds in the atmosphere.

When a cloud is dissolved into rain, and pours down thousands of gallons of water, the question may well be asked how it is possible for such a weight of water to have remained suspended. The cause lies in its extreme divisibility. Left to themselves, the vesicles would fall. Calculation shows that it would take them more than half an hour to fall a little more than one mile in the atmosphere—that is to say, that the rapidity of their descent is about one yard per second; it is often less. But during the day the air is constantly traversed by warm *ascending* currents, which rise with a speed of several yards per second. Thus the clouds can not descend during day-time unless the circumstances be exceptional. It is not necessary to suppose that their vesicles are filled with dilated and lighter air, as if they were so many small balloons. Nevertheless, as Fresnel has remarked, the solar heat absorbed by the cloud must contribute to its remaining suspended. At night the clouds are nearer to the ground. But we have seen that the conditions, under which the vapor of water becomes visible, depend upon the temperature and the degree of saturation. It follows that the lower surface of the clouds dissolves as they descend into a warmer air, and frequently, too, the upper surface dissolves when exposed to the action of the sun; so that, as a matter of fact, they are constantly changing in thickness, shape, and even substance. The clouds being but water in a special state, seem to us motionless even when the particles which compose them are incessantly descending from their upper to their lower surface, below which they become dissolved. They rest, moreover, upon the zone of invisible vapor which I have already spoken of. The horizontal march of the currents represents a somewhat considerable effort to maintain the clouds at the same elevation, even when all the aqueous particles are full.

Having dealt with the formation of clouds, and their position in the air, let us consider their varied and characteristic shapes.

The forms of the clouds are of infinite diversity, from the thick fog which bathes the surface of the soil to the luminous detached filaments which hover in the heights of the atmosphere. A methodical nomenclature of clouds, to enable observers to record with precision observations of their various forms, became a necessity. Howard first gave names to the principal types in order to have a means of recognizing each, and his classification has been generally adopted, so much so that his figures have become, so to speak, classic. His description alone I shall use as a basis for my remarks on this subject.

In our climates the clouds are, in most cases, rather oval in shape; they seem to be piled one upon another, and their clearly-defined edges trace curves upon the azure of the sky. This class of clouds have received the name of *cumulus*, and it is in summer that their shape is the most marked. Sailors call them *bales of cotton*. They rise and augment in size during the morning; reach their greatest elevation when the temperature is highest; from which time they descend, and ultimately disappear, when they are not numerous. Their thickness varies from 1300 feet to 1700 feet; their height from 1500 feet to 10,000 feet.

Sometimes these half-spheres become heaped one upon the other, and form those large, accumulated clouds near the horizon which, seen from a distance, resemble mountains covered with snow. These are the clouds which lend themselves most readily to the play of the imagination, for their lightness and the extreme variability of their shape give rise to incessant metamorphoses. It is not difficult to see in them the forms of men, animals, dragons, trees, and mountains. Ossian has utilized them for some of his finest imageries. The popular legends of mountainous regions are filled with strange events, in which these clouds play an important part.

This frequently-occurring shape is coincident with the warm wind from the S. and S.W.—that is to say, with the equatorial current. When this moist current prevails for some time, *cumuli* become more numerous, more dense, and spread in beds over the sky. This second form is seen almost as often in our variable climates as the first, and it is characteristic of winter as the latter is of summer, the principal difference being that condensation, or rain, takes place more rapidly when the sky is in this state than it does during the summer phase. This kind of cloud is termed *cumulo-stratus*. The fleecy clouds, the dappled sky, represent it in well-known aspects.

When the clouds, instead of being detached, form one vast sheet extending to the horizon, the term *stratus* is given.

When a cloud is about to dissolve in rain, it acquires a greater density, becomes more sombre, and, except in the case of hail or partial storms, extends over a vast space. The water which is discharged from it would fall vertically if the atmosphere were calm, and the drops of water heavy enough; but two causes, of which one at least is always in existence—the wind, and the lightness of the rain-drops—cause the water which falls from the cloud to follow an oblique course, generally preceded by the cloud, which the wind drives at a greater rate of speed. The special state of the cloud resolving itself into rain is termed *nimbus*.

All these clouds are formed of aqueous vesicles, more or less considerable in size, and more or less compact. But the clouds do not only reside in the strata, the temperature of which is above 32°; they also float in the regions where the temperature is below the freezing-point. In this state the vesicular water becomes congealed into minute filaments of ice, and the clouds formed in this way are clouds of ice or snow, which have already served to explain such optical phenomena as halos, parhelia, etc. These clouds of ice are those which reach the loftiest regions. No matter the height to which the balloon may rise, these clouds always appear so far above that they seem no nearer than when viewed from the earth; whereas it is a work of scarcely any time to travel through *cumuli* and the other forms of clouds which I have mentioned. Mr. Glaisher found that at 37,000 feet above the soil of England, he was still far below them.

They are composed of loose filaments, the ensemble of which is sometimes like the sweep of a broom, sometimes like a bunch of feathers, sometimes like a mass of hair, or a light and irregular piece of net-work. Their mean height is from twenty to twenty-three thousand feet.

By reason of their very constitution, they remain in the ethereal regions of eternal snow. But, as I have said, the zone of 32° varies in height according to climates and season, whence it follows that these clouds may make their appearance in the lower regions of the atmosphere in the frosty latitudes of the polar regions, and even in our latitudes during a severe frost.

These clouds are designated *cirrus*. With a little practice it is easy to recognize them, and what is most striking in them is that they are nearly always divided into long and narrow strips, quite straight, and white in color, which correspond with the upper currents that direct, mold, or dissolve them.

Sometimes their whitish hue gets bedimmed, their *striæ* interlace each other, and they become denser because the upper air is moist. In this case they look like carded cotton, and this change generally foretells rain. When in this state of excessive density, they are called *cirro-stratus*.

Sometimes, too, they become transformed into light transparent clouds of vesicular vapor—so transparent that the stars and the spots on the moon can be seen through them. These are clouds which give rise to the *coronæ;* when they are in receipt of abundant light, they seem to be well rounded and fleecy; when the sky is covered with them, it is said to be dappled; their mean height is from ten to thirteen thousand feet; they are termed *cirro-cumulus*. The cumulus and the cirro-cumulus are those which impart the most beautiful hues to sunset; their transparency and their distant reflection refracting and coloring its rays. The beautiful sunsets seen in Paris are partially due to the fact that these clouds, situated above Havre for the horizon of Paris, give us a softened reflection of the luminous effects that are produced by the sea.

Such are the principal shapes which clouds take, and which are due to the difference in their constitution and their elevation. These varieties do not constitute, in reality, more than two great categories—the cumulus, formed of liquid vesicles, and the cirrus, formed of frozen particles.

M. A. Poey gives the following "scientific and popular classification" of the various shapes of clouds:

1st Type.—Cirrus. Curly cloud...................... } Frozen clouds. Height, 26,000 to 40,000 feet.
                { Cirro-stratus. Streaky cloud.............. }
Derivatives. { Cirro-cumulus. Dappled cloud............ } Snow clouds. Height, 13,000 to 26,000 feet.
                { Pallio-cirrus. Cloud in strata............ }

2d Type.—Cumulus. Mountainous cloud............ } Rain clouds, vesicular or of vapor of water.
Derivatives. { Pallio-cumulus. Rain cloud.............. } Average height, 3200 feet.
                { Fracto-cumulus. Wind cloud............ }

Among the clouds composed of liquid vesicles, we must now consider the peculiar and characteristic shapes corresponding to the production of aqueous meteors, of which they are either the cause or the forerunner.

My colleague, J. Silbermann, Vice-president of the Meteorological Society, has spent thirty years in studying and making designs of these specially typical shapes. Out of the large number which he has stereotyped and collected in a kind of meteorological museum, I will cite the principal.

Every one is acquainted with the shape of the clouds which usher in a lengthy period of rain; the sky is covered with an immense leaden sheet, and the rain falls continuously from horizontal strata slightly undulated, which are scarcely distinguishable from the sombre mass in its entirety. For days and nights together the sky continues covered with this opaque sheet, the thickness of which is sometimes many thousand yards, there being successive strata by which the light of the autumn sun is entirely absorbed. These are clouds of continental rain, which extend over vast tracts of country, and the contour of which it is impossible to make out.

The *clouds of partial rain* resemble them so far that they are lengthened into horizontal strata; but in this case their shape, less extended, is more definite, as it stands out against the background of the sky, which is no longer darkened by the immensity of the strata that lie one over the other, but is partially covered with *cumuli* that have different densities in different places. The rain issues from the sides of the clouds; it is delineated upon the pale perspective of the sky in oblique streaks of gray, the general tone of which varies with the motion of the wind. These clouds do not always dissolve entirely; certain parts seem, after they have discharged a great quantity of rain, to dry up and fall back into the centre of the cloud, as if attracted by the molecular affinity which gives to clouds their varying contour.

The hail-squall is different; it does not spread out in a large horizontal sheet, but forms a definite mass, which often stands out by itself

in the blue sky. The sun reaches to its edges and sets off its white surface against the rest of the sky; there issues from its open sides a cold
rain, hail, and rime, which a March wind blows into our faces.

The clouds which produce *hail* have the singular aspect of an adhesion of molecule, as if attraction tended to unite them in condensed
masses of a globular form, and their shape has a strange resemblance to
that of a cauliflower. This peculiar adhesion has also been noticed in
thunder-clouds; the lower plane of this species of cloud is horizontal,
and from this kind of table-like base rise projections, the shape of which

Fig. 68.—Formation of a thunder-cloud.

may be compared with enormous balls of wool more or less carded, and
connected the one with the other. These are typical instances which
accentuate rather than attenuate the average appearance of clouds. The
color, the white or the sombre hue of the clouds, can scarcely be taken
as characteristic, for they are dependent upon their position in respect
to the sun, and in regard to the situation of the observer.

If we see a cloud at a great distance, and are standing between it and
the sun, it will seem to us to be white. If, on the contrary, we notice it

as it passes over our heads, we see the lower surface which the light
does not reach, and then it appears black.

The *snow clouds* have not this definite shape. They generally are of
an immense thickness in the atmosphere, and of slight density. The
light sifted athwart their vast extent gives them a yellowish tint, whence
the flakes descend and cover the earth.

Fig. 69.—Above and below the rain cloud.

CHAPTER III.

RAIN: GENERAL CONDITIONS OF THE FORMATION OF RAIN — ITS DIS-
TRIBUTION OVER THE GLOBE—RAIN IN EUROPE.

HAVING treated of the distribution of moisture in the atmosphere, the
manner in which the clouds are formed and remain suspended in space,
their division into two distinct kinds, and the action of temperature
upon the vapor of water, we shall have no difficulty in discovering how
the formation of rain takes place.

Rain is the precipitation of the aqueous vapor which constitutes the
clouds. For this vapor to become precipitate — that is, to form drops,
the weight of which causes them to descend and to produce rain—the
molecular state of the cloud must be modified by some external cause.
This modification may be effected by the influence of upper clouds—
clouds of ice. Under certain circumstances, the least decline of temperature sets them in motion and destroys them. Such is the case with
saturated *cumuli;* the least diminution of temperature precipitates them
in the form of rain.

The ordinary condition of the production of rain consists, therefore,
in the existence of two layers of clouds, one above the other, and it is
the higher which causes the precipitation of the one below it. This is
an observation which any one may verify for himself; and in the course
of many years' observation of the sky when rain is about to fall, I have
never found this condition wanting.

Monck Mason remarked, in his aeronautical voyages, that when rain
falls, the sky being at the time totally covered with clouds, there is always a similar range of clouds situated at a certain height above, and
that when, on the contrary, though it does not rain, the sky presents
the same appearance below, bright sunshine prevails in the space immediately above. Saussure had already noted the same fact in his Alpine explorations. Hatton had noticed that when two masses of air,
saturated or nearly saturated, but of unequal temperature, meet, there is
a precipitation of aqueous vapor. Peltier observed in regard to another
point, that a thunder-storm is always composed of two banks of clouds
which are of opposite electricity. Rozet arrived at the conclusion that

thunder-storms and rain both result from the encounter between the cirrus and the cumulus, between the frozen and the vesicular vapor. Kaemtz and Martius adopt the same theory. M. Renou further adds that water may fall without being frozen at temperatures as low as 27°, 36°, or 45° below the freezing-point of water, in the state of extreme divisibility which constitutes fogs and mists, and that rain and frost are due to the admixture of frozen cirrus with the still liquid cumulus beneath the varying influence of temperature.

Such is the general manner in which rain is formed. It sometimes, however, falls when the sky is clear. On August 9, 1837, at 9 P.M., Wartmann of Geneva noticed that during the space of two minutes large drops of warm rain fell from the sky, then studded with stars. The edges of the horizon were covered with broken patches of black clouds.

On the 31st of May, 1838, at 7 P.M., M. Wartmann again remarked an analogous phenomenon, which this time lasted for six minutes. The warm drops, which were at first very large and thick, gradually decreased in size. On the 11th of May, 1844, at 10 A.M. and 3 P.M., he noticed the same occurrence, and during the time the air being quite calm.

The transit of masses of clouds is an important factor in their dissolution, and in the abundance and the distribution of rain. This has been already pointed out when we were considering how the various directions of the wind corresponded with the amount of rain that fell. The south-west wind, which prevails in our country, brings the greatest amount of rain, because it is accompanied by the cloudy strata formed over the ocean, these strata of humidity being, moreover, sometimes invisible.

Thus we can form an idea of the immense evaporation which daily takes place from the surface of the ocean, and see in it the origin of clouds and rain. The trade-winds, which blow over the surface of the sea in the tropics, carry this vapor of water as far as the regions of equatorial calm, where they rise into the higher and colder part of the atmosphere, and from thence pass to the temperate countries laden with moisture. As they rise through the atmosphere in the equatorial regions, a portion of vapor is condensed; and as this occurs every day, there is a constant zone of clouds and rain. It is what English sailors term the *cloud-ring*, and French sailors the *Pot au Noir*.

The oceanic clouds from the south and the south-west distribute the water which they contain according to their course, height, and temperature; the more or less thick, and more or less cold, strata of clouds which weigh down upon them varying with the accidental winds which may affect them, and influenced by the undulations of the ground which alter their course. All other conditions being unchanged, the proportion of rain decreases from the equator to the poles, since, on the one hand, evaporation takes place almost entirely in the warm latitudes; and, on the other, the quantity of vapor which the air is capable of dissolving augments rapidly as the temperature increases. Thus, for instance, there is an annual rain-fall of more than six and a half feet at Guiana and Panama, while it is only seven and three-quarter inches at Archangel.

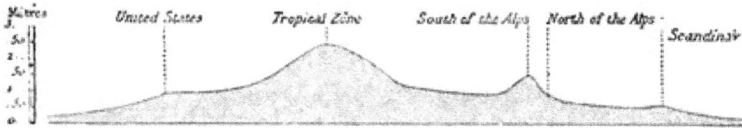

Fig. 70.—Diminution in the rain-fall from the tropics to the poles.

There is also a second law in regard to the proportion of rain, viz., that it diminishes in amount according to the distance from the sea, measured in the direction of the prevailing winds. It is easy to understand that clouds, being unable to reform in the interior of continents, yield less rain in proportion as they pass farther from the ocean. The evaporation that proceeds from rivers, lakes, pools, and moist plains, does indeed give rise to clouds; but this is a very insignificant source of rain compared to that of the ocean. There falls forty-nine inches nearly at Bayonne; forty-seven inches at Gibraltar; fifty-one inches at Nantes; only sixteen and a half inches at Frankfort; seventeen and three-quarter inches at St. Petersburg and Vienna. In Siberia the rain-fall is but seven and three-quarter inches, and less still farther east. At Algiers there is a mean of seven and three-quarter inches, and at Oran and Mostaganem of less than four inches. Farther south, the quantity of rain diminishes rapidly; and at Biskra, on the borders of the desert, there falls two-tenths of an inch in the course of the year.

Numerous observations enable us to establish a third law. The undulating nature of the ground causes a variation in the two distributing elements which we have just been considering. If a mass of air, saturated with moisture, encounters a mountain chain, it will be partially stopped by this protuberance of the soil. But the check is not a long

one.   The currents of air which ascend the slopes of mountains will
elevate them at the same time; they will become colder at the rate of
one degree to 200, 250, or 330 feet; according to the season and temperature, they will consequently undergo a progressive condensation, so
that when they reach the summit they will be able to pass above it; a
great part of the water they contained will already have fallen, and the
remainder will descend upon the summit of the mountain.   The lessened speed of the air also deprives them of their water, much in the same
way as the diminishing rapidity of a stream facilitates the fall of the
deposits which it keeps suspended.   There falls, therefore, more rain in
a mountainous than in a level region; there is also more rain upon the
slope that faces the sea-wind than upon the opposite.   Thus, clouds
which, as they pass over Lisbon, give but an annual rain-fall of twenty-

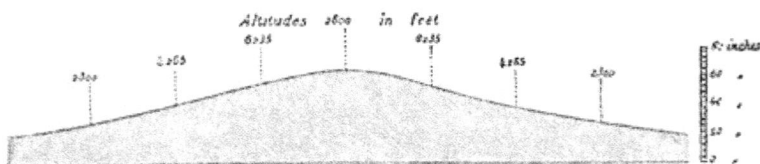

Fig. 71.—Increase of rain, according to the undulations of the soil.

seven and a half inches, are soon arrested by the cold-tipped mountains of Portugal and Spain, there being a rain-fall of 118 inches at Coim-
bra.   The clouds which pass at the zenith of Paris yield nineteen and
three-quarter inches of rain in a year.   As the altitude augments, so
does the rain.   Thus, taking merely the basin of the Seine, we have
three and a quarter feet of rain-water upon the plateau of Langres,
and six feet nearly at the higher point of Morvan, in the Nièvre.   At
Geneva, at the foot of the Alps, the annual quantity of rain is thirty-
two and a half inches, and at the Great St. Bernard ridge it is six and
a half feet in the year.

There are regions in which these conditions are so complete, that the
rain stops as if attracted there permanently.   Thus the Great Himalaya
chain stops the clouds which come from the immense evaporation of
the Indian Ocean.   At Cherra-Poejen, upon the Garrows Mountains, at
a height of 4500 feet, and to the south of the Brahmapootra Valley,
the quantity of rain which the clouds pour down is forty-eight and a
half feet.   In these mountainous regions near the tropics, the maximum
rain-fall is probably to be found; they are also the great reservoirs of
the Asiatic rivers.   In these same lower slopes of the Himalayas, upon

the eastern side of the Ghauts, an average annual rain-fall of twenty-five feet nearly has been recorded, after observations extending over a period of fourteen years. A downfall lasting only four hours has been known to cover the ground to a depth of thirty inches—more than falls at Paris in a whole year. It is certain that in no other part of the torrid zone is the precipitation of the rain so much facilitated by attendant circumstances. The Antilles are not wide enough to prevent the winds and clouds from veering obliquely to the right or to the left; but notwithstanding, certain districts there receive thirty-two and three-quarter feet in the course of the twelvemonth. In the Gulf of Mexico the summer rains also give a depth of more than thirteen feet at Vera Cruz. Farther from the tropical regions we only noticed these remarkable maxima of rain upon the mountain chains which, being in the way

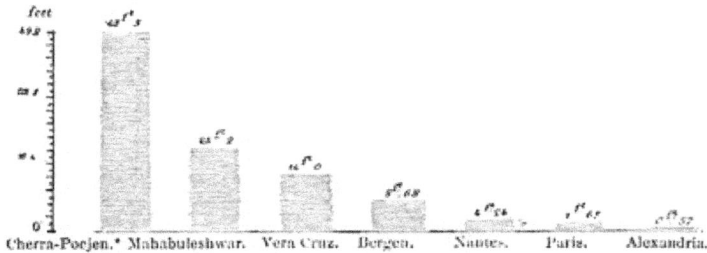

Fig. 72.—Comparative depths of rain-fall.

of the general current, bring it to a stop. Such, for instance, is the effect produced by the Scandinavian Alps that separate Sweden and Norway, for its western slope receives much more rain than the eastern side, there being an annual rain-fall of eight and three-quarter feet at Bergen, which exceeds that of any other town in Europe. Moreover, several points are again specially favored in respect to their frontage to the south-west current; as Nantes, for instance, where there is a mean annual rain-fall of four and a quarter feet.

Collecting and comparing the observations that have been made at a great number of places in different parts of the globe, it has been found possible to register the three predominating causes which we have re-

* [At Cherra-Poejen the fall of rain in April is 22 inches; in May, 62 inches; in June, 195 inches; in July, 121 inches; in August, 104 inches; in September, 75 inches; and in October, 29 inches; making a total fall in seven months of 608 inches. No rain falls either in November or December, and less than five inches in the months of January, February, and March. See my "Report on the Meteorology of India," in relation to the health of the troops stationed there, 1863.—Ed.]

viewed, to lay down upon a diagram the depths of the rain-fall that have been observed, and to make a map exhibiting the comparative depth of rain all over the globe. The heaviest rain takes place to the north of the equator in the Atlantic, in the Pacific, and to the east of America. In these regions, the maximum falls exceed six and a half feet in depth; in Asia, in the islands of Borneo, Sumatra, and Java, along the Himalaya and Ghauts Mountains; in Africa, along the table-lands of the eastern coast; in the Atlantic, between Guinea and Guiana; in South America, upon the Andes in Chili, at Cape Horn, and upon the summit, above Peru, which, by contrast, is a country where no rain falls. Lastly, the mountain chain which runs eastward along the borders of North America, from fifty to sixty degrees longitude, yields an annual maximum of more than six and a half feet.

The rainless regions extend along the desert of Sahara, Egypt, Arabia, and Persia, reaching as far as Mongolia, and even to Siberia, with the exception of the region of Central Asia, upon which the monsoons and the winter rains yield some little moisture.

If we consider Europe in particular, we find, relatively, abundant rain, ranging from three and a quarter to six and a half feet, in the marine zones of Portugal, Brittany, Ireland, and Sweden. The proportion of rain gradually diminishes toward the east, with the zones of condensation produced by the undulating nature of the soil. There are certain points where rain is very rare, as in Greece, for instance. The climate of Attica is dry, and the sky is generally clear, the air having always been considered the purest in Greece. As an instance of this, I may mention that M. Lusieri exposed a piece of paper to the air all night, and that he was able to write upon it the next morning. To this remarkable dryness of the air has been attributed the excellent state of preservation of the Athenian monuments.

The northern hemisphere receives more rain than the southern by about one-fourth. This excess of rain is especially due to the northern equatorial zone of rains and monsoons. Nevertheless, there is much more dry land in the former than in the latter, and evaporation proceeds on a much larger scale in the southern hemisphere, which is nearly all sea. Thus, our clouds, our rain, our rivers, and our streams are chiefly fed by the ocean in the hemisphere of our antipodes.

As the distribution of rain has for its origin both the variations of temperature and the prevailing winds, it can be easily seen that in different countries it is more or less abundant according to the time of year.

The countries in which there is what is termed *a rainy season* are those situated in the tropics, where the sun, which twice a year passes perpendicularly over them, occasions at those epochs an excessive heat, which must, of course, be succeeded both by a great rarefaction of the strata next to the ground; as these latter, becoming too light to bear the weight of the upper strata, rise, and afterward by the diminution of temperature and fall of rain, which always follow, no matter what may have been the producing cause. It is impossible to form an idea of the mass of water which, during the rainy season, falls into the basins of the Amazon and the Orinoco. After these streams and their tributaries have overflowed their banks to a height of several feet, a tract of country as large as Europe becomes a fresh-water sea, the outflow of which into the ocean destroys the salt for some distance from the shore, and in comparison with which the North American lakes are mere mill-ponds. The scientific study of this great display of physical forces, in which Nature, whose action is irresistible, commands the attention of us whose existence is menaced, is making rapid progress, and none are better qualified to throw light upon the subject than the inhabitants themselves, whose life depends upon their being familiar with the vicissitudes of the seasons.

Thus, in the United States, upon the Atlantic, from the twenty-fourth and as far as the fortieth degree of latitude, in Spain, in the south of France, in Italy, Greece, Turkey, Asia, China, Japan, and in the Pacific, in the same latitudes, nearly all the rain falls in winter, excepting the region of periodical monsoons and in certain southern countries, where, during the summer months, no cloud appears in the sky. It is the same between the twenty-fifth and fortieth degrees of south latitude, at Buenos Ayres, the Cape, and at Melbourne.

Over a zone extending from twelve to fifteen degrees of south latitude, over nearly all the globe, it is in summer that most rain falls.

Over a zone extending from forty to sixty degrees of north latitude, and which reaches as far as seventy-five degrees, beyond Iceland and Sweden, and within a limited zone in Asia, rain falls at all times of the year.

Nevertheless, even in our variable regions, there are well-defined proportions for each particular season. Thus, taking France in particular, we find that it may be divided into two parts. The western region has the maximum of rain in summer, and the minimum in winter. Such is also the case in England, while in Germany it is the reverse, under even more marked conditions. The same holds good with regard to Russia.

We have said that there is an annual rain-fall of seven and a quarter feet at Bergen, in Norway. This town is, in this respect, a remarkable exception in the meteorology of the globe. It is, in all Europe, the town where there is the most rain. It is situated in the centre of a deep bay, exposed to westerly winds, which are stopped by the mountains, so that the rain is, to use Kaemtz's expression, mechanically pressed out.

The following table gives the rain-falls throughout Europe, and is the result of many years' observations:

QUANTITY OF RAIN IN EUROPE BY SEASONS.

Names of Places.	Winter.	Spring.	Summer.	Autumn.	Year.	Number of Observations.	Height.	Latitude.
	In.	In.	In.	In.	In.		Feet.	° ′
Breslau	2·2	3·0	3·5	3·2	13·9	56	400	51 6
Prague	2·2	3·7	6·3	3·1	15·3	52	627	50 5
Upsal	2·7	2·9	5·6	4·5	15·7	102	—	59 52
Vienna	3·3	3·9	6·5	4·0	17·6	15	512	48 13
St. Petersburg	2·9	2·9	6·7	5·1	17·6	16	—	59 56
London	5·4	5·5	6·9	7·5	25·2	55	—	51 31
Berlin	4·4	4·3	7·1	4·6	20·4	12	128	52 34
Paris	4·1	4·6	5·4	5·6	19·7	140	285	48 50
Stockholm	3·0	3·3	7·6	6·7	20·6	36	135	59 21
Palermo	8·4	5·2	1·3	8·9	22·8	24	—	38 8
Copenhagen	5·0	4·6	7·1	6·3	23·0	12	—	55 41
Abo	4·7	3·9	7·2	7·9	23·7	48	—	60 27
Stuttgart	4·2	5·7	8·5	5·9	24·3	31	814	48 46
Toulouse	5·2	7·0	5·9	6·6	24·7	25	499	43 36
Metz	5·6	5·7	7·2	7·4	25·9	22	—	49 7
Dijon	5·7	6·1	7·0	8·5	27·3	30	—	47 19
Edinburgh	5·8	5·0	6·7	7·4	24·9	27	289	55 57
Brussels	6·4	6·2	8·3	7·6	28·5	21	—	50 51
Rouen	7·6	6·8	7·1	8·9	30·4	26	190	49 26
Ghent	6·5	6·5	9·5	8·4	30·9	16	36	51 3
Dublin	6·8	5·9	8·1	8·4	29·2	16	—	53 23
Rome	9·3	7·3	3·4	10·9	31·0	40	174	51 54
Geneva	5·2	7·2	9·0	11·0	32·4	29	1299	46 12
Montpellier	9·2	7·2	4·1	11·9	32·4	26	—	43 36
Padua	7·0	7·4	9·0	10·6	34·0	48	—	45 24
Manchester	8·2	7·0	9·9	10·7	35·7	47	154	53 29
Florence	10·2	8·6	5·3	12·7	36·8	16	210	43 47
Turin	5·6	11·3	11·2	9·5	37·6	15	915	45 4
Milan	8·1	9·1	9·2	11·7	38·1	68	479	45 28
Lausanne	6·1	8·1	14·9	11·2	40·3	6	1663	46 31
Nicolaïef	14·5	9·1	24·8	14·6	63·0	6	—	46 5

The quantity of rain at Breslau, Prague, Upsal, Vienna, and St. Petersburg, shows how little falls in these places, as the mean is less than 15⅞ inches.

In the Netherlands, Belgium, France, Germany, and Poland, the average is $19\frac{3}{4}$, $23\frac{1}{2}$, and $27\frac{1}{4}$ inches. It is easy to see that there is a diminution as one recedes from the sea inland. Thus, in the Belgian cities, there is more than $27\frac{1}{2}$ inches of rain, while in the same latitudes at the German towns and those nearer to Asia the quantity is much smaller. Upon the other hand, it is evident that the two most rainy seasons are summer and autumn, no matter what the distance of the locality from the sea. England is very peculiarly situated in this respect, as, being surrounded by the sea, she receives more rain than her latitude would lead one *a priori* to expect.

CHAPTER IV.

HAIL: PRODUCTION OF HAIL — COURSE OF HAILSTORMS — VARYING
DISTRIBUTION OF HAILSTORMS IN DIFFERENT PARTS OF THE COUN-
TRY — HEAVIEST HAILSTORMS KNOWN — NATURE, SIZE, AND SHAPE
OF HAILSTONES — PERIODS OF THEIR OCCURRENCE.

WHEN several strata of black and grayish clouds are flying through
the atmosphere, and when the thunder-storm has burst forth, millions
of pounds of hailstones are launched from the clouds as if precipitated
from the opened cataracts of a vast reservoir. For several minutes the
hail drives through space, pelting trees and gardens; it then ceases as
the wind blows it off in some other direction, and the close and sultry
temperature which had preceded it gives place to the fresh odor of
refreshed plants, light returns, the rainbow appears, and the blue sky
emerges from the banks of clouds. What is the force which produces
in the clouds these lumps of ice (often very large), what bears them up
in space, and then launches them upon the earth? While studying the
production of rain, we saw that it does not, as a rule, occur except when
there are two or more strata of clouds one over the other. Such is also
the case with the formation of hail, though there is a difference in the
respective physical conditions of the clouds.

Hail occurs during a thunder-storm, when the temperature is very
high upon the surface of the ground, but decreases rapidly with elevation. This rapid decrease is the principal element in the formation
of hail, and it has been known to be as much as one degree in a little
more than 100 feet. What then takes place in the region of clouds?
Those above, from 10,000 to 20,000 and 25,000 feet high, contain, the
highest of them, ice at $-22°$ or at $-40°$ Fahr.; the lowest of them,
vesicular water at $+14°$ and at $-4°$. The lower clouds contain vesicular water above $32°$. As a rule, these clouds travel in different directions, and hail is formed when there is a collision and admixture of
winds that are opposed to currents and clouds the temperatures of
which are different. The vapor, which then resolves into rain, freezes
instantaneously in so cold a temperature. Carried off by the wind, and
even exposed to the influence of opposite electricities of the diverse

strata of cloud, these frozen drops do not fall at once, notwithstanding their weight, and they have time to become enlarged by the addition of a considerable quantity of water which they collect during their passage through the air.

The extreme cold that prevails in the clouds below the region of perpetual snow is due, in a great measure, to evaporation, which has itself a double cause—the action of the sun and of electricity—it having been remarked that after every electric discharge the rain or hail falls in great quantities, and the reaction produces a dilatation which gives rise to rapid evaporation.

The formation of hailstones is always a very speedy process. Volta was of opinion that the upper cloud was formed by the condensation of vapor from the lower strata, and that it contained positive electricity, while the latter retained negative electricity. Just as pith-balls placed between two copper plates laden with opposite electricity are seen to bob up and down under the influences of this double attraction, in the same way he thought that hail was formed by a like movement of the corpuscles of ice or snow, becoming successively enlarged by condensed vapors. This theory is not now considered admissible, and it is, indeed, far simpler to suppose that hail is formed like rain, but amidst an atmospheric cold which freezes the globules of water at the very moment of their formation.

It appears that this formation, or the shock of hailstones that are borne along by the wind, sometimes produces a noise audible upon the surface of the ground. Aristotle and Lucretius, of ancient writers, record this fact; and the meteorologists Kalm and Tessier assert that they heard it, the former in France on July 13, 1788, the latter at Moscow on the 30th of April, 1744. Peltier states that at Ham the approach of a hailstorm was preceded by a sound like that of a cavalry squadron at full gallop. In 1871, M. Pessot, corresponding member of the Montsouris Observatory, reported from Doulevant-le-Château (Haute-Marne) a hailstorm which was preceded by this same phenomenon.

The surfaces of hail-clouds show here and there immense irregular protuberances. Seen from underneath, they are generally dark in color, because of their opaqueness, which the solar light is scarcely able to traverse. Arago pointed out that they seem to be thick, and to be distinguishable from other storm-clouds by their ashen hue. Their edges are indented; but they very soon are lost in the general mass of the *nimbi* which discharge rain.

To what height do they soar? From what elevation do hailstones fall? Saussure noticed a hailstorm upon the *Col du Géant* at a height of 11,246 feet, Balmat upon the summit of Mont Blanc itself, and Paccard discovered hailstones beneath the snow which forms its peak. Hail often falls upon the high slopes of the Alps. Thus the phenomenon of hail occurs at all elevations. But when the height at which it commences to fall is very great, the hailstones melt during their passage through the thousands of feet of air above the temperature of 32° which cover the surface of the globe. In the case of our hailstorms, on the contrary, the clouds which emit them are at a less height, and seem to be between 5000 and 6500 feet above the ground. Below them extend the storm and the rain clouds, at a height of about 3300 feet, or even lower. The clouds which discharge hail are never very large. Borne along by the wind, they cover a narrow strip of land, which is often only three-fifths of a mile in breadth, and rarely more than ten miles long; but the length is sometimes as much as 500 miles.

One of the most curious and remarkable hailstorms in the annals of meteorology is that of July 13, 1788. It was divided into two bands: that on the left, or the western one, began at Touraine, near Loches, at 6·30 A.M.; passed over Chartres at 7·30 A.M.; over Rambouillet at 8 A.M.; Pontoise, 8·30 A.M.; Clermont (Oise) at 9 A.M.; Douai, 11 A.M.: whence it entered Belgium, passing over Courtrai at 12·30 P.M.; and finally dying out beyond Flushing at 1·30 P.M. The total length was 420 miles, and it extended over a width of ten miles.

The right, or eastern, branch began at Orleans at 7·30 A.M., passing over Arthenay and Andonville, reached the Faubourg St. Antoine in Paris at 8·30 A.M., Crepy-en-Valois at 9·30 A.M., Cateau-Cambrésis at 11 A.M., and Utrecht at 2·30 P.M., the length being near 500 miles, and the width only five miles. There was a mean interval of twelve miles of ground between the two bands, and rain fell in this space. The passage of the hailstorm was preceded on each line by a profound darkness. The speed of the storm was thirty-two miles per hour on both lines, the hail not falling for more than seven or eight minutes in the same place, but with so much violence that the crops were cut to pieces. This is the greatest hailstorm known. No less than 1039 communes in France suffered from its ravages; the destruction of property was found to amount to no less than £1,000,000. The hailstones were not all of the same shape; some were round, others long and pointed; and some were found to weigh 3900 grains, or more than half a pound.

It is seldom that the same hailstorm extends over such a length of country, and in so regular a line. It is probable that the clouds which produced this hail were more than half a mile high. Generally they are at a less height than this, and are influenced by the undulations of the soil. Certain storms, without having extended over so much ground, are remarkable for their abundant quantity. On May 9, 1865, for instance, a storm began at 8·30 A.M. over Bordeaux and proceeded in a N.N.E. direction, passing over Périgueux at 10 A.M., Limoges at noon, Bourges at 2 P.M., Orleans at 5·30 P.M., Paris at 7·45 P.M., Laon at 11 P.M., and collapsing a little after midnight in Belgium and the North Sea. Its mean breadth was from fifteen to twenty leagues. The hail fell only in certain places: to the left of Périgueux, over the arrondissement of Limoges, to the right of Châteauroux, to the south-east of Paris, from Corbeil to Lagny, and in the arrondissements of Soissons and Saint Quentin. At this latter point it was of a formidable character. The crystal mass which fell from the sky upon the Catelet meadows formed a bed a mile and a quarter long and 2000 feet broad, estimated to amount altogether to 21,000,000 of cubic feet. The hailstones did not disappear for more than four days afterward. These hailstones sometimes destroy all the crops, as, for instance, that which occurred in the neighborhood of Angoulême on August 3, 1813. The day had been fine, and the wind was due north until 3 P.M., when it suddenly veered right round; the sky gradually became covered with clouds, which, collecting one on the top of the other, offered a terrible spectacle. The wind, which from noon until 5 P.M. had been rather violent, suddenly dropped. Thunder was heard in the distance, and gradually became louder; the sky, at last, became totally obscured, and at 6 P.M. there was a tremendous fall of hail, the stones being as large as eggs. Several persons were severely wounded, and a child was killed near Barbezieux. The next day the ground looked as it might do in midwinter: the hailstones had accumulated in the hollows and the roads to a height of thirty to forty inches; trees were entirely stripped of their leaves; vines were cut into pieces, the crops crushed, the cattle, sheep and pigs especially, were severely injured. The whole neighborhood was deprived of game, and some few young wolves were found dead. The effects of the storm were still visible in 1818, the vines, in particular, not having recovered their productive powers.

The storm which burst over Chaumont, in the Haute Marne, on July 17. 1852, spread over a district nearly sixty miles long by five miles

broad; wheat, vines, and nearly every tree, were destroyed by hailstones of abnormal dimensions. The same hurricane swept violently
over the department of the Aisne, uprooting trees, blowing down cot-
tages, and killing several persons. In a few seconds all trace of the
crops had disappeared from the fields.

On July 17, 1868, at about 8 P.M., a heavy hailstorm devastated the
neighborhood of Rheims; the stones were as large as Barcelona nuts,
and the downfall lasted three parts of an hour. In some of the hollows,
where the ground was sandy, there were remarked impressions like
those which might be made by a cannon-ball. These cavities, into
which the hailstones were first driven, constitute regular physical impressions of the hail, which seemed, in regard to the construction placed
by geologists upon similar marks, to possess a special importance.

Disastrous hailstorms are, fortunately, rare in our climates, though
they do, from time to time, remind us of their existence. A heavy
storm began at Brussels on June 18, 1839, about 7 P.M.; thick clouds
drove from the S.S.W., while at the same time the vane indicated a
lower current from the N.W. Until 7·30 there was a continuous rolling sound, during which the flashes of lightning succeeded each other
with astonishing rapidity. Soon after, a large cloud of very ashen
hue, and the direction of which was from W.N.W. to S.E., veiled the
city in the most complete obscurity, and burst in a shower of hail
which did immense damage. Most of the hailstones were from a half
to three-quarters of an inch in size, some of them as much as one inch.
In shape some were spherical, but the greater number were more or less
flat. The depth of water that fell during the storm was an inch and a
half. The temperature rose as high as 92° Fahr.—the maximum recorded at Brussels; the barometer reading was 29·70 inches at 4 P.M.

Hailstorms have a tendency to follow the direction of the valleys and
the rivers when the clouds are not high; for, as is shown by the cases
cited above, the storms then become regular currents, which come from
the Atlantic and, following the ordinary course of the currents which
reach us, continue their progress from the south-westerly regions toward those of the north-east. But in all partial secondary storms (which
are the most frequent, and are generally confined to a limited area) there
is an evident deviation from the valleys. It seems, too, that they keep
away from forests. Since meteorological facts have been registered by
the French *Écoles Normales*, there has been plenty of evidence collected
as to the influence of the ground in regard to the distribution of storms

and of hail. One district may be visited by hailstorms every year, another not once in ten years. It has even been found possible to compose statistical maps showing the damage done by the hail in each department, by aid of the documents appertaining to insurance companies. These maps are scarcely reliable from a meteorological point of view, as they are based on pecuniary losses; and the same quantity of hail would cause ten times as much damage were it to fall over a tobacco plantation of the Lower Rhine, as it would if it were to rage over an uncultivated or even a wooded district. It is true that the intrinsic quantity of hail differs in neighboring countries, according to their geological, orographical, and climatological situation.

Hailstorms are those in which the development of electrity attains the largest proportions. The thick clouds in which the meteor becomes elaborated are laden with a large quantity of the electrical fluid, part of which becomes exhausted within themselves or in reciprocal discharges with neighboring clouds.

The thunder is then not merely a report following the flash; it is a continuous rolling sound, during which it is not unusual for no lightning to be perceptible, either because the flashes are of very small dimensions or because they take place entirely within the interior of the clouds. Thus, on the 4th of September, 1871, I noticed, in the hailstorm which took place in Paris at 3·36 P.M., that when the hail had passed over the district in which the Observatory is situated, and when it was over Ménilmontant, there was a continuous rolling of thunder, *unaccompanied by lightning*, which lasted six minutes, and recommenced again after several short intervals. On the 7th of May, 1865, a violent storm burst over the department of the Aisne, causing damages amounting to several million francs. Above the strata of clouds there was visible a thick cumulus, of a livid white hue, from which there was a continuous flashing of lightning; the rolling of the thunder was uninterrupted, though not very loud; there was an unintermittent crepitation of the lightning, and the explosions seemed to be confined to the interior of the largest cloud. When the cloud had slowly ascended the heights of Roussay, upon the apex of the basins of the Somme and the Scheldt, it swept down with tremendous rapidity into the valley of this latter stream, pelting Vend'huile, Câtelet, and Beaurevoir, with so many hailstones that they lay five yards deep upon the ground. They were still visible five days after, and, at some places, formed such a solid mass that they acted as a dike to keep back the water. When it was attempted to sweep them away they slipped along like fields of ice! M. Quételet remarked, during a severe storm that occurred at Brussels on June 18. 1839, a continuous rolling of thunder, during which time the flashes of lightning succeeded each other with marvelous rapidity. Soon after, a thick ashen cloud plunged the whole city into profound darkness, and burst in a heavy fall of hail.

It is interesting to ascertain what is the greatest dimension which a hailstone can attain. I am able to give some very curious comparisons on this subject from a number of well-authenticated documents.

After the great hailstorm of July 13, 1788, alluded to above, the geologist Tessier cut pieces of ice which seemed to him to be of the consistency of hail, into the shape and size of pigeons', hens', and turkeys' eggs, in order that meteorologists might be enabled to calculate approximately the weight of hailstones according to their size. The first weighed 169 grains; the second, 254; and the third, 1065 grains.

The most ordinary size of a hailstone is that of a small nut: some, indeed, are not larger than a good-sized pea. In ordinary storms, the stones weigh from 46 to 120 grains.

The three weights above often occur in the annals of meteorology. There is nothing absolutely abnormal in a fall of hailstones weighing from a quarter of an ounce to two and a quarter ounces.

Some extraordinary facts are the following, which are, however, perfectly authenticated and certified by well-known savants: In a disastrous hailstorm near the Rhine, a hailstone was picked up by Voget at Heinsberg weighing 1400 grains. At Randerath they weighed twice as much. During a storm that occurred at Morbihan, and which lasted three-quarters of an hour, on June 21, 1846, the hailstones were of all dimensions, from the size of a nut to that of a turkey's egg. One was eight and three-quarter inches in circumference. Muncke weighed some hailstones in Hainault that exceeded three and three-quarter ounces in weight. Halley relates that some hailstones were picked up on April 29, 1697, in Flintshire, the weight of which exceeded four ounces; and on May 4, in the same year, Taylor found that the circumference of some that fell in Staffordshire was eleven and three-quarter inches.

Volney tells us how, during the storm of July 13, 1788, he was staying at Pontchartrain, ten miles from Versailles. The sun's rays were almost unbearable; the air still and suffocating: the sky was cloudless, and claps of thunder were from time to time audible. Toward 7·15 P.M. a cloud appeared in the south-west, followed by a very sharp wind.

" A few minutes afterward the cloud filled the horizon and sped toward our zenith, accompanied by a wind which had become quite cool; hail began to fall obliquely at an angle of 45°, the stones being as large as pieces of plaster thrown down from the top of a house. I could scarcely believe my eyes; several of the stones were as large as a man's fist, and some of these were but pieces that had been broken off stones still larger. When I ventured to put out my hand beyond the door of the house where I had taken refuge, I picked up one and found that it weighed more than five ounces. It was very irregular in shape, there being three protuberances, thick as the thumb and nearly as long, which projected from the main body of the stone!"

Volta states that, during the night of April 19–20, 1787, among the enormous hailstones which fell in Como and the neighborhood, there was one which weighed nearly nine ounces. Parent, member of the Academy of Sciences, relates that hailstones as big as a man's fist, and weighing from nine and a half ounces to twelve and three-quarter ounces, fell in Le Perche on May 15, 1703. Montignot and Tressan picked up some at Toul on July 11, 1753, which had the shape of an irregular polyhedron, with a diameter of three inches.

During a hailstorm at Constantinople on October 5, 1831, there fell stones weighing more than one pound, and larger than a man's fist. Analogous stones are said to have been picked up in May, 1821, at Palestrina (Italy).

The following are, however, even more remarkable instances: On June 15, 1829, there was a hailstorm at Cazorta, in Spain, which crushed in houses; some of the blocks of ice weighed four and a half pounds. For hailstones to attain such proportions, several must have become agglomerated together, either when they reached the ground or during their descent. This is, in fact, in accordance with experience. And this explanation is, therefore, specially applicable to the following cases, if, indeed, they be authentic: During the latter part of October, 1844, during a terrible hurricane which devastated the south of France, there fell hailstones weighing eleven pounds; the town of Cette, in particular, was severely damaged; men were struck to the ground as if they had been stoned, partition walls were blown down, and vessels sunk.

It seems that there was a very singular hailstorm on May 8, 1802, a piece of ice having been picked up which measured more than three feet both in length and in width, with a thickness of two and a quarter feet. Dr. Foissac, who cites this fact, does not consider it to be an exaggeration; and he adds, "M. Huc, a Catholic missionary in Tartary, relates that hailstones of a remarkable size often fall in Mongolia, and that some of them have been found to weigh twelve pounds. During a heavy storm in 1843 the noise as of a terrible wind was heard in the air, and soon after there fell in a field not far from our house a *piece of ice larger than a millstone.* It was broken up with a hatchet; and though the weather was very warm, it took three days to melt completely."

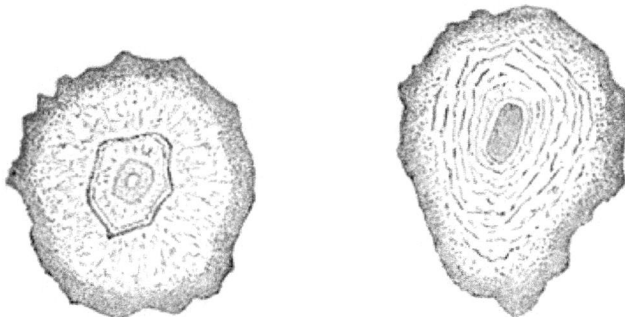

Fig. 73.—Section of hailstones, showing their ordinary interior structure.

If this be true, there is nothing improbable in the chronicle dating from Charlemagne, which relates that there fell hailstones fifteen feet wide by six long and eleven thick, nor in that of Tippoo Sahib, which speaks of a hailstone as big as an elephant.

The shape of hailstones differs very much. They are, as a rule, round, spherical, more or less irregular, like peas, grapes, or nuts. Several are more elongated, like a grain of wheat, cornelian cherries, or olives. When very large, they are formed by the juxtaposition of crystallized particles. On July 4, 1819, during a nocturnal storm which spread over a large portion of Western France, Deleros picked up several entire spherical hailstones, in which was visible a first spherical nucleus of a somewhat opaque, whitish hue, offering the traces of concentric strata. Around this nucleus was an envelope of compact ice, radiated from the centre to the circumference, and terminating upon the exterior with twelve large pyramids, between which were intercalated smaller pyramids. The whole formed a spherical mass nearly three and a half inches in diameter.

Some hailstones picked up on September 12, 1863, in a road to the south-west of Tiflis, drawings of which were exhibited to the Academy of Sciences at St. Petersburg, were ellipsoidal in shape, and their sur-

face was covered with a large number of small prominences. The poly-
hedric tissue, examined through
a glass, had the aspect of a series
of six-fronted pyramids; and a
section of the interior revealed
the existence of a hexagonal net-
work of meshes, which is represented in Fig. 74.

On July 29, 1871, at 6 P.M., the
sun shining brightly, and there
being hardly any clouds, a sound
was heard at Auxerre, like that
of a heavy luggage-train. A few
flashes of lightning preceded the
fall of the hail, which came down

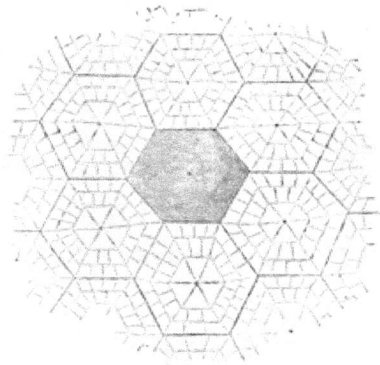

Fig. 74.—Section of a hailstone, enlarged.

unaccompanied by any tempest or atmospheric disturbance. The hailstones preserved their shapes when they reached the ground, which are
represented in the four corners of Fig. 74, after the designs of M.
Daudin. The two stones in the centre are those to which I alluded in
connection with the Academy of St. Petersburg, and the remainder have
been added as illustrative of the smaller and more usual size of hailstones. During the same storm M. Parent remarked at Montargis that
there was a heavy fall of hail at 6·45 P.M., the pieces of ice being from
one to two inches in length, oval in shape, and transparent as crystal.

During the storm of May 22, 1870, in Paris, M. Trécul, of the Institute, noticed that several of the hailstones were conical, or rather pyri-
form—that is, larger at the base than at the top, some of them being
about three-quarters of an inch long by half an inch wide. One of
them, carefully examined, presented characteristics worthy of notice.
The third part of it, at the top (the narrowest portion of the hailstone),
was opaque and white; while the lower, or the broadest, part was perfectly translucid, like the purest ice. In addition, this hailstone, when
looked at from its broadest end—that is, when the narrowest diameter
was placed crosswise in respect to the visual axis—presented the shape
of an obtuse-angled rhombus; and from the sides there started oblique
facets which converged and died away toward the obtuse summit of the
hailstone.

As to the epochs of hailstorms, it is generally known that they occur
in summer and in the afternoon—that is, when the meteorological conditions mentioned above happen together—viz., great heat upon the

Fig. 75.—Different forms of hail.

surface of the ground, which diminishes rapidly with increase of elevation, and which is accompanied by a considerable evaporation from the
clouds under the action of the sun. As, however, the mere collision of
a very cold upper wind with a very warm wind at the same altitude may
produce hail, it occasionally falls in winter and at night; but this is of
rare occurrence.

Meteorologists often class together hoar-frost and hail, and hence assert that these aqueous meteors occur oftener in winter and spring than
in summer and autumn. But hoar-frost differs from hail, not only
from being divided into so much smaller particles, but in its mode of
formation, for it does not spring from the bosom of the clouds, nor does
it necessitate great atmospheric movements. It is merely frozen rain,
or a rough-grained and dense snow.

CHAPTER V.

PRODIGIES: SHOWERS OF BLOOD — OF EARTH — OF SULPHUR — OF PLANTS—OF FROGS—OF FISH—OF VARIOUS KINDS OF ANIMALS.

APART from the ordinary showers, more or less heavy, of rain, snow, or hail, which we have been considering above, the history of meteors is supplemented by certain extraordinary showers which have often inspired the ignorant and credulous with terror, who have seen in them direct manifestations of God's anger.

I do not refer to stones falling from the sky, the aerolites, which Greek philosophers looked upon as fragments detached from the celestial vault, but which are, as we have seen, cosmical corpuscles circulating in space. Nor will we deal with the showers of stones, bricks, planks, and earthenware, which are caused by whirlwinds. But we will just glance at certain phenomena which we have not yet taken notice of. We will begin by the Showers of Blood.

Homer relates how a shower of blood fell upon the heroes of Greece, as a presage of death for many of their number. Obsequens cites the following: After the capture of Fidènes, in the year 14 of the Romish era, drops of blood fell from the sky, to the great surprise of all men. In 538 a heavy shower of blood fell over the Aventine Hill and at Aricia. In 570 and 572 it rained blood for two days upon the Squares of Vulcan and Concordia; in 585 during one day. In 587 this prodigy occurred in several districts of the Campagna, upon the territory of Præneste; in 626 at Ceres, in 648 at Rome, in 650 at Duna, in 652 in the neighborhood of the Anio. There was a shower of blood when Tatius was murdered. Plutarch speaks of showers of blood after great battles—in the Cimbric war, for instance, after the massacre of so many thousand Cimbri upon the plains of Marseilles. He admits that the bloody vapors distilled from the corpses and diluted in the clouds would lend to these their crimson inge. The following are the showers of blood which, principally by aid of the researches made by M. Grellois, I have succeeded in collecting as having occurred since the commencement of the Christian era down to the close of the last century. In the first instance, Gregory of Tours relates that in the year

582 A.D. "a shower of blood fell over the district about Paris. Many persons had their clothes stained with it, and cast them off in terror." An analogous shower is said to have taken place at Constantinople in 652. In 654 the sky seemed on fire in Gaul, blood descending from the clouds in large quantities. In 787 Fritsch mentions a shower of blood in Hungary, followed by the plague. Others were witnessed at Brixen in 869, and at Bagdad in 929. In 1117 there occurred strange phenomena, showers of blood, and subterraneous noises, which scattered terror throughout Lombardy during the struggle for freedom there, and a meeting of Bishops took place at Milan to consider their origin. The same phenomenon was remarked at Brescia for three days and three nights before the death of the Pope, Adrian II. In 1144 there were several showers of blood in Germany; in 1163 at La Rochelle. In 1181, during the month of March, there was a constant rain of blood for three days in France and Germany: a luminous cross was visible in the skies. Toward the end of 1543 blood fell at the castle of Sassemburg, near Barendorf, in Westphalia; in 1580 at Louvain. In 1571 there fell near Einden, during the night, so much blood that over a space of five or six miles the grass and clothes exposed had assumed a dark purple hue. Many persons preserved some of it in vessels. It was attempted, but unsuccessfully, to show that this prodigy was due to the rising into the air of the vapor from the blood of oxen that had been killed. No other explanation was found more deserving of credit among natural causes. These phenomena were also noticed at Strasbourg in 1623, at Tournay in 1638, and at Brussels in 1640.

We learn from the records of the Academy of Sciences that on March 17, 1669, at 4 A.M., there fell in several parts of the town of Châtillon-sur-Seine a kind of rain or reddish liquor, thick, viscous, and putrid, which resembled a shower of blood. Large drops were seen imprinted against walls, and one wall was even splashed all over on both sides, "which would lead one to believe that this rain was composed of stagnant and muddy waters, carried into the air by a hurricane out of some neighboring marshes." There was a shower of blood at Venice in 1689.

In 1744 there fell a red rain in the Faubourg of St. Peter d'Arena, at Genoa, which, on account of the war then going on in the territory of the Republic, terrified the inhabitants very much; but it was subsequently ascertained that this tint was due to some red earth which a strong wind had carried into the air from a neighboring mountain.

History speaks of showers of blood at Cleves in 1763, in Picardy in 1765, and in Italy in 1803. Rain of a red color has been observed often enough in our own day to prevent there being any doubt as to the reality of the phenomenon, and the only mistake of our forefathers was in assigning it a supernatural origin. Bede was of opinion that a rain thicker and warmer than usual might become blood-red, and so deceive the uninstructed. Kaswini, El Hazen, and other savans of the Middle Ages, relate that about the middle of the ninth century there fell a red powder and a matter resembling coagulated blood. These philosophers were thus on the road to a reasonable explanation; they saw in it only a resemblance which might be correct, and not a reality which is repugnant to the simplest logic. "What the vulgar call a shower of blood," says G. Schott, "is generally a mere fall of vapors tinted with vermilion or red chalk. But when blood actually does fall, which it would be difficult to deny takes place, it is a miracle due to the will of God." Eustathius, the commentator of Homer, says that in Armenia the clouds discharge showers of blood because this country contains the Cinabrian mines, the dust of which, mixed with water, colors the drops of rain.

Conrad Lycosthenes, in his "Book upon Prodigies," represents the showers of blood and the showers of crosses in the shape of childish figures, which give us an idea of the simple-mindedness prevalent in those days.

In the early part of July, 1608, one of these pretended showers of blood fell in the outskirts of Aix (Provence), and this shower extended to the distance of half a league from the town. Some priests, either being themselves deceived or wishing to work upon the credulity of the people, at once attributed it to diabolic influence. Fortunately, a person of education, M. de Peiresc, examined very minutely into this apparent prodigy, studying in particular some drops that fell upon the wall of the cemetery attached to the principal church in Aix. He soon discovered that they were in reality the excrements of some butterflies which had been noticed in large numbers during the early part of July. There were no spots of the kind in the centre of the town, where the butterflies had not made their appearance, and, moreover, none were noticed upon the higher parts of the houses, above the level to which they flew. Besides, the presence of these drops in places protected from the air rendered it impossible that they could have their origin in the atmosphere. He at once pointed this out to those who regarded the occurrence as miraculous; but, in despite of the proofs which he adduced, the inhabitants persisted in attributing these drops to a supernatural cause.

Fig. 76.—Rain of blood in Provence, July, 1608.

Réaumur gives the butterfly known as "the great turtle" as being the most capable of depositing these drops. "There are thousands of others," he says, "which turn into chrysalises toward the end of May or the beginning of June. When this transformation is about to take place, they leave the trees and often take refuge upon walls, entering houses, hanging on to the arch of a door-way or a plank. If the butterflies which emerge from them at the end of June or the beginning of July flew in masses together, they would be numerous enough to form small clouds, and consequently to cover the stones in certain places with spots of a blood-red color, and thus to make the timid believe that they were spectators of a supernatural occurrence." Gen-

erally speaking, showers of blood are not only red spots produced by
certain insects, but *regular showers*, colored by the dust which the wind
carries into the air. This general origin was not ascertained until the
present century. On March 14, 1813, one of these strange red showers
fell in the kingdom of Naples and the Two Calabrias. Sementina examined and analyzed it, rendering the following account to the Naples
Academy of Sciences: "An east wind had been blowing for two days,
when the inhabitants of Gerace noticed a dense cloud moving toward
the sea. At 2 P.M. the sea became calm, but the cloud already covered
the neighboring mountains and began to intercept the light of the sun.
Its color, originally a pale red, soon became deep as fire. The town
was then plunged into such profound darkness that, about 4 P.M., it
was necessary to light candles in the houses. The inhabitants, alarmed
by the obscurity and the color of the cloud, rushed in crowds to the
cathedral to pray. The obscurity increased, and *the whole sky seemed
red as fire;* thunder began to growl; and the sea, though six miles distant, added to the general alarm by the roar of its waves. There then
began to fall large drops of reddish rain, which many persons took for
blood, and others for fire. At last, as night advanced, the air became
clear, the thunder and lightning ceased, and the inhabitants regained
their self-possession."

With the exception of there being no popular alarm, the same phenomenon of a shower of reddish dust occurred not only in the Two
Calabrias, but also at the opposite extremity of the Abruzzes. This
dust was of a yellowish hue, like cinnamon, and had a slight earthy
taste; it was unctuous to the touch, and, seen through a glass, contained small and hard bodies resembling pyroxene. Heat at first
embrowned it, then made it black, and finally gave it a reddish tint.
After the action of the heat, this dust displayed, even to the naked eye,
an immense number of small and brilliant points, which were of yellow
mica. Its specific gravity, when deprived of hard substances, was 2·07:
it was composed of silica, 33·0; aluminium, 15·5; lime, 11·5; chrome,
1·0; iron, 14·5; and carbonic acid, 9·0.

Whence came this dust? This it was found impossible to ascertain
at that time. It was not until 1846 that a general examination of these
rains was made, and their origin found by following them up into
space. On May 16 in that year an earthy rain fouled all the water at
Syam (Jura). In the autumn of the same year there was a similar fall,
accompanied by lightning, diluvian rain, very disastrous hurricanes,

etc., which occurred alternately, or nearly so, over a large circular tract
of country, in such a way as to be only explicable by some great disturbance in the system of the trade-winds. The cyclones also swept
over the Atlantic; amidst fearful squalls, whirlwinds, and hailstorms,
vessels were dismasted and their decks swept clean. Then also occurred severe tempests in France, Italy, and at Constantinople; while,
farther eastward, the typhoons spent their fury in the China seas. The
winds were sufficiently intense to detach a stratum of land in districts
where the surface of the ground was sandy or of some other soft substance. This earth, carried into the air, was, of course, certain to be
deposited somewhere. This took place in the south of France, between
Puy and Mont Cenis, in the direction of the prevailing wind, and crosswise from Bourg to Drôme. The quantity of earth precipitated varied,
however, according to the locality; at Lyons, in fact, it was scarcely
apparent, though it occurred in the shape of a reddish slime which was
popularly converted into a *shower of blood*. But at Meximieux a bat-
talion of soldiers marching toward the Swiss frontier were covered with
the mud, and their uniforms impregnated with it. The Château de
Chamagnieu was bespattered in such a way that it could scarcely be
recognized, and there was such a thick layer at Valence that the inhabitants were compelled to clean water-shoots and gutters. Fournet gives
a calculation which shows that in the department of the Drôme the
clouds must have taken up from and again discharged upon the ground
the enormous weight of 720 tons, which represent 180 four-horse wag-
on-loads. Ehrenberg, who analyzed samples of this earth, found in
them seventy-three organic formations, some of which were peculiar to
Southern America. This earth must, therefore, have come from the
New World. The interval of time between their leaving America, October 13, and their arrival in France, October 17, was about four days,
which gives a speed of eighteen and three-quarter yards per second.

Subsequent to that date we have had a remarkable fall of colored
rain in the neighborhood of Chambéry, on March 31, 1847. It was
imbued with a milky matter, which seemed like thin clay suspended in
the air. The clothes of persons exposed to this rain were bespattered
with whitish spots. Information from Savoy and the Great St. Bernard came to hand soon after this, stating that there had been a fall of
earthy red snow, coming from the south-west, and covering the ground
to the depth of several inches.

This coloring of the snow by the dust must not be confounded with a hue which it often derives from a small insect which lives in it—*uredo nivalis*—a kind of microscopic infusory often extraordinarily numerous in the Alps and the Polar regions.

At the period of the red rain in 1847 cited above, the falls of snow extended over a large portion of France—at Orleans, at Paris, in the Vosges, and La Bresse ; and there were hurricanes at Havana, Bahama, the Azores, Newfoundland, the Sorlingues, Portugal, and Spain. There were numerous atmospheric whirlwinds in the north and the west, at Le Havre, Paris, and at Grignan, no less than twenty-four storks falling dead at this place. At Nantua, a whirlwind, which carried a sentry-box ten feet into the air, covered the streets with débris of tiles, chimneys, and windows. The numbers given by Fournet show a very rapid and marked depression of the barometer on March 31, followed by a still greater decrease on April 2.

There was also a very remarkable shower of earth on March 27, 1862. The residue, when moist, was, like that of 1846, so far red in hue as to revive the popular belief about a shower of blood ; when dry, the earth was fine and yellowish. Ehrenberg discovered in it forty-four organic forms, among which were those microscopical *galionelles*, a cubic inch of which may contain 466,000.

The shower which fell at Beauvais in May, 1863, from 5 to 11 A.M., was also very remarkable, the spots which it left upon clothes being as marked as in the preceding cases.

About 3 A.M. on the morning of May 1, a violent thunder-storm broke over Perpignan, and afterward a reddish dust was noticed in several parts of the town, which, it was subsequently ascertained, must have fallen during the storm. The same storm extended to the level district in the department of the Eastern Pyrenees ; but here the phenomenon witnessed was a fall of red snow, and the appearance of these flakes alarmed the inhabitants. The occurrence was also noticed on many coast-towns of the Mediterranean. There was discovered in them a dust of marshy and ferruginous clay, mixed up with fine sand, which, as it passed through the atmosphere, deprived it of a portion of the organic matters in suspension there. In this way these rains serve a fertilizing purpose, being in fact *showers of manure.* Each heavy gust of wind raises clouds of dust, as may especially be remarked when, animated by a gyratory movement, it possesses a certain force of aspiration which enables it to form those small whirlwinds of dust which may be seen upon the high-roads.

The whole extent of the vast zone of deserts which reaches over the intertropical and the subtropical countries of the Old as of the New World contains earthy elements, which the wind drives to an immense distance. Europe, like Asia, Africa, and America, furnishes the wind with a supply of this kind.

We have already pointed out the powers of whirlwinds. To cite but that of 1780: it developed its force near Carcassonne, upon the banks of the Aude, raised high into the air immense quantities of sand, unroofed eighty houses, and blew in all directions stacks of wheat standing in fields. Large ash-trees were uprooted, and their biggest branches carried to a distance of forty yards. Such a power amply explains the fact of earth and sand being taken so much farther. The shower of blood which fell at Sienna on December 28–31, 1860, analyzed by D. Campani, seemed to be of organic origin.

One of the latest showers of blood recorded is that which occurred on March 10, 1869. On this day the sirocco was blowing at Naples, and its squalls were accompanied by that nebulosity which is peculiar to it, and which resembles a slight mist; the barometer had fallen considerably; the weather was very warm, and from time to time there fell sharp but short showers, either of very fine rain or in large drops; each drop of this rain left a muddy spot behind it.

These spots, when examined carefully, had a marked yellowish brown tint, and resembled spots left by water containing iron. A sheet of white paper, first damped and then exposed to the wind, was soon covered with a number of small and reddish grains, nearly spherical in shape, the diameter of which varied from 0·004 inch to 0·0004 inch. There can be no doubt, considering the direction of the wind at the time, that these grains of sand came direct from the desert of Sahara.

M. Breton, of Grenoble, noticed that this residue was exactly analogous to that which was picked up at Valence in September, 1846, after the red rain spoken of above. As was imagined, this sand came from Sahara. It appears from another account that Algeria was the theatre of a very violent hurricane on March 3, 1869.

French soldiers were overtaken by the wind, near El-Outaia, in the midst of a sea of sand. It took them four hours to travel six and three-quarter miles. "During the seventeen years that I have been in Algeria," says an eye-witness, "I have never seen such a whirlwind. Our little column was compelled to stop and to take precautions against being killed. At the second halt we turned our backs to the squall, and for an hour and a half we could see neither the sun nor the sky, although just before there had been scarcely any clouds. For more than a quarter of an hour together we could not see a distance of two or three yards in front of us."

The red rain which fell at Naples had undoubtedly been brought from the desert of Sahara, itself exposed to a tempest which in fact extended over all Europe, the Mediterranean, and Africa.

These phenomena are intimately connected with the great movements of the atmosphere, as M. Tarry has judiciously pointed out.

Ten days after the red rain mentioned above, on the 20th of March, a violent tempest, coming from England, swept over the north coast of France. There was a very marked centre of atmospheric depression (28.90 inches) at Boulogne on the 20th; by the next day it had reached Lesina, upon the Adriatic. For several days a violent north-west wind raged over France, and afterward over Italy. On the 22d the cyclone had reached Africa, where it raised into the air the sands of Sahara; a retrograde movement then took place; a fresh decrease of the barometer reading occurred in the south of Europe, where the pressure had risen after the passage of the cyclone. On the 24th the barometer fell to 29·13 inches at Palermo, and 29·21 inches at Rome: the wind grew very violent; the instrument of Father Secchi, in the latter city, indicating a speed of 640 miles in the twenty-four hours—the greatest of the year.

The atmosphere in Sicily was noticed, on the 23d, to be laden with thick clouds and a yellowish dust, which lent to the sky an unusual appearance. Rain falling, each drop left a yellow residuum, which it needed two or three filterings to remove. This substance, analyzed by Professor Silvestre, at Catania, contained the following elements: clay, chalky sand, peroxide of hydrate of iron, nitrogenized sodium, silica, and organic matter.

The same phenomenon was remarked at Subiaco, near Rome, and at Lesina, in Illyria. Thus the prodigies spoken of by Livy are now registered at the Paris Observatory.

The last remarkable red rain was that of February 13, 1870. On February 7 a great barometrical depression occurred in England; the barometer marked 29.33 inches at Penzance; on the 9th it had reached the Mediterranean; on the 10th, Sicily, where the barometer reading was lower than at Rome. This fall of the barometer was accompanied by a violent tempest; at Rome there was a violent north wind for three days—the 8th, the 9th, and the 10th. It superinduced a severe frost in France and Italy, snow falling in Rome on the nights of the 8th and the 9th. On the 11th and 12th the weather was calmer, and the barometer reading increased again, the cyclone raging over the desert of Sahara. The retrograde movement alluded to above soon made itself manifest. On the 12th the barometer fell to 29·45 inches in the south of Spain; a violent wind from the south blew over Spain and Italy on the 13th and 14th; and from Africa the cyclone, accompanied by the hurricane, again made its way back to Europe, with the sand swept up from Sahara. As a matter of fact, at 2 P.M. on the 13th of February, a reddish sand was remarked in the rain that fell at Subiaco, near Rome, by M. Alvarez; at Tivoli, by Father Ciampri; and at Mondragone, by Father Lavaggi. In the night of the 13th to the 14th there fell at Genoa an earthy and reddish substance; and at Moncalieri, Father Denza, Director of the Observatory, picked up some *red snow* which contained the same kind of sand.

This recital of the showers of blood shows us—1st, that they are a reality; 2d, that they are mostly due to dust taken up by the wind into very distant regions; 3d, that they are not so infrequent as they appear to be. Thus there are no less than twenty-one occasions upon which they have been known to occur during the present century in Europe and Algeria, as the following table will show:

1803.	February	Italy.
1813.	February	Calabria.
1814.	October	Oneglia, between Nice and Genoa.
1819.	September	Studein, Moravia.
1821.	May	Giessen.
1839.	April	Philippeville, Algeria.
1841.	February	Genoa, Parma, Canigou.
1842.	March	Greece.
1846.	May	Syam, Chambéry.
1846.	October	Dauphiné, Savoy, Vivarais.
1847.	March	Chambéry.
1852.	March	Lyons.
1854.	May	Horbourg, near Colmar.
1860.	31st December	Sienna.
1862.	March	Beauman, near Lyons.
1863.	March	Rhodes.
1863.	April	Between Lyons and Aragon.
1868.	26th April	Toulouse.
1869.	10th March	Naples.
1869.	23d March	Sicily.
1870.	13th February	Rome.

It will be noticed that these remarkable showers mostly take place in the spring and the autumn, at the epoch of the equinoctial gales. We have seen that they may be due to the traces left by certain kinds of butterfly. A third cause must also be noticed—viz., volcanoes, the ashes of which are sometimes conveyed by the winds to an immense distance. Several cases in proof of this might be adduced.

We now come to another series of remarkable showers spoken of in ancient legends, exaggerated and interpreted in different ways, and the true explanations of which it is not always easy to give.

Showers of milk are often spoken of as having taken place. Thus Obsequens relates that upon the territory of Veies there fell a shower of milk and oil in 629. The absence of all definite information upon facts of this kind prevents one from doing more than hazard a few conjectures borrowed from volcanic eruptions or the carrying into the air of white or chalky earth by some hurricane. In 620 streams of milk are said to have flowed into the Roman lake. In 643 milk is reported to have flowed for three days in some place not mentioned; numerous victims were immolated when this prodigy took place. These so-called streams of milk are a common phenomenon in some countries; the washing of the rain over a white soil suffices to cause this illusion, which, however, the most cursory analysis would dispel.

Dion Cassius speaks of a rain that looked like milk, and which, falling on coins or copper vessels, made them retain the appearance of silver for three days. If this fact be true, it is clear that it must have arisen from a downfall of sublimated mercury which had become condensed, and consequently had fallen to the ground. But in what way this sublimation and condensation was brought about, it is first necessary to ascertain, before believing in the occurrence of this prodigy.

Glycas also speaks of a shower of mercury, which might be the same as the above, though it is stated to have taken place during the reign of Aurelian.

We may compare with these showers a phenomenon which has been observed too often to permit of its reality being questioned. I allude to the appearance of *crosses* upon men's clothes, a few instances of which I append:

In 764 the misbehavior of the monks of St. Martin drew down the anger of God. Blood fell from the heavens on to the earth, and crosses appeared upon men's garments.—*Gregory of Tours.*

Fritsch speaks of the same phenomenon as occurring in 783. In

1094 crosses fell from heaven on to the garments of priests, for the purpose, no doubt, of warning them of their impiety, says G. Schott. In 1534 there fell in Sweden a shower which left the mark of a red cross upon men's garments. Cardan explains this phenomenon by the statement that red dust was diluted in the rain-water, and that the crosses were formed by the drops falling in the woof of the cloth. Fromond and Schott do not accept this explanation, because, according to them, these crosses were formed not only upon certain parts of the garment, but all over it, and that when drops of blood fall upon a piece of cloth they never take this shape. The pious of that date considered it to be a direct intervention of the Deity. But this is not all. It is related that in 1501 crosses fell in Germany and Belgium, not only upon the garments, *even when inclosed in boxes,* and especially upon the garments of women, but that they left a mark upon the skin, and upon bread. This prodigy lasted three years, recurring during Passion-week and Easter; no doubt, adds the chronicler, to inspire the respect too often forgotten to the blood and cross of the Lord.

John of Horn, Prince of Liége, told the Emperor Maximilian I. of a young woman of that town, twenty-two years of age, whose garments were perpetually covered with blood-red crosses, although she continually changed her clothes.

It must, at the same time, be mentioned that many instances are cited in which nutritious substances have descended in a shower. Thus in 1824 and 1828 there was so abundant a shower of this kind in one of the districts of Persia that it covered the ground to the depth of five or six inches. It was a kind of lichen, of a sort already known; cattle and sheep devoured it greedily, and some bread was even made from it.

We may also class with the preceding the descent of a soft substance which Muschenbroeck states to have occurred in Ireland in 1675. This was a glutinous and fat substance, which softened when held in the hand, and emitted an unpleasant smell when exposed to the action of fire.

On the 10th of March, 1695, at about 7 P.M., a heavy storm burst over Châtillon-sur-Seine: the front part of the cloud appeared inflamed, the air to be on fire, and the spectators who saw it believed that the neighboring villages were being burned, as sparks of flame fell to the ground in all directions. This shower lasted a quarter of an hour, and extended over a large tract of country, where it caused no conflagration; immediately after the storm there was a heavy fall of large snow-flakes.

In 828 there fell from the sky a number of grains like those of wheat, but much smaller.

This fact may easily be credited, as also the following, which is told by Johnston: There fell for the space of two hours, over a tract of country two miles in extent, in Carinthia, a shower of wheat with which bread was afterward made.

We may also accept the statement of Cassiodorus, that there fell in 371 a shower of rain, in the country of the Atrebates, in which there was a plentiful admixture of wool.

The showers of sulphur, which are often spoken of, are, as a rule, nothing more than the pollen of certain plants, pine and nut trees in particular, which may be carried by the wind to an immense distance. Without going so far back as the storm of sulphur which destroyed Sodom and Gomorrah, there are certain storms of the kind which appear well authenticated. Olaus Wormius states that on May 16, 1646, there fell a heavy shower at Copenhagen which inundated the whole city, and contained a dust exactly like sulphur, both in regard to color and smell. Simon Paulli states that on May 19, 1665, there raged in Norway a fearful tempest, with a dust so like sulphur that, when thrown into the fire, it produced the same smell, and that, when mixed with spirits of turpentine, it produced a liquor the odor of which was just like that of balm of sulphur. The close neighborhood of the Iceland volcanoes is sufficient to explain this occurrence. Phenomena of the same kind are not infrequent in Naples. Sigesbek, in the "Breslau Memoirs," speaks of a shower of sulphur which fell in Brunswick, and which was a *regular mineral sulphur*. This fact can not be accepted without further proof: as to the showers of pollen, flowers, leaves, etc., they are well authenticated.

At Autrèche (Indre et Loire), at 12·10 P.M. on April 9, 1869, the air was very still, and the sky cloudless. M. Jallois relates that one of his correspondents remarked a shower of dry oak-leaves falling from the higher regions of the atmosphere. Being gifted with excellent sight, he saw them appear like bright specks upon the azure of the sky, at a very great height, and fall about him, after having descended almost vertically, with a trifling inclination eastward. This continued for ten minutes; but the shower of leaves had probably commenced previously. There was at least one to each square yard upon a piece of water close by.

This phenomenon seems to have resulted from a great squall which occurred on April 3; the oak leaves carried up by a hurricane into the higher regions of the atmosphere were kept there by the wind for six days, and fell again when the weather became calm.

This shower of leaves reminds me of a shower of oranges. On July 8, 1833, a water-spout, which took place at Pausilippus, near Naples, burst upon the shore and swept off two large baskets of oranges; a few minutes afterward they descended to the ground at some distance.

After the vegetable showers we come to a series even more remarkable, and perfectly well authenticated. I refer to the *showers of live animals*.

In the chapter on water-spouts we have seen that fish are sometimes taken up in this way out of a pond. Peltier relates that frogs once fell upon his head from a water-spout. This was at Ham, in 1835, and the fact was duly certified. I may cite another, still more recent.

In the morning of January 30, 1869, toward 4·30 A.M., after a violent gust of wind, there began a fall of snow which lasted until daylight, at Arache, in Upper Savoy; and in the morning a large number of live larvæ were found in the snow. They could not have been hatched in the neighborhood, for, during the days preceding, the temperature had been very low. On January 24 the thermometer had marked 60·8°, and upon the following days a temperature of 41° at 7 A.M. They seemed to be mostly the *Trogosita mauritanica*, which is common in the forests of Southern France. There were also found among them a few caterpillars of a small butterfly belonging to the *noctuelian* tribe, probably the *Sibia stagnicola*. This caterpillar reaches its full size in the course of February, and is indigenous to the centre and the south of France.

This shower of insects at Arache, at an altitude of from 1000 to 1200 yards, can only be explained by a violent wind which must have brought them from some locality in the south of France.

M. Tissot, the village school-master, who observed this phenomenon, adds, that in the course of November, 1854, the wind being very violent, thousands of insects, most of them alive, alighted upon a plantation near Turin; some of them were larvæ, and others had attained their full growth, while all belonged to an order of hemiptera which are nowhere seen except in the island of Sardinia. Ancient authors have related several instances of falls of insects.

Phanias, cited by Porta, states that there fell a shower of fish for three days in the Chersonesus.

In Athens, Philareus asserts that he saw large quantities of fish and frogs fall from the sky in many different places. Heraclides Lembus, in Book XXI. of his "Histories," says that God sent showers of frogs upon Pœnia and Dardania in such large quantities that the houses and roads were covered with them. They were found mixed up in the food, and were consumed with it. The water was filled with them; it was impossible to walk without treading upon them. The decomposition of their bodies produced such an odor that it was found necessary to quit the country.

Varro declares that all the inhabitants of a certain town in Gaul were driven from their houses on account of the countless frogs which fell from the sky.

Scaliger states that the town of Mirabel, in Aquitania, was filled with half-formed frogs which fell from the sky. Johnston relates that, in the island of Auckland (Friesland), "in which there are no frogs," a number fell in a shower of rain. Olaus Magnus also states that frogs, worms, and fish fall from the clouds in the north oftener than in the south, "on account of the viscosity of the clouds and the heat which they derive from the sulphurous principle!"

Fromond relates that, while standing with several friends at one of the gates of Tournai, in 1625, a shower of rain suddenly fell, and produced so many frogs, all of the same size and color, that the ground was covered with them.

Porta says that he often saw, between Naples and Pouzzoles, a quantity of frogs suddenly emerge from the dust upon which a heavy shower of rain had just fallen. This peculiarity, he adds, is well known to many inhabitants of these two towns.

These sudden appearances of frogs and toads are generally due to the fact that these animals mostly issue from the mud after a thunderstorm, and are in the habit of crossing frequented routes. It is excessively rare for whirlwinds to carry up into the air either fishes or frogs.

The showers of locusts are due to flying masses of orthoptera, the nomad cricket in particular. These insects are a scourge to agriculture. They are brought by the wind; and when they alight, they transform a fertile region into a desert. Seen from a distance, their countless swarms present the appearance of thunder-clouds. These dark masses hide the sun. As far and as high as the eye can reach, the sky is black and the ground covered with them. The sound of their million wings is like the noise of a cataract. As they reach the ground, they break the branches of the trees. In a few hours all signs of vegetation have disappeared over an extent of several leagues. The wheat is gnawed to its roots, the trees are stripped of their leaves. Every thing is destroyed, sawn, cut to pieces, and devoured. When nothing is left, the terrible swarm rises, as if at a given signal, and flies off, leaving famine and desolation behind it.

It often happens that, after having consumed every thing, they die of starvation before depositing their eggs. Their bodies, heated by the sun, soon become putrefied, and emit exhalations which breed terrible epidemics in the district.

In 1690 locusts arrived in Poland and Lithuania from three different points, and in three distinct masses. The Abbé de Ussans, who saw them, says, "At certain places where they had died in large quantities they lay four feet deep. Those which were still alive, and which had settled upon the trees, made the boughs bend beneath their weight."

In 1749 locusts arrested the march of Charles XII.'s army when it was retreating through Bessarabia after the defeat of Pultowa. The king thought that it was a hailstorm which was thus swooping down upon his army. The arrival of the locusts was announced by a hissing sound like that which precedes a tempest, and the rustling of their flight drowned the sound of the waves of the Black Sea. All the country in their track was laid bare.

In the south of France locusts sometimes multiply at such a prodigious rate that they soon produce enough eggs to fill several barrels. They have at times caused terrible damages; notably so in the years 1805, 1820, 1822, 1824, 1825, 1832, and 1834.

Mezeray states that in January, 1613, during the reign of Louis XIII., locusts invaded the district round Arles. In seven or eight hours all the wheat and forage were devoured to the very roots over 20,000 acres of ground. They then crossed the Rhône and visited Tarascon and Beaucaire, where they consumed the garden produce and the lucerne. They went from thence to Aramon, Monfrin, Valebrégues, etc., where most of them were, fortunately, destroyed by starlings and other insect-eating birds which had been attracted thither by the prospect of such a banquet. The consuls of Arles and Marseilles had their eggs picked up. It cost the former town 25,000 and the latter 20,000 francs, and 3000 cwt. of eggs were thrown into the Rhône. Counting 1,750,000 eggs to the cwt., 5,250,000,000 of locusts, as they would afterward have become, must have been destroyed.

Fig. 77.—Shower of locusts.

In 1825, in the territory of Saintes-Maries, not far from Aigues-Mortes, upon the shores of the Mediterranean, 1518 wheat-sacks were filled with dead locusts, the weight of which was nearly sixty-nine tons; at Arles there were picked up 165 sackfuls, or between six and seven tons.

Locusts are always to be met with in Algeria, in the provinces of Oran, Bone, Algiers, and Bougie; but they are not so numerous as to produce those terrible invasions which change a fertile country into a desert. There are locust years in Algeria, just as in France there are years when beetles, caterpillars, etc., are especially abundant. These scourges are, fortunately, very rare. The most disastrous took place in 1845 and 1866.

Regular showers of beetles have also been known to descend like a thick cloud and cover the fields and the highways.

27

As with the locusts, they swarm from one province into another. Masses of these coleoptera, which are not transported by a whirlwind, but which are generally driven by the wind, emigrate from a district after they have devoured every thing in it.

To give an idea of the prodigious numbers in which cock-chafers sometimes make their appearance, I will quote some few historical instances.

In 1574 these insects so abounded in England that they stopped several mills on the Severn.

In 1688 they formed so dense a cloud in Galway that the sky was darkened to the distance of a league, and the peasants had a difficulty in finding their way about. They destroyed all vegetation, so that the country around had the look of winter. Their voracious jaws made a noise like that caused by the sawing of a thick piece of timber; and

Fig. 78. Shower of cock-chafers.

in the evening the flapping of their wings resembled the distant rolling of a drum. The unhappy Irish were compelled to cook and eat them, for want of other food. In 1804 vast clouds of cock-chafers, precipitated by a violent wind into the Lake of Zürich, formed a thick mass upon the shore, where their bodies were heaped up, the putrid exhalations from which poisoned the atmosphere. On May 18, 1832, at 9 P.M., a legion of beetles encountered a diligence upon the route from Gournay to Gisors (as it was leaving Talmoutiers) with so much violence that the horses, blinded and frightened, were compelled to return.

Such is the series of showers of blood, earth, vegetables, and animals, which the history of meteorology has registered. We will stop here. Just as in the preceding chapter we saw that there were writers who spoke of hailstones as big as elephants, so too, in this case, there has been considerable exaggeration. Fabulous as may be the force which the wind sometimes acquires, we may assign to the domain of romance the story told by Avicenne, that prince of Arab doctors, as to his having seen the body of a *calf* fall from the skies. Nevertheless, Xavier de Maistre declares that a young girl was carried off by a whirlwind in 1820; but it is not said to what height. Cabeus, in the seventeenth century, declared that a violent wind had blown away a woman who was washing linen in the lake. In regard to large animals, the most exaggerated story is the one which is also the oldest—viz., as to the Nemæan Lion falling from the moon on to the Peloponnesus. . . . It is true that stones to the weight of hundreds of pounds sometimes fall from the sky, as we saw in regard to aerolites. But hitherto the other worlds have sent us nothing more valuable than stones. The animals, fish, insects, grains, and leaves, which fall from the sky come originally from the earth, not from any of the planets.

BOOK SIXTH.

ELECTRICITY, THUNDER-STORMS, AND LIGHTNING.

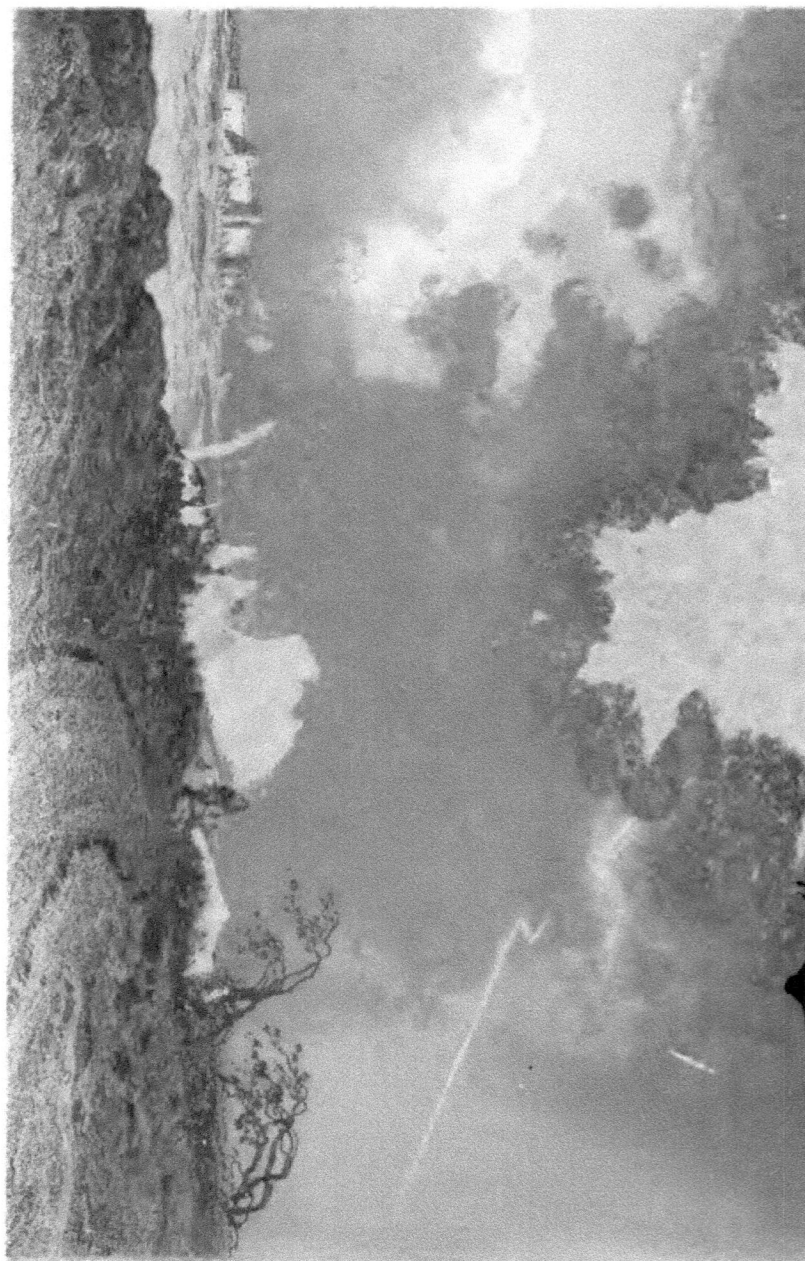

CHAPTER I.

ELECTRICITY UPON THE EARTH AND IN THE ATMOSPHERE: ELECTRIC CONDITION OF THE TERRESTRIAL GLOBE — DISCOVERY OF ATMOS-PHERIC ELECTRICITY—EXPERIMENTS OF OTTO DE GUÉRICKE, WALL, NOLLET, FRANKLIN, ROMAS, RICHMANN, SAUSSURE, ETC.—ELECTRICITY OF THE SOIL, OF THE CLOUDS, OF THE AIR—FORMATION OF THUN-DER-STORMS.

WE now come to the most marvelous and singular agent that exists, the study of which will complete and close the immense panorama developed in this work—viz., electricity, thunder-storms, and lightning. The study of them is by no means devoid of complications; but our close attention will be amply repaid by the wonderful spectacles which it will reveal. Following our general plan, we will see how it is distributed over the earth and in the atmosphere. It is, however, first necessary to obtain an idea of its history, which is somewhat remarkable.

Otto de Guéricke, burgomaster of Magdeburg, the celebrated inventor of the pneumatic machine, first discovered (about 1650) some signs of electric light. Dr. Wall, at about the same epoch, by applying friction along a cylinder of amber, saw a bright spark emitted, and heard a sharp noise; and, curiously enough, this first electric spark produced by the hand of man was at once compared to the lightning's flash. This light and this sound, says Dr. Wall, in his "Memoirs" (see "Phil. Trans."), seem to represent, in a certain measure, the lightning and the thunder. The analogy was striking, and needed only an effort of the imagination to be understood; but to demonstrate its truth, to discover in so insignificant a phenomenon the causes and the laws of the greatest phenomena in nature, required a series of proofs which could only be expected from a great genius. Nevertheless, many physical philosophers endeavored to obtain them by comparisons of a more or less ingenious kind: some remarked that the spark is zigzag, like lightning: others opined that thunder in the hands of nature is the same as electricity in the hands of man. "I confess," said Abbé Nollet, "that I should look upon this idea with great complacency if it could be well

sustained ; and there are many specious reasons by which it might be."
Still, this was nothing more than a train of reasoning which could not
be conclusive, inasmuch as in physics experiment alone is absolutely
decisive.   While Europe and the whole of the Old World were thus
reasoning, America was conducting experiments in special reference to
the subject of lightning.   Franklin succeeded in bringing electricity
down from the sky, in order to investigate it by direct examination.
After having made several discoveries in respect to electricity, especially in regard to the Leyden jar and the attractive power of fine points,
Franklin went in search of electricity into the very midst of the clouds.
He had concluded, as the result of certain experiments, that a stem of
pointed metal, placed at a great height, upon the summit of a building,

Fig. 79.—Experiments of Franklin and Romas.

formed a receptacle for the electricity of thunder-clouds.   He was await-
ing, with no little impatience, the construction of a steeple then being
built at Philadelphia ; but unwilling to remain so long in doubt, he
had recourse to a more expeditious and not less certain method for ascertaining what he desired to know.   As all that was necessary was to
raise a substance of some kind into the region of the thunder—that is
to say, high enough into the air—he thought that an ordinary kite
would serve his purpose as well as any steeple.   He accordingly arranged two pieces of stick, laid crosswise and covered with a silk handkerchief, which he took into the fields upon the occasion of the first
thunder-storm.   Fearing the ridicule which failure would entail, he was
accompanied only by his son.   The kite remained some time in the air

without any perceptible effect being produced; but at last the fibres of the rope were somewhat agitated. Encouraged by this, Franklin placed his finger upon the end of the rope, a motion which immediately led to the appearance of a bright spark, which was soon followed by several others. Thus, for the first time, the genius of man succeeded in playing with the lightning and discovering the secret of its existence.

Franklin's experiment took place in June, 1752, and was shortly afterward repeated in every civilized country with the same success. A French magistrate, De Romas, assessor to the Nérac Tribunal, profiting by the ideas of Franklin, which had been made public in France, conceived the idea of using the kite with raised bars; and in the month of June, 1753, before the result of Franklin's experiments was known, he had obtained very strong electric signs, because he had prudently attached a metal wire to the cord along its whole length, which measured 850 feet. A little later, in 1757, Romas repeated these experiments during a thunder-storm; and this time he elicited sparks of an enormous size. He says, "Imagine tongues of fire nine or ten feet in length, and an inch thick, which made as loud a report as a pistol. In less than an hour I had obtained at least thirty sparks of these dimensions, to say nothing of a thousand others of seven feet or less." A great number of persons, including several ladies, were present at these experiments.

As may be imagined, these experiments were not unattended with danger. Romas was on one occasion knocked down by an excessively heavy discharge, but, fortunately, escaped severe injury. Richmann, a member of the St. Petersburg Academy of Sciences, was not so fortunate, as one of his experiments cost him his life. He had erected an iron rod, which conducted the atmospheric electricity from the roof of the house to his study, so that he could measure its intensity every day. On the 6th of August, 1753, in the midst of a violent storm, and while standing at some distance from the rod, in order to avoid the large sparks, he incautiously approached too near the conductor. A globe of bluish fire struck him on the forehead, and killed him on the spot.

For the last hundred years the study of electricity has been pursued both by experiments made in the laboratory and in the atmosphere—with what splendid results it is needless to relate. The electric telegraph, which enables us to carry on a whispered conversation with our neighbors across the ocean, and the process which effects a faithful re-

Fig. 80.—Richmann, of St. Petersburg, struck by lightning during an electrical experiment.

production of the *chefs-d'œuvre* of statuary and engraving, are but two of the most important applications of the first. The experiments upon the electricity of the atmosphere, devoted to more complex and potent phenomena, have enabled us to acquire a more exact notion concerning the conditions of this electricity and its various manifestations.

Electricity is a power the inner nature of which, like that of heat, light, and attraction, remains unknown to us. This power produces certain effects; and it is the study of these effects which constitutes the science. To explain them, it is admitted—first, that electricity is a subtle fluid, capable of becoming amassed, condensed, and rarefied; of discharging itself from one body into another; of traversing immense distances more rapidly even than light, which itself travels at the rate of about 185,000 miles per second; secondly, that this fluid has two modes of existence—two modes of manifesting itself—which are distinguished, the one from the other, by the terms *positive* and *negative.* These distinctions do not exist in nature, and are only perceptible to human sense by relative variations in intensity. Be this as it may, it has been ascertained that *opposite electricities attract*, whereas *like electricities repel each other.* The union of equal quantities of fluids of an opposite denomination forms a *neutral*, or natural, fluid, which, it is believed, exists in inexhaustible quantities throughout all bodies. Under many influences, among which must be cited that of friction, the neutral fluid becomes decomposed into one or the other of these two elements. The terrestrial globe and the atmosphere are two vast reservoirs of electricity, between which there is a constant exchange by decomposition and reconstitution, which plays a complementary part to the action of heat and moisture in the life of plants and of animals.

The general result of the researches into the conditions of electricity upon the surface of the globe and in the atmosphere is, that in a normal condition the globe is charged with *negative* and the atmosphere with *positive* electricity. At the surface of the soil, where continual exchanges are taking place, electricity is in a neutral state, as also in the lower stratum of air, which is in contact with the surface, upon the sea as well as upon land. Positive electricity increases in the atmosphere in proportion to height.

The large amount of evaporation which takes place from the surface of the sea in the regions of the equator loads the clouds with positive electricity, and these, carried by the upper currents, travel toward the polar regions, and charge the atmosphere there with an accumulation of this electricity. Its influence causes in the soil of the polar regions an opposite condensation of negative electricity. The auroræ boreales are, in chief, caused by these two conflicting tensions; it is a silent but visible reconstitution of the natural fluid by the two opposite tensions of the atmosphere and the soil. Thus the appearance of an aurora borealis is accompanied by electric currents, which circulate upon the soil at a distance sufficiently great to permit of the movements of the magnetic needle, indicating, at the Paris Observatory, for instance, an aurora which may be visible in Sweden or Norway.

Clouds are generally charged with positive electricity; nevertheless, negative clouds are sometimes met with. It is not unusual to see upon the summit of a mountain clouds adhering to it, as if they were attracted thither, making a halt there, and then following the general movements of the wind. It often happens that in this case the clouds lose their positive electricity by their contact with the mountain, and assume the negative electricity of the latter, which, far from serving to attract, has, on the contrary, a tendency to repel and drive them away. On the other hand, a stratum of clouds situated between the ground, negative, and an upper stratum, positive, is almost neutral; its positive electricity becomes accumulated upon its inside surface, and the first drops of rain cause it to disappear altogether.

The electricity of the atmosphere is subject, like heat and atmospheric pressure, to a double annual and diurnal oscillation, and to accidental oscillations greater than those which are fixed and regular. The maximum occurs from 6 to 7 A.M. in summer, and from 10 A.M. to noon in winter; the minimum is between 5 and 6 P.M. in summer, and between 2 and 3 P.M. in winter. A second maximum is also noticeable at sunset, followed by a diminution during the night until sunrise. This oscillation is connected with that of the hygrometrical condition of the air. In the annual variation the maximum occurs in January and the minimum in July: it is due to the great atmospheric circulation. Winter is the period when the equatorial currents are most active in our hemisphere, and it is then that the auroræ boreales are most numerous.

As the positive or negative conditions of electricity, as determined by apparatus constructed for measuring their intensities, are but a comparison more or less between two different *charges*, it follows that, when an electric cloud passes over our heads and dissolves itself into rain, the air may manifest negative electricity both before and after the rain, and even during its fall, according to the intensity of the charge contained in the cloud. M. Quételet demonstrates this state of things in the following manner:

Let A, B, C, D, E, be five positions on the earth in a straight line, which we suppose to be neutral. The stratum of air above and parallel with the positions on the earth—A' B' C' D' E'—is in a state of positive electricity in the absence of clouds, and to an equal extent throughout.

The stratum A" B" C" D" E", still higher and parallel, is also in a state of positive and more intense electricity. There comes suddenly a cloud at the three central positions B′ C′ D′, in a state of positive electricity, greater than that of the circumambient air. It follows that, relatively to it, the air which is around will display negative electricity.

To an observer situated at A, the electricity above the earth will be positive. As the cloud approaches, these indications will become gradually less until they vanish altogether, and even become negative on the passage of the cloud. But the rain will bring back positive electricity. A corresponding variation will be manifest when the rain stops, and the cloud moves off. At D, the indications will be negative; at E, they will again become positive.

We saw in the conflicts of the great atmospheric currents in the tropical regions, where the node of the circuit accomplished from the equator to the poles takes place, that the evaporation of the seas, caused by solar heat in these foci of condensation, the variation of atmospheric pressure, etc., engender cyclones, hurricanes, and tempests, the whirling march of which reaches as far as our temperate latitudes. These violent movements develop electricity in immense proportions, and it is rarely that these phenomena are not accompanied by thunder-storms, lightning, and thunder. The formation of the clouds upon sea and land, the fogs which occur in our regions, the course of the clouds along our valleys and mountains, all emit varying quantities of electricity. There is a storm when this electricity of the clouds, instead of effecting a mutual and tranquil interchange, collects at certain points, and, becoming condensed, saturates them, so to speak, and finally bursts, afterward uniting itself to the negative electricity which has been instantaneously amassed, either upon the ground or in other clouds.

The great storms reach us from the Atlantic. They arise from the cyclones, and the clouds which convey them are generally more than 3000 or 5000 feet high, traveling from S.W. to N.E., without being apparently affected by the undulations of the ground in France. The secondary storms, which are formed in France itself, are conveyed by clouds of a less elevation than the above, and which sometimes just skim the ground, being, in fact, influenced by it, scarcely reaching over the mountains, following the valleys, amidst which they distribute in large quantities lightning and hailstorms.

The formation of storms is preceded by a slow but steady decline in the reading of the barometer. The calm of the air and a stifling heat, due to the absence of evaporation from the surface of our bodies, are specially characteristic circumstances. The variations in the electric condition of the soil and the atmosphere, added to the above, have a great effect upon our organic system. A peculiar nervous feeling, with no visible cause for it, takes possession of many persons, in spite of all their efforts to shake it off. It is under these circumstances that one is especially enabled to see how intimate is the connection between man's physical and moral condition.

CHAPTER II.

LIGHTNING AND THUNDER.

WHEN electricity is discharged from a cloud by which it is overloaded, and is precipitated either into another cloud or to the ground with opposite electricity, electric light is produced—a rapid spark such as we display on a small scale in our experiments in physics. This spark traverses in an instant the distance, whatever it may be, which separates the two electrized points. It has been ascertained that it does not last $\frac{1}{10000}$ of a second. It is this electric spark which constitutes lightning; it is by it that lightning is made manifest during a storm.

As a general rule, these flashes appear in the shape of a sudden diffused light which illuminates the clouds, the sky, and the earth, and is followed by a darkness which seems more intense than it was before by the force of contrast. Whether in this case the exchange of electricity between the clouds takes place simultaneously over a large surface which is lighted up and which dies away instantaneously, or whether there be a spark as in lightning concealed by the clouds, in either event one only sees—which is of the most frequent occurrence—a sudden diffused light, upon which are momentarily displayed the more or less marked contours of the clouds.

These diffused lightnings are the most frequent. Hundreds of flashes are seen during a stormy day, or rather night, to one flash of linear lightning. The latter is, however, characteristic lightning. It is but a strong electric spark, a small ball of fire which darts from an overcharged cloud to the earth, or from one cloud to another, or which even rises from the earth to the clouds; the rapidity of its progress produces the effect of a narrow and luminous line. It is rare that it darts in a straight line, in spite of the axiom as to "the nearest road;" whether because of the varied distribution of moisture in the air, which causes it to be a more or less better conductor, or because of the varying excess of electricity in different parts of the soil and of the clouds, the lightning is nearly always zigzag. The subtle fluid shows, by the way in which it traverses our dwelling-places, that it leaps suddenly from one point to another as if by caprice, but being evidently obedient to the laws of the distribution and conductibility of electricity. Generally speaking, linear lightning darts in obtuse-angled zigzags, or else is curled like a snake. Sometimes it splits into two or more branches. Nicholson and the Abbé Richard observed forked flashes. Occasionally, though more rarely, it splits into three branches; Arago cites several instances of this, especially in the volcanic thunder - storms; Kaemtz noticed it once. At times, too, the flashes have four or five ramifications, or, it may be, the branches which issue from the original flash become ramified into several small lateral branches. M. Liais observed and sketched flashes with five branches.

The flashes are not always of a shining white hue, but have at times a yellow, red, blue, and even a violet or purple tint; this color depends upon the quantity of electricity which traverses the air, upon the density of the latter, upon its moisture, and upon the substances suspended in it. The violet flashes generally indicate that the cloud from which they are emitted is at a great height, and the air which they travel through an air so rarefied as to call to mind that of the Geissler tubes.

It is rarely that a correct idea as to the length of lightning-flashes is formed. While we produce with the greatest difficulty in our laboratories an electric spark of a few inches, Nature shoots forth sparks as much as ten miles long. F. Petit measured at Toulouse some flashes which were ten and a half miles long—the extreme length with which I am acquainted. Arago found that a series of flashes which he measured were seven or eight miles in length.

In reply to the question as to the height of thunder-clouds, it is evident that they are of different elevations. De l'Isle measured one on June 6, 1712, which was 26,250 feet above Paris; Chappe, on July 13, 1761, remarked one that was situated 10,400 feet over Tobolsk; and Kaemtz noticed another 10,200 feet above Halle. These observations give a decreasing series of elevations which gradually decline until they almost reach the ground. Haidinger measured thunder-clouds which were only 230 feet above Gratz, on June 15, 1826, while upon another occasion he remarked some only ninety-two feet above the ground at Admont. This refers to a level country. In the mountains, Saussure observed some of these clouds over Mont Blanc; Bouguer and La Condamine, over the Pichincha, at 16,000 feet; Ramond, upon Mont Perdu, at 11,100 feet, and upon the Peak du Midi at 9630 feet, and indeed at all heights. They are generally from 2950 to 3280 feet high over the sea.

Whether the flash takes place horizontally between two groups of clouds, or obliquely either between clouds of different strata or between the clouds and the ground, it is generally several miles long. It is this length which is the primary cause of the rolling of thunder. Thunder is, in reality, but the sound of the electric spark effecting an exchange of electricity, a neutralization, between two points more or less distant from each other.

The noise of the thunder may be due to several different causes. The spark itself, as it traverses in an instant the atmospheric air, forces back the molecules upon its passage, and produces a momentary void into which the circumambient air at once rushes, and so on for a certain distance. Pouillet met this rather natural explanation by the objection that if the sound of thunder was produced in this way, the passage of a cannon-ball would produce an analogous noise. The objection is not well founded, for the cannon-ball is but a tortoise in comparison to the dart of the lightning. In the second place, the sound of thunder may be due to the fact that clouds become dilated under the influence of the electric tension which swells them in a certain measure, lengthens them, and stretches them with so much force at certain points that, if a spark causes the cloud to discharge, the outer air, being no longer retained by the expansive force of the electric fluid in equilibrium with it, rushes from all directions toward the clouds. To this may be attributed the cause of thunder, and of the fall of rain which follows. The electric conditions of the various clouds which compose a storm being dependent the one upon the other, the discharge of one must lead to that of several others more or less distant. In the one case as in the other, the sound is always caused by the expansion of the air at the spot where the more or less partial void has just been made, as happens with fire-arms, the bursting of a bladder, etc. When situated at the point where the lightning terminates—where the thunder-bolt falls, according to the vulgar expression—this noise is never very long, and is exactly like the report of a cannon, a fowling-piece, or a pistol, according to the intensity. But one of the special characteristics of thunder consists in the *rolling*, as its name imitates it in every language— *thunder, tonnerre, tonitruum, bronté, donner.*

It is frequently asked to what this rolling, often very prolonged, can be due. There are several causes for it. The first is due to the length of the flash, and to the difference in the speed of sound and of light. Let us imagine, for instance, a horizontal flash 35,000 feet long and

28

3000 feet high. An observer, placed beneath one extremity of the flash, will see this flash in its full length for an instant; the sound will be formed at the same moment along the whole line of the flash; but the sound-waves will reach his ears at different times. That which starts from the nearest point will arrive in three seconds, as sound travels at the rate of about 1100 feet per second. That which was formed at the same instant at a point 6000 feet distant takes twice the time to arrive. That which proceeds from a point at 13,200 feet will take twelve seconds. The sound formed at a distance of 35,000 feet would take thirty-three seconds to travel; thus the rolling will continue half a minute, gradually becoming fainter, until it dies away altogether.

If, as most frequently happens, the observer is not situated exactly at one of the extremities of the flash, but at a certain point in its line of passage, he hears first the report, which gradually grows louder, and then diminishes. In this case, the sound starting from a point situated over his head, and at a height of 1000 yards, reaches him in three seconds; but the sounds formed on either side at equal distances arrive at the same time during several seconds, and sound ceases in less than thirty-two seconds.

To this cause of the prolonged rolling must be added the numerous discharges which often take place very rapidly among thunder-clouds —the zigzags and ramifications of the lightning, caused by the hygrometrical diversity of the various strata of air—the echoes repeated by mountains, the soil, waters, and the clouds themselves, to which must further be added the interference produced by the encounter of different systems of sound-waves.

The duration of the rolling of thunder varies very much, as every one may have remarked. The greatest length recorded for a single flash is forty-five seconds, by De l'Isle, at Paris, on June 17, 1712. Upon the same day he remarked another, which lasted forty-one seconds; and on July 8, in the same year, one of thirty-nine seconds. The intervals, included between the commencement of the thunder and the different phases of intensity in its rolling, were as follows upon this last occasion (July 8):

> At 0 seconds, flash ;
> At 11 seconds, slight thunder ;
> At 12 seconds, it bursts ;
> At 32 seconds, the explosions cease ;
> At 50 seconds, the sound dies gently away.

The intensity of thunder varies to an enormous extent. In certain cases it has been compared to the report *of a hundred pieces of artillery discharged at the same time.* In other instances the report is no louder than that of a pistol, followed by a rolling sound more or less dull. At times the explosions remind one of the tearing of a piece of silk, at others of the noise made by a cart loaded with bars of iron sent loose down a steep paved street.

The longest interval ever remarked between the flash and the report was seventy-two seconds. This was at Paris, and the same interval was also noticed to elapse by the astronomer De l'Isle on April 30, 1712. In these two cases the cloud must have been six leagues off. Next to these exceptional cases, the longest interval was forty-nine seconds, which represents ten miles' distance. Direct researches have shown that a storm is never heard at a greater distance than thirteen miles, rarely at more than seven to ten; the flashes are visible, but the sound does not travel so far. The fact is the more curious as cannon are heard at a much greater distance, as much as twenty-five miles; and when very large, they may be heard at double that distance.

Continued cannonading, as during a siege or a pitched battle, has been heard at a distance of thirty leagues. During the winter of 1870, the Krupp guns, exhibited in Paris in 1867, were heard at Dieppe, a distance of eighty-four miles, during the bombardment of Paris. The cannonade of March 30, 1814, was heard at Casson, a village between Lisieux and Caen, at a distance of forty-four leagues from Paris. Arago relates that the firing at Waterloo was audible at Creil, 120 miles distant. Thus the thunder manufactured by man reaches much farther than the thunder produced by nature. If thunder is not audible at more than six leagues, it follows that, if thunder is heard with the sky clear, the report must be produced by clouds below the visible horizon, as we can not see beyond six leagues. A person of five feet five inches in height is able to see, when the horizon is clear, an object placed upon the ground at a distance of 13,000 feet. If the object in question is eighty feet high in the air, it may be seen at five and a half leagues. If it is 1600 feet high, as in the case of an isolated mountain, it will be visible at a distance of fifty miles. If the object be 3300 feet high, as *cumulus* clouds are, as a rule, in our climates, it can be seen at a distance of seventy miles.

For a thunder-clap which takes place when the sky is clear, to be produced by a cloud, we must consequently suppose the cloud to be less than 100 feet above the ground—a state of things never witnessed. Thus electricity may be emitted from certain regions of the air, from invisible clouds, and may produce flashes and thunder-claps during fine clear weather. Observation has proved this to be a fact, but one of very rare occurrence.

To these statements bearing upon the general action of thunder and lightning, I may add that, notwithstanding the extreme rapidity of the flash, it has been found possible to measure its duration, which does not exceed $\frac{1}{100000}$ of a second. To effect this, a round piece of card-board, divided from the centre to the circumference into black and white sections, is made use of. This circle is made to turn like a wheel, with a speed equal to that of the wind. It is well known that luminous impressions remain for one-tenth of a second upon the retina. Thus, if a hot coal is turned round, and if the revolution is made in one-tenth of a second, and as each successive position of the coal remains impressed for the same length of time upon the retina, a continuous circle becomes visible. If the circular piece of card-board, with its black and white stripes, is made to revolve, the sectors cease to be visible, and we can only see a grayish circle, if each stripe passes before our eyes in less than the tenth of a second. But it is possible to make the card-board revolve more than a hundred times in a second. This being the case, if the card-board circle is exposed to a continuous light, we shall be unable to distinguish the lines, inasmuch as they come before our eye much more rapidly than the impression which they produce remains. But if the circle is made to revolve in a dark place, and an instantaneous flash of light suddenly falls upon it, and as suddenly disappears, the impression produced upon our eye by each of the sectors will last less than one-tenth of a second, it will be almost instantaneous, and the circle *will seem to us to be motionless*. By giving this apparatus a fixed rate of rotation, it has been ascertained that a flash lasts but $\frac{1}{1000}$ of a second.

Light, traveling a distance of 185,000 miles in a second, takes but an instant, too short to be reckoned, to come from the spot, never more than a few miles off, at which the flashes are produced. Thus we see the flash at *the very moment* at which it occurs. But sound travels, as we have seen above, less rapidly—at the rate of 1100 feet a second. It follows that the thunder-clap, which takes place at the same time as the flash, will only be audible to us ten seconds afterward if we are 11,000 feet away from the storm; and any one can, therefore, calculate how far off the storm is by the interval between the flash and the thunder.

Fig. 81.—Harvesters killed by lightning.

½ second interval corresponds to............................	550 feet.
1 " " "	1,100 "
2 " " "	2,200 "
3 " " "	3,300 "
4 " " "	4,400 "
5 " " "	5,500 "
6 " " "	6,600 "
7 " " "	7,700 "
8 " " "	8,800 "
9 " " "	9,900 "
10 " " "	11,000 "
11 " " "	12,100 "
12 " " "	13,200 "

There are about twelve beatings of the pulse to a league. When the flash extends over a length of several miles, the spot struck by the thunder may be very distant, although the report is heard immediately after the flash, because it is the sound which starts from the nearest extremity of the flash which is heard first. For instance, in a storm on the 27th of June, 1866, M. Hirn remarked that the report followed immediately upon the flash, although this same flash had struck down two persons beneath a tree three miles distant.

Many are the marvelous freaks and jests played by electricity, sometimes ending in tragedy. Among the most remarkable is that of striking a person dead, and leaving him in the exact position occupied at the moment the shock was given, just as if he were still alive, and yet so thoroughly consumed as to be nothing but a mass of cinders. Thus we are told that at Vic-sur-Aisne, France, in 1838, three soldiers sought refuge from a violent thunder-storm under a linden-tree. Some peasants, seeing them stand motionless long after the storm had passed, and receiving no response to a pleasant salutation, touched them on the shoulder. The bodies instantly crumbled to fine ashes! Yet the moment before there was no evidence that the lightning had touched them. Their clothing was not torn, and their faces wore a natural appearance. The following remarkable circumstance was witnessed by Pastor Butler: On the 27th of July, 1691, ten harvesters took refuge under a hedge on the approach of a thunder-storm. The lightning struck and killed four of them, who remained as if suddenly petrified. One of them was just putting a bit of tobacco in his mouth, another was fondling a little dog on his knee with one hand and feeding him with the other. M. Cardan relates that eight harvesters, tak-

ing their noonday repast under a maple-tree during a thunder-storm,
were killed by one stroke of lightning. When approached by their
companions, after the storm had cleared away, they seemed to be still
at their repast. One was raising a glass to drink, another was in the
act of taking a bit of bread, a third was reaching out his hand to a
plate. There they sat as if petrified, in the exact position in which
death surprised them.

Fig. 82.—Curious freak of lightning.

On the 10th of September, 1845, during a violent thunder-storm, a
house in the village of Salagnac, France, was struck by lightning. A
large ball of fire descended the chimney, and rolled across the floor of a
room in which sat a child and three women. No one was hurt. It
then rolled out through the centre of the kitchen, passing close to the
feet of a young peasant, and disappeared through a crevice in the wall.
Its erratic course ended in the pig-sty, the harmless occupant of which
it despitefully slew, without setting on fire the straw on which the crea-
ture lay.

CHAPTER III.

THE SAINT ELMO FIRES AND THE JACK-O'-LANTERNS.

THE Saint Elmo fires are a slow manifestation of electricity, a quiet and steady outflow (like that of the hydrogen in a gas-burner), which radiates gently over the topmost points of lightning conductors, of buildings and vessels, during thunder weather, when the terrestrial electric tension is strongly attracted by that of the clouds.

The Saint Elmo fires are generally seen as a light resting on the masts of ships. The following are some of the most recent observations made :

On December 23, 1869, in latitude 46° 53' north, and longitude 9° 55' west, the barometer reading 29·61 inches, thermometer 49° 1', the log of the packet *Impératrice-Eugénie* records the occurrence of very violent squalls. Sharp and numerous flashes of lightning were visible in all parts of the horizon, without being followed by a single clap of thunder. During the night these squalls were accompanied by heavy hailstorms, and, when they passed over the vessel, they presented the phenomenon known under the name of *the Saint Elmo fire.*

Luminous tufts, blue in color and about a foot and a half high, appeared above the tips of the conductors upon each mast. The masts and the rigging looked phosphorescent, and the tips of the waves also seemed decked with tufts, but less showy than those that appeared above the masts. These glimmerings were visible whenever the squall reached the vessel. Very brilliant when the wind was blowing with its full force, they became less bright as it fell, and disappeared when it dropped altogether. Only those parts of the masts and the rigging which were exposed to the direct action of the squall presented this luminous appearance. They looked as if they had been rubbed with phosphorus. The phenomenon did not take place upon the parts which were at all sheltered from the wind, nor did it come down lower than the top-yards, about ninety feet above the level of the sea. The phenomenon repeated itself several times during the night, but only when the squalls were accompanied by hail. The Saint Elmo fires are also seen over steeples. The following is one of the most recent instances:

On March 2, 1869, these flames appeared over the church at Sainte-Catherine-de-Fierbois, in the canton of Sainte-Maure and the arrondissement of Chinon; no thunder was audible during the storm, and the steeple disarmed the thunder-clouds. A correspondent of the French Scientific Association wrote as follows: "Toward the end of the tempest, when the wind had somewhat abated and the rain was not so heavy, several persons remarked a crown of fire around the cross that surmounted the steeple of the church, about 130 feet high. One of the eye-witnesses saw it for at least five minutes (he did not perceive it begin); the light was so bright that the steeple and cross were as plain to the eye as in full daylight; the light finally died away like that of a burned-out candle, without the least change of position."

Fig. 85.—Saint Elmo fire over the spire of Notre-Dame, Paris.

Luminous tufts of electricity have often been seen above the spire of Notre-Dame during certain violent thunderstorms of a summer evening.

The Saint Elmo fires are occasionally seen playing over man himself, over his clothes, or any object that he has in his hand.

Julius Cæsar relates how in the month of February, about the second watch of the night, a thick cloud suddenly arose, followed by a shower of stones; and that during the same night the pike-heads of the fifth legion seemed to be on fire.

According to Procopius, a similar phenomenon was seen over the pikes and lances of Belisarius's army in the war with the Vandals.

Livy states that the pikes of some soldiers in Sicily, and a whip which a horseman in Sardinia had in his hand, seemed as if on fire. Even the coats of mail were luminous and bright with numerous flames of fire.

When in 1769, in the midst of a violent storm, bright tufts appeared over the cross upon the steeple at Hohen-Gebrachim, two persons, who had come to put out the conflagration, as they thought, were at once surprised and terrified to see their heads covered with fire and light.

On May 8, 1831, after sunset, the whole atmosphere was on fire, presaging a violent storm. At the extremity of the flag-staff at Algiers there appeared a white light in the shape of a brush which lasted for half an hour. Some artillery and engineer officers were walking upon the terrace of Fort Bab-Azoun, and each noticed, to his surprise, that the heads of his companions were tipped by small luminous tufts. When they raised their hands, brushes of light formed at the tips of their fingers.

In some instances the Saint Elmo fires have been noticed in the shape of flames; at other times a man's whole body has been seen radiant with light. Peytier and Hossard were frequently enveloped, in the Pyrenees, in centres of storms, which seemed so formidable as seen from the plains below, that the spectators believed they must have perished in them. On several occasions their hair and the tassels of their caps stood upright and emitted a bright light, accompanied by a loud hissing noise.

Letestu, in 1786, remained for three hours of the night in his balloon during a storm; he heard a deafening noise; the ear was filled with snow and hail, and the gilding upon his flag emitted scintillations.

The discharge of electricity from the soil into the atmosphere is sometimes accompanied by remarkable phenomena—by a kind of electric *hum* upon the summits of mountains.

These various phenomena are due solely to disengagements of electricity. We must not confound with the Saint Elmo fires gleams of light which resemble them very much, viz., the ignes-fatui. These latter are not caused by electricity.

The ignes-fatui, or will-o'-the-wisp, is a wandering and shadowy fire, produced by the emanations of phosphureted hydrogen gas, which rises out of places where vegetable and animal substances are in process of decomposition, such as cemeteries, manure-heaps, or marshes, and which become spontaneously inflamed when combined with the oxygen of the air.

These vacillating lights have appealed to the superstitious feelings of the people. The frightened imagination has often looked upon them as wandering spirits, and they have often terrified those who have seen them gliding between the graves of a church-yard during the silence of night.

They are sometimes emitted suddenly when an old burying-vault is opened; and as in former days lighted lamps were placed in the graves, the credulous believed they were inextinguishable.

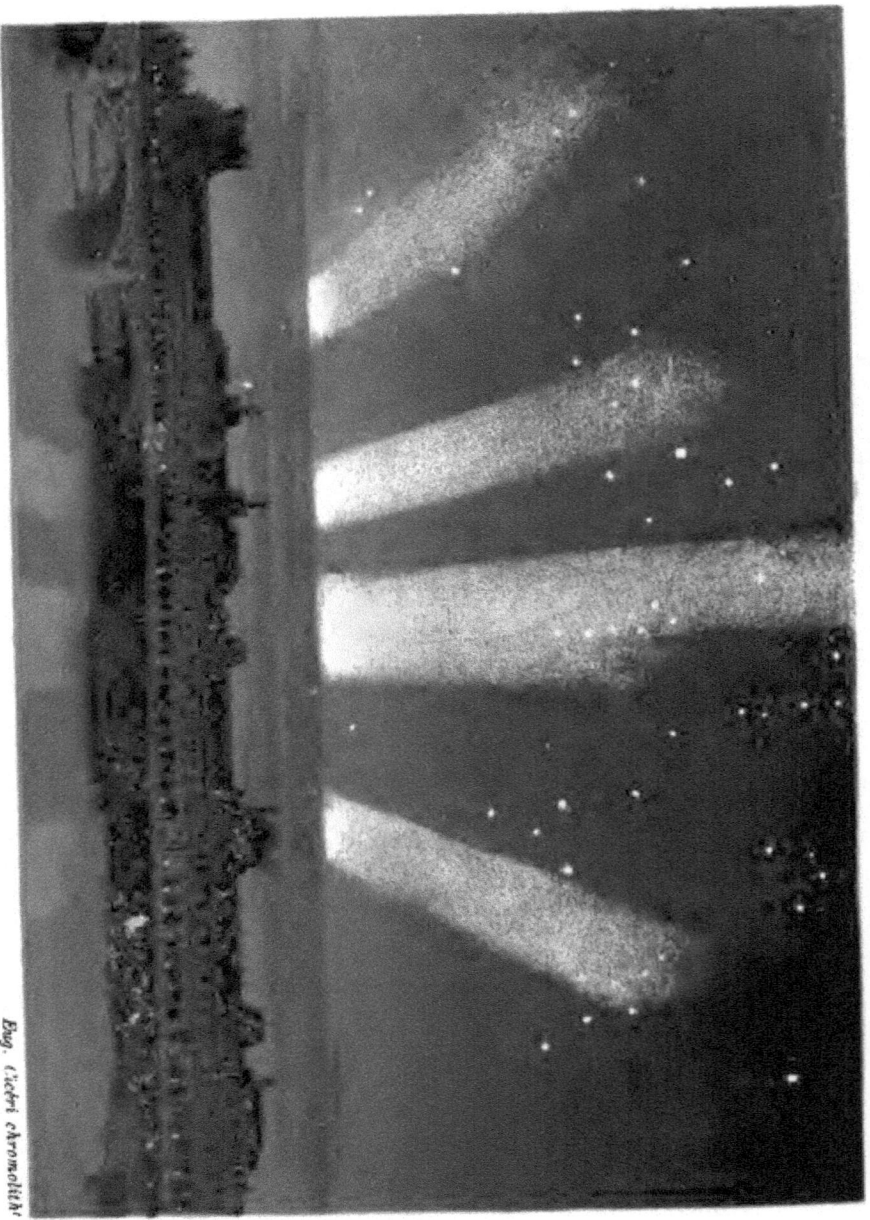

CHAPTER IV.

AURORÆ BOREALES.

WE now come to the most curious and the grandest of the various manifestations of electricity in the atmosphere. As we have seen, the globe is one vast reservoir for this subtle fluid, which exists in all the worlds appertaining to our system, and of which the radiating focus is in the sun itself. Like attraction, light, and heat, electricity is a general power in nature. Its palpitations sustain the life of the universe, and even upon our planet currents of it are in constant circulation from the equator to the poles, and from the poles to the equator. The delicate magnetic needle and the sea-compass indicate this perpetual circulation as moving northward. The magnetic needle oscillates and becomes agitated when these disturbances become violent and there are great changes in its position. The lightning which falls upon a ship often exercises an ineffaceable influence upon the compass; and while the pilot assumes that the needle is still pointing north, he runs the risk of being driven on to the rocks of some unknown shore. If a bright aurora borealis is shining over Stockholm or Reikjavik, the compass in the Paris Observatory, hundreds of leagues off, is affected by it, and seems as if it were asking the editor of the *Bulletin International* to see what was the matter.

The aurora borealis is one of the grand results of atmospheric electricity. Instead of a furious and violent storm limited to a few leagues, it is a gentle and gradual recomposition of the negative fluid of the earth with the positive fluid of the atmosphere, taking place in the aërial heights, in the upper hydrogenous atmosphere. This disengagement of electricity in a vast sheet is only visible at night, and assumes every imaginable kind of shape, according to the way in which it takes place, and to the perspective caused by the distance of the observer. At one time the eye may scarcely have time to catch its rapid undulations, alternately rose-colored and white in hue, as they dart across the sky. Now it takes the shape of a cloth of gold and purple, which seems to fall from the celestial heights; now it is a fiery dew, accompanied by a strange, rustling sound, or it may appear in the form of sheaves of flame, darting from the north to the various points of the compass. It is principally in the neighborhood of the polar circles, where thunder-storms are rare, that these manifestations of terrestrial electricity are seen to the fullest advantage. Michelet, who describes so graphically the great phenomena of nature, speaks of the aurora borealis in this way:

"The pole seems a kingdom of death. But, in reality, general life is triumphant there. The two spirits of the globe (magnetic and electric) make their nightly rejoicing in this desert."

"The aërial currents, and the currents of the sea, are their vehicles. The two torrents of heated waters which, from Java and Cuba, travel northward, where they cool and freeze, and then return refreshed to the centre whence they started, both assist in keeping up the magnetic and electric correspondence between the equator and the pole. Their storms are dependent upon each other. In summer, when the melted ice from the poles and the northern currents make their cooling influence felt, the magnetic element seems to extend in the direction of the central electricity; hence the violent storms, especially those near to this centre."

Spitzbergen is a very favorable region for witnessing an aurora bo-realis. In a voyage undertaken in 1839, M. Ch. Martins observed and analyzed a large number, which he describes thus (see "Le Tour du Monde," 1865, vol. ii., p. 10):

"At times they are simple diffused gleams or luminous patches; at others quivering rays of pure white which run across the sky, starting from the horizon as if an invisible pencil were being drawn over the celestial vault; at times it stops in its course, the incomplete rays do not reach the zenith, but the aurora continues at some other point; a bouquet of rays darts forth, spreads out into a fan, then becomes pale and dies out. At other times long golden draperies float above the head of the spectator, and take a thousand folds and undulations, as if agitated by the wind. They appear to be but at a slight elevation in the atmosphere, and it seems strange that the rustling of the folds, as they double back on to each other, is not audible. Generally a lumi-nous bow is seen in the north: a black segment separates them from the horizon, its dark color forming a contrast with the pure white or bright red of the bow which darts forth the rays, extends, becomes di-vided, and soon presents the appearance of a luminous fan, which fills the northern sky, mounts nearly to the zenith, where the rays, uniting, form a crown, which, in its turn, darts forth luminous jets in all directions. The sky then looks like a cupola of fire: the blue, the green, the yellow, the red, and the white vibrate in the palpitating rays of the aurora. But this brilliant spectacle lasts only a few minutes; the crown first ceases to emit luminous jets, and then gradually dies out; a diffuse light fills the sky; here and there a few luminous patches, resembling light clouds, open and close with an incredible rapidity, like a heart that is beating fast. They soon get pale in their turn; every

Fig. 84.—An aurora borealis over the Polar Sea.

thing fades away and becomes confused; the aurora seems to be in its death-throes; the stars, which its light had obscured, shine with a renewed brightness; and the long polar night, sombre and profound, again assumes its sway over the icy solitudes of earth and ocean." In presence of such phenomena, the poet and the artist are compelled to confess their littleness—the savant alone does not despair. After having admired the spectacle, he studies, analyzes, compares, and discusses it; he succeeds in proving that these auroræ are due to electric radiations from the poles of the earth, which is a colossal magnet, the northern pole of which is situated to the north of North America, not far from the pole of our hemisphere, while its southern pole is in the sea to the south of Australia, near Victoria."

A few instances will suffice to prove the electro-magnetic nature of the aurora borealis. At Spitzbergen, a magnetic needle suspended horizontally by an untwisted piece of silk-thread is turned toward the west. As soon as the aurora begins, the person observing this needle remarks that, instead of being sensibly motionless, it is agitated, passing to and fro from right to left, and from left to right. In proportion as the aurora becomes more brilliant, the agitation of the needle increases, and the observer is able to judge of the intensity of the aurora by the motions of the needle without leaving his study. Lastly, when the corona is formed in the sky, its centre will be found exactly in the direction to which a magnetic needle, hanging freely, points. The auroræ boreales are therefore intimately connected with the magnetic phenomena of the terrestrial globe.

What a strange world is that of the poles! Nearly every night there is a more or less brilliant display of these auroral lights; from the middle of January, there is an hour's twilight at noon; the aurora, announcing the return of the sun, becomes grander as it mounts toward the zenith. Lastly, on the 16th of February, a segment of the solar disk, resembling a luminous point, shines brightly for an instant, and as rapidly disappears; but every day at noon the segment increases, until the whole orb rises above the sea: it is the end of the long winter night; after that, day and night follow each other for sixty-five days, until the 21st of April, when begins day-time, lasting four months, during which period the sun revolves above the horizon, gradually becoming lower, and finally disappearing.

In North America, to the east of Behring's Straits, there is a large tract of territory little known to Frenchmen—*Alaska*—which is traversed by the Arctic Circle. It formed part of Russian America a few years ago, and was 45,000 square leagues in extent. It was purchased by the United States in October, 1867. In a curious account of a voyage which Frederick Whymper made there, in 1865 (see "Le Tour du Monde," 1869, vol. ii., p. 247), there is recorded the observation of that very rare phenomenon, viz., an aurora borealis in the shape of a ribbon, extending in undulating folds in the heights of the air.

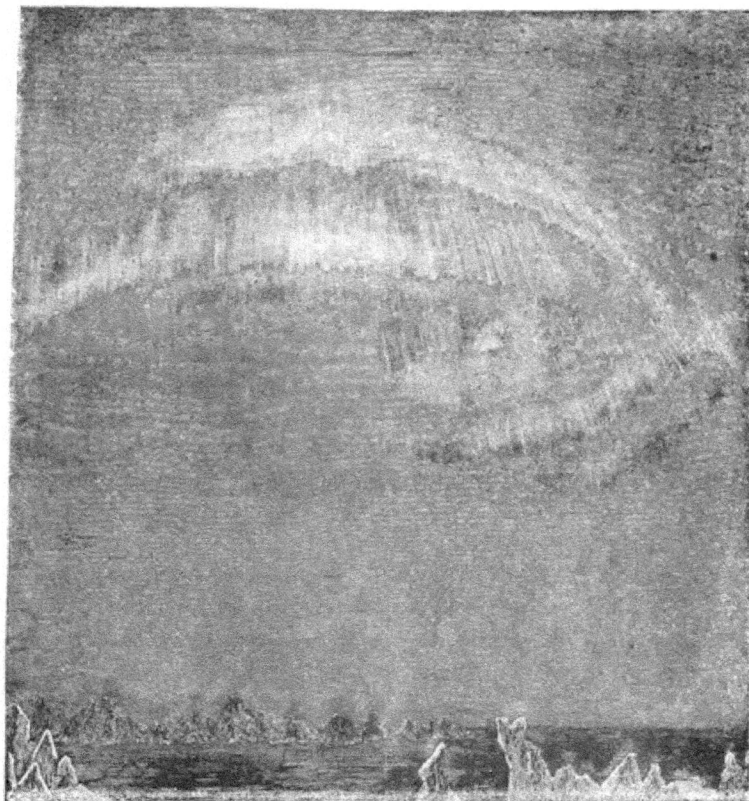

Fig. 85.—Aurora borealis observed at Bossekop (Spitzbergen) January 6, 1839.

To use the traveler's own words, "It was on the 27th of December, as we were about to retire for the night, that we were informed that an aurora borealis was visible in the west. We at once climbed the roof of the highest building in the fort, in order to contemplate this splendid phenomenon. It was not in the form of an arch, as often is the case, but the light was serpentine-shaped and undulating, the form and the color varying every instant, being at one moment of a pale and soft tint like moonbeams, while at the next long bands of blue, rose, and violet stood out upon the silvery background. The scintillations extended from the lower extremity upward, and their brightness became fused with that of the stars, the brilliancy of which was visible through the spiral vapor."

from New York to Siberia, and from both hemispheres, at the Cape of Good Hope, in Australia, Salvador, Philadelphia, and Edinburgh! This was the first time that the eye verified what theory had advanced, viz., that auroræ boreales and the southern auroræ occur at the same time in the two hemispheres under the influence of the same current. The extremities of the globe are brought into intimate relation with each other by the fluid which circulates incessantly in the air and upon the soil. At certain solemn moments, magnetism augments in intensity, and seems to reanimate the life of our planet.

The production of auroræ boreales is, in Humboldt's opinion, one of the most striking proofs of the faculty which our planet possesses of emitting light. He says, "It results from the phenomenon of auroræ, that the earth is endowed with the property of emitting a light distinct from that of the sun. The intensity of this light is rather greater than that of the moon in its first quarter. It is at times (January 7, 1831) strong enough to admit of one's reading printed characters without difficulty. This light of the earth, the emission of which toward the poles is almost continuous, reminds us of the light of Venus, the part of which, not lighted by the sun, often glimmers with a dim phosphorescent light. Other planets may also possess a light evolved out of their own substance. There are other instances in our atmosphere of this production of terrestrial light, such as the celebrated fogs of 1783 and 1831, which emitted a perceptible light during the night. Such, too, are those large clouds which are brilliant with a steady and motionless light; and such, too, as Arago has truly remarked, is that diffuse light which guides our steps during the nights of spring and autumn, when the clouds intercept all celestial light, and snow does not cover the ground."

I may further remark that auroræ boreales are more or less periodical. They were very numerous in Belgium and Western Europe during the last half of the eighteenth century. They were very rare in the seventeenth, and very frequent in the sixteenth century. This secular periodicity seems to be of about a century and a half. There is a monthly variation, more accurately ascertained. They are most frequent about the time of the equinoxes, and seem to be seven times more numerous in March and October than in June.

Such are the last and the grandest of the phenomena which we have to contemplate in this gallery of the works of the Atmosphere.

THE END.

VALUABLE & INTERESTING WORKS

FOR PUBLIC AND PRIVATE LIBRARIES,

PUBLISHED BY HARPER & BROTHERS, NEW YORK.

☞ *For a full List of Books suitable for Libraries, see* HARPER & BROTHERS' TRADE-LIST *and* CATALOGUE, *which may be had gratuitously on application to the Publishers personally, or by letter enclosing Six Cents.*

☞ HARPER & BROTHERS *will send any of the following works by mail, postage prepaid, to any part of the United States, on receipt of the price.*

FLAMMARION'S ATMOSPHERE. The Atmosphere. Translated from the French of CAMILLE FLAMMARION. Edited by JAMES GLAISHER, F.R.S., Superintendent of the Magnetical and Meteorological Department of the Royal Observatory at Greenwich. With 10 Chromo-Lithographs and 86 Woodcuts. 8vo, Cloth, $6 00.

HUDSON'S HISTORY OF JOURNALISM. Journalism in the United States, from 1690 to 1872. By FREDERICK HUDSON. Crown 8vo, Cloth, $5 00.

PIKE'S SUB-TROPICAL RAMBLES. Sub-Tropical Rambles in the Land of the Aphanapteryx. By NICHOLAS PIKE, U. S. Consul, Port Louis, Mauritius. Profusely Illustrated from the Author's own Sketches; containing also Maps and Valuable Meteorological Charts. 8vo, Cloth, $3 50.

TYERMAN'S OXFORD METHODISTS. The Oxford Methodists: Memoirs of the Rev. Messrs. Clayton, Ingham, Gambold, Hervey, and Broughton, with Biographical Notices of others. By the Rev. L. TYERMAN, Author of "Life and Times of the Rev. John Wesley," &c. Crown 8vo, Cloth, $2 50.

TRISTRAM'S THE LAND OF MOAB. The Result of Travels and Discoveries on the East Side of the Dead Sea and the Jordan. By H. B. TRISTRAM, M.A., LL.D., F.R.S., Master of the Greatham Hospital, and Honorary Canon of Durham. With New Map and Illustrations. Crown 8vo, Cloth, $2 50.

SANTO DOMINGO, Past and Present; with a Glance at Hayti. By SAMUEL HAZARD. Maps and Illustrations. Crown 8vo, Cloth, $3 50.

LIFE OF ALFRED COOKMAN. The Life of the Rev. Alfred Cookman; with some Account of his Father, the Rev. George Grimston Cookman. By HENRY B. RIDGAWAY, D.D. With an Introduction by Bishop FOSTER, LL.D. Portrait on Steel. 12mo, Cloth, $2 00.

HERVEY'S CHRISTIAN RHETORIC. A System of Christian Rhetoric, for the Use of Preachers and Other Speakers. By GEORGE WINFRED HERVEY, M.A., Author of "Rhetoric of Conversation," &c. 8vo, Cloth, $3 50.

CASTELAR'S OLD ROME AND NEW ITALY. Old Rome and New Italy. By EMILIO CASTELAR. Translated by Mrs. ARTHUR ARNOLD. 12mo, Cloth, $1 75.

THE TREATY OF WASHINGTON: Its Negotiation, Execution, and the Discussions Relating Thereto. By CALEB CUSHING. Crown 8vo, Cloth, $2 00.

PRIME'S I GO A-FISHING. I Go a-Fishing. By W. C. PRIME. Crown 8vo, Cloth, $2 50.

HALLOCK'S FISHING TOURIST. The Fishing Tourist: Angler's Guide and Reference Book. By CHARLES HALLOCK, Secretary of the "Blooming-Grove Park Association." Illustrations. Crown 8vo, Cloth, $2 00.

SCOTT'S AMERICAN FISHING. Fishing in American Waters. By GENIO C. SCOTT. With 170 Illustrations. Crown 8vo, Cloth, $3 50.

ANNUAL RECORD OF SCIENCE AND INDUSTRY FOR 1872. Edited by Prof. SPENCER F. BAIRD, of the Smithsonian Institution, with the Assistance of Eminent Men of Science. 12mo, over 700 pp., Cloth, $2 00. (Uniform with the Annual Record of Science and Industry for 1871. 12mo, Cloth, $2 00.)

COL. FORNEY'S ANECDOTES OF PUBLIC MEN. Anecdotes of Public Men. By JOHN W. FORNEY. 12mo, Cloth, $2 00.

MISS BEECHER'S HOUSEKEEPER AND HEALTHKEEPER: Containing Five Hundred Recipes for Economical and Healthful Cooking; also, many Directions for securing Health and Happiness. Approved by Physicians of all Classes. Illustrations. 12mo, Cloth, $1 50.

FARM BALLADS. By WILL CARLETON. Handsomely Illustrated. Square 8vo, Ornamental Cloth, $2 00; Gilt Edges, $2 50.

POETS OF THE NINETEENTH CENTURY. The Poets of the Nineteenth Century. Selected and Edited by the Rev. ROBERT ARIS WILLMOTT. With English and American Additions, arranged by EVERT A. DUYCKINCK, Editor of "Cyclopædia of American Literature." Comprising Selections from the Greatest Authors of the Age. Superbly Illustrated with 141 Engravings from Designs by the most Eminent Artists. In elegant small 4to form, printed on Superfine Tinted Paper, richly bound in extra Cloth, Beveled, Gilt Edges, $5 00; Half Calf, $5 50; Full Turkey Morocco, $9 00.

THE REVISION OF THE ENGLISH VERSION OF THE NEW TESTAMENT. With an Introduction by the Rev. P. SCHAFF, D.D. 615 pp., Crown 8vo, Cloth, $3 00.

This work embraces in one volume:
I. ON A FRESH REVISION OF THE ENGLISH NEW TESTAMENT. By J. B. LIGHTFOOT, D.D., Canon of St. Paul's, and Hulsean Professor of Divinity, Cambridge. Second Edition, Revised. 196 pp.
II. ON THE AUTHORIZED VERSION OF THE NEW TESTAMENT in Connection with some Recent Proposals for its Revision. By RICHARD CHENEVIX TRENCH, D.D., Archbishop of Dublin. 194 pp.
III. CONSIDERATIONS ON THE REVISION OF THE ENGLISH VERSION OF THE NEW TESTAMENT. By J. C. ELLICOTT, D.D., Bishop of Gloucester and Bristol. 178 pp.

NORDHOFF'S CALIFORNIA. California: For Health, Pleasure, and Residence. A Book for Travelers and Settlers. Illustrated. 8vo, Paper, $2 00; Cloth, $2 50.

MOTLEY'S DUTCH REPUBLIC. The Rise of the Dutch Republic. By JOHN LOTHROP MOTLEY, LL.D., D.C.L. With a Portrait of William of Orange. 3 vols., 8vo, Cloth, $10 50.

MOTLEY'S UNITED NETHERLANDS. History of the United Netherlands: from the Death of William the Silent to the Twelve Years' Truce—1609. With a full View of the English-Dutch Struggle against Spain, and of the Origin and Destruction of the Spanish Armada. By JOHN LOTHROP MOTLEY, LL.D., D.C.L. Portraits. 4 vols., 8vo, Cloth, $14 00.

NAPOLEON'S LIFE OF CÆSAR. The History of Julius Cæsar. By His late Imperial Majesty NAPOLEON III. Two Volumes ready. Library Edition, 8vo, Cloth, $3 50 per vol.

Maps to Vols. I. and II. sold separately. Price $1 50 each, NET.

HAYDN'S DICTIONARY OF DATES, relating to all Ages and Nations. For Universal Reference. Edited by BENJAMIN VINCENT, Assistant Secretary and Keeper of the Library of the Royal Institution of Great Britain; and Revised for the Use of American Readers. 8vo, Cloth, $5 00; Sheep, $6 00.

MACGREGOR'S ROB ROY ON THE JORDAN. The Rob Roy on the Jordan, Nile, Red Sea, and Gennesareth, &c. A Canoe Cruise in Palestine and Egypt, and the Waters of Damascus. By J. MACGREGOR, M.A. With Maps and Illustrations. Crown 8vo, Cloth, $2 50.

WALLACE'S MALAY ARCHIPELAGO. The Malay Archipelago: the Land of the Orang-Utan and the Bird of Paradise. A Narrative of Travel, 1854–1862. With Studies of Man and Nature. By ALFRED RUSSEL WALLACE. With Ten Maps and Fifty-one Elegant Illustrations. Crown 8vo, Cloth, $2 50.

WHYMPER'S ALASKA. Travel and Adventure in the Territory of Alaska, formerly Russian America—now Ceded to the United States—and in various other parts of the North Pacific. By FREDERICK WHYMPER. With Map and Illustrations. Crown 8vo, Cloth, $2 50.

ORTON'S ANDES AND THE AMAZON. The Andes and the Amazon; or, Across the Continent of South America. By JAMES ORTON, M.A., Professor of Natural History in Vassar College, Poughkeepsie, N. Y., and Corresponding Member of the Academy of Natural Sciences, Philadelphia. With a New Map of Equatorial America and numerous Illustrations. Crown 8vo, Cloth, $2 00.

WINCHELL'S SKETCHES OF CREATION. Sketches of Creation: a Popular View of some of the Grand Conclusions of the Sciences in reference to the History of Matter and of Life. Together with a Statement of the Intimations of Science respecting the Primordial Condition and the Ultimate Destiny of the Earth and the Solar System. By ALEXANDER WINCHELL, LL.D., Professor of Geology, Zoology, and Botany in the University of Michigan, and Director of the State Geological Survey. With Illustrations. 12mo, Cloth, $2 00.

WHITE'S MASSACRE OF ST. BARTHOLOMEW. The Massacre of St. Bartholomew: Preceded by a History of the Religious Wars in the Reign of Charles IX. By HENRY WHITE, M.A. With Illustrations. 8vo, Cloth, $1 75.

LOSSING'S FIELD-BOOK OF THE REVOLUTION. Pictorial Field-Book of the Revolution; or, Illustrations, by Pen and Pencil, of the History, Biography, Scenery, Relics, and Traditions of the War for Independence. By BENSON J. LOSSING. 2 vols., 8vo, Cloth, $14 00; Sheep, $15 00; Half Calf, $18 00; Full Turkey Morocco, $22 00.

LOSSING'S FIELD-BOOK OF THE WAR OF 1812. Pictorial Field-Book of the War of 1812; or, Illustrations, by Pen and Pencil, of the History, Biography, Scenery, Relics, and Traditions of the Last War for American Independence. By BENSON J. LOSSING. With several hundred Engravings on Wood, by Lossing and Barritt, chiefly from Original Sketches by the Author. 1088 pages, 8vo, Cloth, $7 00; Sheep, $8 50; Half Calf, $10 00.

ALFORD'S GREEK TESTAMENT. The Greek Testament: with a critically revised Text; a Digest of Various Readings; Marginal References to Verbal and Idiomatic Usage; Prolegomena; and a Critical and Exegetical Commentary. For the Use of Theological Students and Ministers. By HENRY ALFORD, D.D., Dean of Canterbury. Vol. I., containing the Four Gospels. 944 pages, 8vo, Cloth, $6 00; Sheep, $6 50.

ABBOTT'S FREDERICK THE GREAT. The History of Frederick the Second, called Frederick the Great. By JOHN S. C. ABBOTT. Elegantly Illustrated. 8vo, Cloth, $5 00.

ABBOTT'S HISTORY OF THE FRENCH REVOLUTION. The French Revolution of 1789, as viewed in the Light of Republican Institutions. By JOHN S. C. AB-BOTT. With 100 Engravings. 8vo, Cloth, $5 00.

ABBOTT'S NAPOLEON BONAPARTE. The History of Napoleon Bonaparte. By JOHN S. C. ABBOTT. With Maps, Woodcuts, and Portraits on Steel. 2 vols., 8vo, Cloth, $10 00.

ABBOTT'S NAPOLEON AT ST. HELENA; or, Interesting Anecdotes and Remarkable Conversations of the Emperor during the Five and a Half Years of his Captivity. Collected from the Memorials of Las Casas, O'Meara, Montholon, Antommarchi, and others. By JOHN S. C. ABBOTT. With Illustrations. 8vo, Cloth, $5 00.

ADDISON'S COMPLETE WORKS. The Works of Joseph Addison, embracing the whole of the "Spectator." Complete in 3 vols., 8vo, Cloth, $6 00.

ALCOCK'S JAPAN. The Capital of the Tycoon: a Narrative of a Three Years' Residence in Japan. By Sir RUTHERFORD ALCOCK, K.C.B., Her Majesty's Envoy Extraordinary and Minister Plenipotentiary in Japan. With Maps and Engravings. 2 vols., 12mo, Cloth, $3 50.

ALISON'S HISTORY OF EUROPE. FIRST SERIES: From the Commencement of the French Revolution, in 1789, to the Restoration of the Bourbons, in 1815. [In addition to the Notes on Chapter LXXVI., which correct the errors of the original work concerning the United States, a copious Analytical Index has been appended to this American edition.] SECOND SERIES: From the Fall of Napoleon, in 1815, to the Accession of Louis Napoleon, in 1852. 8 vols., 8vo, Cloth, $16 00.

BALDWIN'S PRE-HISTORIC NATIONS. Pre-Historic Nations; or, Inquiries concerning some of the Great Peoples and Civilizations of Antiquity, and their Probable Relation to a still Older Civilization of the Ethiopians or Cushites of Arabia. By JOHN D. BALDWIN, Member of the American Oriental Society. 12mo, Cloth, $1 75.

BARTH'S NORTH AND CENTRAL AFRICA. Travels and Discoveries in North and Central Africa: being a Journal of an Expedition undertaken under the Auspices of H. B. M.'s Government, in the Years 1849-1855. By HENRY BARTH, Ph.D., D.C.L. Illustrated. 3 vols., 8vo, Cloth, $12 00.

HENRY WARD BEECHER'S SERMONS. Sermons by HENRY WARD BEECHER, Plymouth Church, Brooklyn. Selected from Published and Unpublished Discourses, and Revised by their Author. With Steel Portrait. Complete in 2 vols., 8vo, Cloth, $5 00.

LYMAN BEECHER'S AUTOBIOGRAPHY, &c. Autobiography, Correspondence, &c., of Lyman Beecher, D.D. Edited by his Son, CHARLES BEECHER. With Three Steel Portraits, and Engravings on Wood. In 2 vols., 12mo, Cloth, $5 00.

BOSWELL'S JOHNSON. The Life of Samuel Johnson, LL.D. Including a Journey to the Hebrides. By JAMES BOSWELL, Esq. A New Edition, with numerous Additions and Notes. By JOHN WILSON CROKER, LL.D., F.R.S. Portrait of Boswell. 2 vols., 8vo, Cloth, $4 00.

DRAPER'S CIVIL WAR. History of the American Civil War. By JOHN W. DRAPER, M.D., LL.D., Professor of Chemistry and Physiology in the University of New York. In Three Vols. 8vo, Cloth, $3 50 per vol.

DRAPER'S INTELLECTUAL DEVELOPMENT OF EUROPE. A History of the Intellectual Development of Europe. By JOHN W. DRAPER, M.D., LL.D., Professor of Chemistry and Physiology in the University of New York. 8vo, Cloth, $5 00

DRAPER'S AMERICAN CIVIL POLICY. Thoughts on the Future Civil Policy of America. By JOHN W. DRAPER, M.D., LL.D., Professor of Chemistry and Physiology in the University of New York. Crown 8vo, Cloth, $2 50.

DU CHAILLU'S AFRICA. Explorations and Adventures in Equatorial Africa with Accounts of the Manners and Customs of the People, and of the Chase of the Gorilla, the Crocodile, Leopard, Elephant, Hippopotamus, and other Animals. By PAUL B. DU CHAILLU. Numerous Illustrations. 8vo, Cloth, $5 00.

BELLOWS'S OLD WORLD. The Old World in its New Face: Impressions of Europe in 1867-1868. By HENRY W. BELLOWS. 2 vols., 12mo, Cloth, $3 50.

BRODHEAD'S HISTORY OF NEW YORK. History of the State of New York. By JOHN ROMEYN BRODHEAD. 1609-1691. 2 vols. 8vo, Cloth, $3 00 per vol.

BROUGHAM'S AUTOBIOGRAPHY. Life and Times of HENRY, LORD BROUGHAM. Written by Himself. In Three Volumes. 12mo, Cloth, $2 00 per vol.

BULWER'S PROSE WORKS. Miscellaneous Prose Works of Edward Bulwer. Lord Lytton. 2 vols., 12mo, Cloth, $3 50.

BULWER'S HORACE. The Odes and Epodes of Horace. A Metrical Translation into English. With Introduction and Commentaries. By LORD LYTTON. With Latin Text from the Editions of Orelli, Macleane, and Yonge. 12mo, Cloth, $1 75.

BULWER'S KING ARTHUR. A Poem. By EARL LYTTON. New Edition. 12mo, Cloth, $1 75.

BURNS'S LIFE AND WORKS. The Life and Works of Robert Burns. Edited by ROBERT CHAMBERS. 4 vols., 12mo, Cloth, $6 00.

REINDEER, DOGS, AND SNOW-SHOES. A Journal of Siberian Travel and Explorations made in the Years 1865-'67. By RICHARD J. BUSH. late of the Russo-American Telegraph Expedition. Illustrated. Crown 8vo, Cloth, $3 00.

CARLYLE'S FREDERICK THE GREAT. History of Friedrich II., called Frederick the Great. By THOMAS CARLYLE. Portraits, Maps, Plans, &c. 6 vols., 12mo, Cloth, $12 00.

CARLYLE'S FRENCH REVOLUTION. History of the French Revolution. Newly Revised by the Author, with Index, &c. 2 vols., 12mo, Cloth, $3 50.

CARLYLE'S OLIVER CROMWELL. Letters and Speeches of Oliver Cromwell. With Elucidations and Connecting Narrative. 2 vols., 12mo, Cloth, $3 50.

CHALMERS'S POSTHUMOUS WORKS. The Posthumous Works of Dr. Chalmers. Edited by his Son-in-Law, Rev. WILLIAM HANNA, LL.D. Complete in 9 vols., 12mo, Cloth, $13 50.

COLERIDGE'S COMPLETE WORKS. The Complete Works of Samuel Taylor Coleridge. With an Introductory Essay upon his Philosophical and Theological Opinions. Edited by Professor SHEDD. Complete in Seven Vols. With a fine Portrait. Small 8vo, Cloth, $10 50.

DOOLITTLE'S CHINA. Social Life of the Chinese: with some Account of their Religious, Governmental, Educational, and Business Customs and Opinions. With special but not exclusive Reference to Fuhchau. By Rev. JUSTUS DOOLITTLE. Fourteen Years Member of the Fuhchau Mission of the American Board. Illustrated with more than 150 characteristic Engravings on Wood. 2 vols., 12mo, Cloth, $5 00.

GIBBON'S ROME. History of the Decline and Fall of the Roman Empire. By EDWARD GIBBON. With Notes by Rev. H. H. MILMAN and M. GUIZOT. A new cheap Edition. To which is added a complete Index of the whole Work, and a Portrait of the Author. 6 vols., 12mo, Cloth, $9 00.

HAZEN'S SCHOOL AND ARMY IN GERMANY AND FRANCE. The School and the Army in Germany and France, with a Diary of Siege Life at Versailles. By Brevet Major-General W. B. HAZEN, U.S.A., Colonel Sixth Infantry. Crown 8vo, Cloth, $2 50.

HARPER'S NEW CLASSICAL LIBRARY. Literal Translations. The following Volumes are now ready. Portraits. 12mo, Cloth, $1 50 each. CÆSAR.—VIRGIL.—SALLUST.—HORACE.—CICERO'S ORATIONS.—CICERO'S OFFICES, &c.—CICERO ON ORATORY AND ORATORS.—TACITUS (2 vols.).—TERENCE.—SOPHOCLES.—JUVENAL.—XENOPHON.—HOMER'S ILIAD.—HOMER'S ODYSSEY.—HERODOTUS.—DEMOSTHENES.—THUCYDIDES.—ÆSCHYLUS.—EURIPIDES (2 vols.).—LIVY (2 vols.).

DAVIS'S CARTHAGE. Carthage and her Remains: being an Account of the Excavations and Researches on the Site of the Phœnician Metropolis in Africa and other adjacent Places. Conducted under the Auspices of Her Majesty's Government. By Dr. DAVIS, F.R.G.S. Profusely Illustrated with Maps, Woodcuts, Chromo-Lithographs, &c. 8vo, Cloth, $4 00.

EDGEWORTH'S (Miss) NOVELS. With Engravings. 10 vols., 12mo, Cloth, $15 00.

GROTE'S HISTORY OF GREECE. 12 vols., 12mo, Cloth, $18 00.

HELPS'S SPANISH CONQUEST. The Spanish Conquest in America, and its Relation to the History of Slavery and to the Government of Colonies. By ARTHUR HELPS. 4 vols., 12mo, Cloth, $6 00.

HALE'S (Mrs.) WOMAN'S RECORD. Woman's Record; or, Biographical Sketches of all Distinguished Women, from the Creation to the Present Time. Arranged in Four Eras, with Selections from Female Writers of each Era. By Mrs. SARAH JOSEPHA HALE. Illustrated with more than 200 Portraits. 8vo, Cloth, $5 00.

HALL'S ARCTIC RESEARCHES. Arctic Researches and Life among the Esquimaux: being the Narrative of an Expedition in Search of Sir John Franklin, in the Years 1860, 1861, and 1862. By CHARLES FRANCIS HALL. With Maps and 100 Illustrations. The Illustrations are from Original Drawings by Charles Parsons, Henry L. Stephens, Solomon Eytinge, W. S. L. Jewett, and Granville Perkins, after Sketches by Captain Hall. 8vo, Cloth, $5 00.

HALLAM'S CONSTITUTIONAL HISTORY OF ENGLAND, from the Accession of Henry VII. to the Death of George II. 8vo, Cloth, $2 00.

HALLAM'S LITERATURE. Introduction to the Literature of Europe during the Fifteenth, Sixteenth, and Seventeenth Centuries. By HENRY HALLAM. 2 vols., 8vo, Cloth, $4 00.

HALLAM'S MIDDLE AGES. State of Europe during the Middle Ages. By HENRY HALLAM. 8vo, Cloth, $2 00.

HILDRETH'S HISTORY OF THE UNITED STATES. FIRST SERIES: From the First Settlement of the Country to the Adoption of the Federal Constitution. SECOND SERIES: From the Adoption of the Federal Constitution to the End of the Sixteenth Congress. 6 vols., 8vo, Cloth, $18 00.

HUME'S HISTORY OF ENGLAND. History of England, from the Invasion of Julius Cæsar to the Abdication of James II., 1688. By DAVID HUME. A new Edition, with the Author's last Corrections and Improvements. To which is Prefixed a short Account of his Life, written by Himself. With a Portrait of the Author. 6 vols., 12mo, Cloth, $9 00.

JAY'S WORKS. Complete Works of Rev. William Jay: comprising his Sermons, Family Discourses, Morning and Evening Exercises for every Day in the Year, Family Prayers, &c. Author's enlarged Edition, revised. 3 vols., 8vo, Cloth, $6 00.

JEFFERSON'S DOMESTIC LIFE. The Domestic Life of Thomas Jefferson: compiled from Family Letters and Reminiscences by his Great-Granddaughter, SARAH N. RANDOLPH. With Illustrations. Crown 8vo, Illuminated Cloth, Beveled Edges, $2 50.

JOHNSON'S COMPLETE WORKS. The Works of Samuel Johnson, LL.D. With an Essay on his Life and Genius, by ARTHUR MURPHY, Esq. Portrait of Johnson. 2 vols., 8vo, Cloth, $4 00.

KINGLAKE'S CRIMEAN WAR. The Invasion of the Crimea, and an Account of its Progress down to the Death of Lord Raglan. By ALEXANDER WILLIAM KINGLAKE. With Maps and Plans. Two Vols. ready. 12mo, Cloth, $2 00 per vol.

KINGSLEY'S WEST INDIES. At Last: A Christmas in the West Indies. By CHARLES KINGSLEY. Illustrated. 12mo, Cloth, $1 50.

KRUMMACHER'S DAVID, KING OF ISRAEL. David, the King of Israel: a Portrait drawn from Bible History and the Book of Psalms. By FREDERICK WILLIAM KRUMMACHER, D.D., Author of "Elijah the Tishbite," &c. Translated under the express Sanction of the Author by the Rev. M. G. EASTON, M.A. With a Letter from Dr. Krummacher to his American Readers, and a Portrait. 12mo, Cloth, $1 75.

LAMB'S COMPLETE WORKS. The Works of Charles Lamb. Comprising his Letters, Poems, Essays of Elia, Essays upon Shakspeare, Hogarth, &c., and a Sketch of his Life, with the Final Memorials, by T. NOON TALFOURD. Portrait. 2 vols., 12mo, Cloth, $3 00.

LIVINGSTONE'S SOUTH AFRICA. Missionary Travels and Researches in South Africa; including a Sketch of Sixteen Years' Residence in the Interior of Africa, and a Journey from the Cape of Good Hope to Loando on the West Coast; thence across the Continent, down the River Zambesi, to the Eastern Ocean. By DAVID LIVINGSTONE, LL.D., D.C.L. With Portrait, Maps by Arrowsmith, and numerous Illustrations. 8vo, Cloth, $4 50.

LIVINGSTONES' ZAMBESI. Narrative of an Expedition to the Zambesi and its Tributaries, and of the Discovery of the Lakes Shirwa and Nyassa. 1858–1864. By DAVID and CHARLES LIVINGSTONE. With Map and Illustrations. 8vo, Cloth, $5 00.

M'CLINTOCK & STRONG'S CYCLOPÆDIA. Cyclopædia of Biblical, Theological, and Ecclesiastical Literature. Prepared by the Rev. JOHN M'CLINTOCK, D.D., and JAMES STRONG, S.T.D. 5 vols. now ready. Royal 8vo. Price per vol., Cloth, $5 00; Sheep, $6 00; Half Morocco, $8 00.

MARCY'S ARMY LIFE ON THE BORDER. Thirty Years of Army Life on the Border. Comprising Descriptions of the Indian Nomads of the Plains; Explorations of New Territory; a Trip across the Rocky Mountains in the Winter; Descriptions of the Habits of Different Animals found in the West, and the Methods of Hunting them; with Incidents in the Life of Different Frontier Men, &c., &c. By Brevet Brigadier-General R. B. MARCY, U.S.A., Author of "The Prairie Traveller." With numerous Illustrations. 8vo, Cloth, Beveled Edges, $3 00.

MACAULAY'S HISTORY OF ENGLAND. The History of England from the Accession of James II. By THOMAS BABINGTON MACAULAY. With an Original Portrait of the Author. 5 vols., 8vo, Cloth, $10 00; 12mo, Cloth, $7 50.

MOSHEIM'S ECCLESIASTICAL HISTORY, Ancient and Modern; in which the Rise, Progress, and Variation of Church Power are considered in their Connection with the State of Learning and Philosophy, and the Political History of Europe during that Period. Translated, with Notes, &c., by A. MACLAINE, D.D. A new Edition, continued to 1826, by C. COOTE, LL.D. 2 vols., 8vo, Cloth, $4 00.

NEVIUS'S CHINA. China and the Chinese: a General Description of the Country and its Inhabitants; its Civilization and Form of Government; its Religious and Social Institutions; its Intercourse with other Nations; and its Present Condition and Prospects. By the Rev. JOHN L. NEVIUS, Ten Years a Missionary in China. With a Map and Illustrations. 12mo, Cloth, $1 75.

THE DESERT OF THE EXODUS. Journeys on Foot in the Wilderness of the Forty Years' Wanderings; undertaken in connection with the Ordnance Survey of Sinai and the Palestine Exploration Fund. By E. H. PALMER, M.A., Lord Almoner's Professor of Arabic, and Fellow of St. John's College, Cambridge. With Maps and numerous Illustrations from Photographs and Drawings taken on the spot by the Sinai Survey Expedition and C. F. Tyrwhitt Drake. Crown 8vo, Cloth, $3 00.

OLIPHANT'S CHINA AND JAPAN. Narrative of the Earl of Elgin's Mission to China and Japan, in the Years 1857, '58, '59. By LAURENCE OLIPHANT, Private Secretary to Lord Elgin. Illustrations. 8vo, Cloth, $3 50.

OLIPHANT'S (MRS.) LIFE OF EDWARD IRVING. The Life of Edward Irving, Minister of the National Scotch Church, London. Illustrated by his Journals and Correspondence. By Mrs. OLIPHANT. Portrait. 8vo, Cloth, $3 50.

RAWLINSON'S MANUAL OF ANCIENT HISTORY. A Manual of Ancient History, from the Earliest Times to the Fall of the Western Empire. Comprising the History of Chaldæa, Assyria, Media, Babylonia, Lydia, Phœnicia, Syria, Judæa, Egypt, Carthage, Persia, Greece, Macedonia, Parthia, and Rome. By GEORGE RAWLINSON, M.A., Camden Professor of Ancient History in the University of Oxford. 12mo, Cloth, $2 50.

RECLUS'S THE EARTH. The Earth: a Descriptive History of the Phenomena and Life of the Globe. By ÉLISÉE RECLUS. Translated by the late B. B. Woodward, and Edited by Henry Woodward. With 234 Maps and Illustrations, and 23 Page Maps printed in Colors. 8vo, Cloth, $5 00.

RECLUS'S OCEAN. The Ocean, Atmosphere, and Life. Being the Second Series of a Descriptive History of the Life of the Globe. By ÉLISÉE RECLUS. Profusely Illustrated with 250 Maps or Figures, and 27 Maps printed in Colors. 8vo, Cloth, $6 00.

SHAKSPEARE. The Dramatic Works of William Shakspeare, with the Corrections and Illustrations of Dr. JOHNSON, G. STEEVENS, and others. Revised by ISAAC REED. Engravings. 6 vols., Royal 12mo, Cloth, $9 00.

SMILES'S LIFE OF THE STEPHENSONS. The Life of George Stephenson, and of his Son, Robert Stephenson; comprising, also, a History of the Invention and Introduction of the Railway Locomotive. By SAMUEL SMILES, Author of "Self-Help," &c. With Steel Portraits and numerous Illustrations. 8vo, Cloth, $3 00.

SMILES'S HISTORY OF THE HUGUENOTS. The Huguenots: their Settlements, Churches, and Industries in England and Ireland. By SAMUEL SMILES. With an Appendix relating to the Huguenots in America. Crown 8vo, Cloth, $1 75.

SPEKE'S AFRICA. Journal of the Discovery of the Source of the Nile. By Captain JOHN HANNING SPEKE, Captain H. M. Indian Army, Fellow and Gold Medalist of the Royal Geographical Society, Hon. Corresponding Member and Gold Medalist of the French Geographical Society, &c. With Maps and Portraits and numerous Illustrations, chiefly from Drawings by Captain GRANT. 8vo, Cloth, uniform with Livingstone, Barth, Burton, &c., $4 00.

STRICKLAND'S (MISS) QUEENS OF SCOTLAND. Lives of the Queens of Scotland and English Princesses connected with the Regal Succession of Great Britain. By AGNES STRICKLAND. 8 vols., 12mo, Cloth, $12 00.

THE STUDENT'S SERIES.
France. Engravings. 12mo, Cloth, $2 00.
Gibbon. Engravings. 12mo, Cloth, $2 00.
Greece. Engravings. 12mo, Cloth, $2 00.
Hume. Engravings. 12mo, Cloth, $2 00.
Rome. By Liddell. Engravings. 12mo, Cloth, $2 00.
Old Testament History. Engravings. 12mo, Cloth, $2 00.
New Testament History. Engravings. 12mo, Cloth, $2 00.
Strickland's Queens of England. Abridged. Engravings. 12mo, Cloth, $2 00.
Ancient History of the East. 12mo, Cloth, $2 00.
Hallam's Middle Ages. 12mo, Cloth, $2 00.
Hallam's Constitutional History of England. 12mo, Cloth, $2 00.
Lyell's Elements of Geology. 12mo, Cloth, $2 00.

TENNYSON'S COMPLETE POEMS. The Complete Poems of Alfred Tennyson, Poet Laureate. With numerous Illustrations by Eminent Artists, and Three Characteristic Portraits. 8vo, Paper, 75 cents; Cloth, $1 25.

THOMSON'S LAND AND THE BOOK. The Land and the Book; or, Biblical Illustrations drawn from the Manners and Customs, the Scenes and the Scenery of the Holy Land. By W. M. THOMSON, D.D., Twenty-five Years a Missionary of the A.B.C.F.M. in Syria and Palestine. With two elaborate Maps of Palestine, an accurate Plan of Jerusalem, and several hundred Engravings, representing the Scenery, Topography, and Productions of the Holy Land, and the Costumes, Manners, and Habits of the People. 2 large 12mo vols., Cloth, $5 00.

TYERMAN'S WESLEY. The Life and Times of the Rev. John Wesley, M.A., Founder of the Methodists. By the Rev. LUKE TYERMAN, Author of "The Life of Rev. Samuel Wesley." Portraits. 3 vols., Crown 8vo, Cloth, $7 50.

VÁMBÉRY'S CENTRAL ASIA. Travels in Central Asia. Being the Account of a Journey from Teheran across the Turkoman Desert, on the Eastern Shore of the Caspian, to Khiva, Bokhara, and Samarcand, performed in the Year 1863. By ARMINIUS VÁMBÉRY, Member of the Hungarian Academy of Pesth, by whom he was sent on this Scientific Mission. With Map and Woodcuts. 8vo, Cloth, $4 50.

WOOD'S HOMES WITHOUT HANDS. Homes Without Hands: being a Description of the Habitations of Animals, classed according to their Principle of Construction. By J. G. WOOD, M.A., F.L.S. With about 140 Illustrations. 8vo, Cloth, Beveled Edges, $4 50.

www.ingramcontent.com/pod-product-compliance
Lightning Source LLC
Chambersburg PA
CBHW020901210326
41598CB00018B/1741